# THE EVOLUTION OF MIDDLE EASTERN LANDSCAPES

CROOM HELM HISTORICAL GEOGRAPHY SERIES
Edited by R.A. Butlin, University of Loughborough

# The Evolution of Middle Eastern Landscapes
## An Outline to A.D. 1840

J.M. Wagstaff

CROOM HELM
London & Sydney

© 1985 J.M. Wagstaff
Croom Helm Ltd, Provident House, Burrell Row,
Beckenham, Kent BR3 1AT

Croom Helm Australia Pty Ltd, First Floor, 139 King Street,
Sydney, NSW 2001, Australia

British Library Cataloguing in Publication Data

Wagstaff, J. Malcolm
    The evolution of Middle Eastern landscapes: an
    outline to AD 1840. — (Croom Helm historical
    geography series)
    1. Near East — Description and travel
    911'.56    DS45

    ISBN 0-85664-812-4

Filmset by Mayhew Typesetting, Bristol, UK

**Printed and bound in Great Britain by
Biddles Ltd, Guildford and King's Lynn**

# CONTENTS

# TABLES

# FIGURES

# PREFACE

Students of the contemporary scene in the Middle East soon discover their need for an historical perspective. Geographers are no exception. This is a long-settled region. Its people are the heirs of ancient civilisations; their attitudes and behaviour are the products of historical accretion, as well as trial and error in the business of gaining a living and getting along with each other. The marks of more than 400 generations of agricultural endeavour are clear in the region's landscapes. The historical legacy is strong in built environments. Yet many geographical studies of the region dealing with the present are diminished by their lack of historical depth. Neophytes, for their part, must fall back all too often on half-remembered biblical studies or vague notions of Europe's 'Eastern Question'. Responsibility for this state of affairs can be attributed, in part at least, to a rigidly Eurocentric education. Part, however, is due to the lack of any study specifically devoted to the historical geography of the region; we tend to read only within our own disciplines. The present study is offered as a move in the direction of a remedy.

Like so many books written by professional academics, this one had its origins in a course of undergraduate lectures. These were first given in the University of Southampton in 1966–67 and I am grateful to Dr L.E. Tavener, then acting head of the Geography Department, for encouraging me to prepare them. The discontinuance of the course increased the need to provide my own pupils with at least some background, some historical depth, to studies essentially focused on the Middle East as it is today, with its problems of development, surplus capital, industrialisation, urbanisation, population growth, inter-communal strife and war. Others have found the same need. Herein lies the rationale for the approach adopted and for the extensive references provided. I hope, none the less, that this book may be more than a university text and appeal to a wider public.

In writing, I have been conscious of many influences and debts in my own intellectual development and my career. My parents set me on the golden road with their interests in armchair travel and history. Teachers were sympathetic and encouraging. I was fortunate as an undergraduate to be part of that humane and outward-looking department founded by Percy Roxby at Liverpool and developed by Wilfrid Smith, H.C. Darby and, in my time, Robert Steel. Work in Greece was a stepping-stone to

Turkey and the further East. Good fortune took me to Durham, then expanding its Centre of Middle Eastern and Islamic Studies and where Geography was acquiring its special Middle Eastern flavour under the leadership of W.B. Fisher. The second edition of Braudel's *Méditerranée* was recently published and impressed me enormously as a way of writing regional historical geography. It would be surprising if some of its excellent influence were not detectable in what follows.

Most of the book was written in the late 1970s and early 1980s, much of it during the turmoil caused by cuts in the finances of British universities. The labour involved was a precious refuge from the storms of academic politics, but my involvement in university administration explains why I am grateful to my editor, Robin Butlin, and to my publishers for their patience over long delays.

Some geographers will find my efforts at writing historical geography idiosyncratic in the extreme. Specialists from other disciplines will probably be irritated by my mistakes and myopia. I crave indulgence. I would like to think, however, that a few geographers and other scholars will find my efforts worthwhile, even stimulating, and that some of them will do a better job.

Highfield,
Ascension

# ACKNOWLEDGEMENTS

In preparing this book I have accumulated a large store of debt which it is my pleasure to discharge here. In addition to those individuals mentioned in the Preface, I should like to thank the library staff of the School of Oriental and African Studies, London, where much of the reading was done; the staff of the inter-library loans office in Southampton University Library who were so diligent in obtaining obscure items; the following for permission to reproduce or use illustrative material: John Bartholomew and Sons Ltd (the Nordic Projection in Figure 1.1), E.J. Brill (*Turkey*, ed. P. Benedict, E. Tümertekin and F. Mansur, 1974, for Fig. 10.3), The University of Chicago (R.M. Adams, *Land Behind Baghdad*, 1965, for Figs. 6.7 and 9.1, and K.W. Butzer, *Early Hydraulic Civilisation in Egypt*, 1976, for Fig. 5.2), the President and Fellows of Harvard College (R.W. Bulliet, *Conversion to Islam in the Medieval Period*, 1979, for Fig. 7.3), Hodder and Stoughton (formerly English Universities Press Ltd, W.C. Brice, *South-West Asia*, 1966, for Fig. 4.11), *The Middle East Journal* (for Fig. 4.2), and the Department of Geography, University of Chicago (for Fig. 4.9); Dr Stephen Mitchell for valuable comments on Chapter 6; Mr A.S. Burn and the Cartographic Unit of Southampton University for drawing the illustrations; and Mrs R. Flint for producing a typescript from the puzzle which was the original manuscript.

JMW

# TRANSLITERATION AND QUOTATION

Consistency is difficult to achieve in a work based on a range of very disparate sources. I have tried to standardise place names on the basis of the versions adopted by *The Times Atlas of the World* (London, 1959), but I have used common English forms for such well-known places as Damascus and Jerusalem. For personal names and technical terms in Arabic and Persian I have generally employed the system of transcription favoured by *The Encyclopaedia of Islam* (new edn, vol. 1, Leiden 1960, xiii), but I am conscious of not being entirely consistent. In particular, I have stuck rigorously to *Qur'ān* and prefer this form to the more normal English, Koran. For the Ottoman period I have followed modern Turkish orthography.

References to the Qur'ān are by *sura* and verse, and to the Bible by book (e.g. Genesis), chapter and verse. For the Qur'ān I have used the Arabic text with an English translation by Muhammud Zafrulla Khan (2nd impression, London 1971), although occasionally I have preferred the Penguin translation by Dawood (Harmondsworth 1959). For the Bible I have used *The Holy Bible*, revised edition (first published 1881–84).

# 1 INTRODUCTION: DATES AND DEFINITIONS

*Middle East* is a convenient geographical expression (Davison 1960; Keddie 1973; Lewis 1964a; Reiss 1969; Smith 1968). It describes a region of vast peninsulas lying about the junction of Africa, Asia and Europe (Fig. 1.1). But its very convenience has made the term somewhat nebulous in application. It probably originated in the British India Office during the 1850s (Koppes 1976), in the early days of the struggle between the expansionist powers — Britain and Russia — for influence in Persia. At that time, 'Middle East' referred to Persia and adjacent territories in central Asia and around the Gulf. The region was thus distinguished from the *Near East* and the *Far East* as part of the wider *East* which, until that time, was viewed from Europe as a largely undifferentiated region stretching from the gates of Vienna and the lagoon of Venice all the way to the China Seas (Steadman 1969). The Near East was virtually synonymous with the Ottoman Empire, that 'sick man of Europe' over whose demise the Eastern Question took a new twist in the 1850s. The Far East consisted principally of China and Japan. They re-emerged into European consciousness with the First Opium War (1839–42) and the arrival of Commodore Perry in Yokhama harbour (1854).

'Middle East' became current in political circles of the English-speaking world around 1900 (Fig. 1.2). It was used to describe 'those regions of Asia which extend to the borders of India, and which are consequently bound up with the problems of Indian defence' (Chirol 1903, 5). This was again the result of great-power rivalry. British politicians not only saw Russian expansion and railway-building in central Asia as putting the northern provinces of Persia at risk and creating a new threat to India through Afghanistan, but also noticed that the new German Empire was extending its influence into the region, conspicuously through the Baghdad Railway project, designed to subvert Britain's dominance of the eastern trade. At the same time, Turkey and Persia converted nominal authority over the shores of the Gulf into actual control of considerable lengths of them, while French treaty rights in Muscat threatened British hegemony over the seaways. Political discourse thus required a term to distinguish an important strategic region from the Near and Far Easts.

In the Far East, the Chinese Empire, already reduced by British and French expansion, seemed about to disintegrate under the antagonistic

Figure 1.1: Location of the Middle East

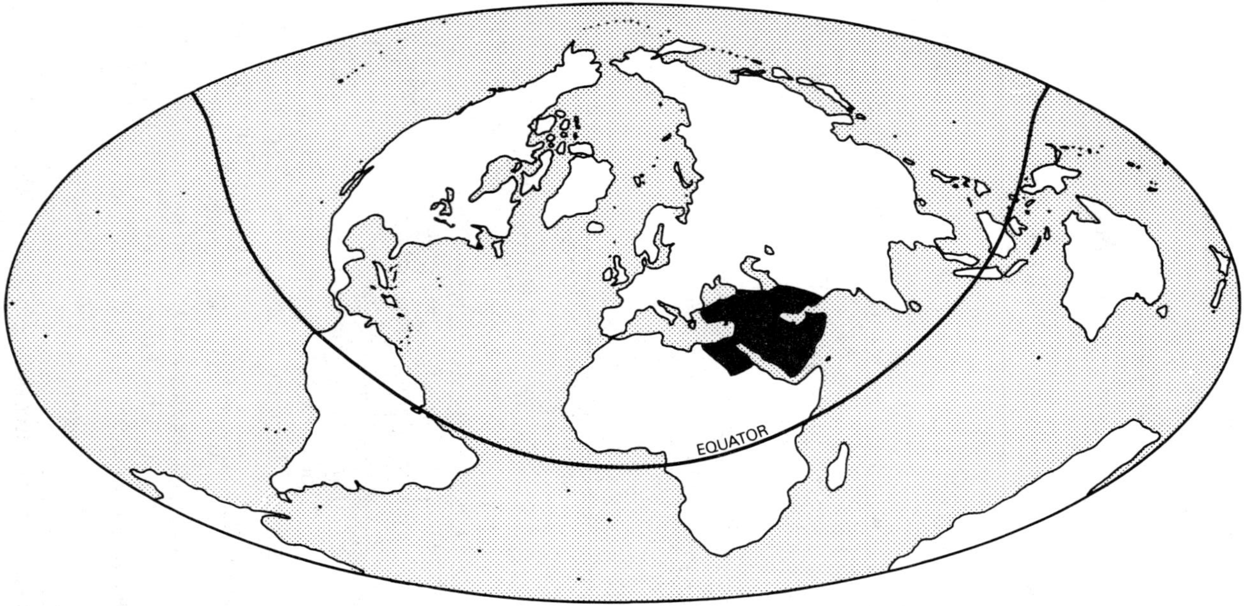

EQUATOR

1:160 m 'Nordic' Projection
(designed by J. Bartholomew)

## Figure 1.2: Middle East Defined in Relation to Europe

Legend:

Middle East as defined by Chirol (1902)

Middle East as studied in this book

IRAQ Present political units

0 — 1000 miles

0 — 1600 km

pressures of an aggressive Japan and an expansionist Russia. Anglo-American rivalry for the trade of the region developed, and the Germans established themselves in the Shantung peninsula. The situation was paralleled in the Near East. The Ottoman Empire continued to lose territory from its European provinces and the Balkans became 'a veritable powder keg which in the end blew up with disastrous consequences' in the summer of 1914 (Stavrianos 1958, 541). Even in the early twentieth century, however, the definition of the Middle East was vague. For Mahan and Chirol in 1902 it included Persia and the Persian Gulf (Chirol 1903; Mahan 1902), as it had for the men of the India Office two generations earlier. But in the same year, Hogarth recognised no Middle East at all. His *Nearer East* covered 'all south-eastern Europe below the long oblique water-parting of the Balkans', included Corfu and Crete and stretched eastwards across the islands, embraced 'all of the north-eastern corner of Africa that is fit for settled human habitation' and stretched to the salt-hollows of central Persia and the shifting sands of Persian Baluchistan which formed, he thought, the 'truly natural boundary with the Further East' (Hogarth 1902, 2–3). In 1911, however, Lord Curzon included not only the whole of Persia and the Persian Gulf territories in his Middle East, but also what was currently called Turkey in Asia (*Parliamentary Debates, Lords*, 1911, col. 575).

During the First World War (1914–18) three, and subsequently two, major army commands operated independently in the region for the British. The Mediterranean Expeditionary Force was concerned with operations in Gallipoli and Macedonia, while the Force in Egypt was confined to that country and the Western Desert. In 1916 the two were amalgamated into a single command, named the Egyptian Expeditionary Force, and made responsible for operations in Palestine and Syria (Macmunn and Falls 1928, 94–7). Meanwhile, the Mesopotamia Expeditionary Force was operating in Iraq, with ultimate authority for its direction divided between New Delhi and London. The existence of two distinct spheres of military operations seems to have fostered the development of Middle East and Near East as popular terms. After the war, English-speaking geographers attempted to give them some precision. Near East was suggested as a collective term to cover the Balkans, Turkey, Cyprus, Syria, Lebanon and Palestine; Egypt was sometimes added. Middle East was restricted to Iran, Iraq, Jordan, Saudi Arabia and the smaller states of the Arabian peninsula (Clark 1944). Despite this degree of clarification, ambiguity persisted. It was compounded by further developments in military nomenclature.

Since the end of the First World War, the RAF had possessed a separate

Middle Eastern command, covering Egypt, Sudan and Kenya. It also had administrative, but not operational, responsibility for Palestine and Transjordan. In April 1938 this command was amalgamated with the then separate commands for Iraq, Aden and Malta. In June 1939 the three separate British army commands in the region, together with the garrisons in Malta and Aden and any troops stationed in Iraq, the Gulf and British Somaliland, were also brought under a unified Middle East command. At the same time, the Committee of Imperial Defence decided to form a joint high command for the three services in the region with its base in Egypt (Playfair 1954, 31–5). The subsequent spread of operations conducted under the joint command during the Second World War (1939–45) extended the application of 'Middle East' across north Africa and into Greece. Persistent use of 'Middle East' in official communiqués and news bulletins did much to establish the westward shift in its definition and to popularise the use of this extended form (Smith 1968). 'Near East' became largely redundant (Lorraine 1943). After the war, various attempts were made to revive and distinguish this term clearly from 'Middle East', but without much success, except perhaps amongst archaeologists. The wider 'Middle East' had taken root in the public consciousness (Martin 1944; Rennell of Rodd 1946). It was even officially defined in the House of Commons in 1951 (*Parliamentary Debates, Commons* 1951, 448–9). Various other attempts at definition have been made, some of wide and others of lesser compass. All seem agreed, though, that nowadays the Middle East includes, as a minimum, the territories of Turkey and Iran, Lebanon, Syria and Iraq, Israel and Jordan, the states of Arabia, and Egypt (Pearcy 1959). This region is the focus for the present book.

A definition in terms of modern political units is consistent with the geopolitical origins and usage of 'Middle East'. It also reflects the very real difficulties faced by any attempt to define the region as homogeneous in its physical or cultural geography (Bacon 1946 and 1953; Patai 1952). In the core area — northern Arabia, Syria and perhaps northern Iraq — common characteristics may be distinguished. Mountain and plain lie side by side. Summer drought alternates with winter precipitation, including snow. Pastoral nomadism and settled agriculture have been carried on side by side for centuries and are symbiotically dependent. Villages and towns have long flourished in a series of man-modified landscapes, in a socio-economic matrix shaped by physical conditions. Civilisation developed early. Three of the world's greatest religions were born here. The area was the heartland of classic Muslim culture developed under the united caliphate.

Despite our ability to recognise such spatio-temporal patterns, definition of their limits is difficult. Any attempt to establish exact limits is
bound to fail. At best, only a series of transitions can be recognised.
Mountains and plains become more extensive away from the vaguely defined core. Drought gets longer or shorter depending on direction; precipitation maxima slip to spring and autumn in the north and move to summer
in southern Arabia. Nomadism sheds its association with the dromedary,
and either nomadism or agriculture became dominant as historical modes
of living. Towns and villages become linked in either tighter or looser
lattices. The historical domain of Muslim culture stretches out through
north Africa to Spain in the west, far into central Asia and India in the
east, and southwards beyond the Sahara deep into Africa. Culture-based
geographical definitions are bound to be time-specific. In the end, we
are left with a broad location and an arbitrary definition.

As a region, the Middle East is analogous to an open system. Its energy
has been generated by the interaction of the various ecological units which
compose it, as well as by its pulsatory relations with neighbouring regions.
These interactions have created social and economic regions, which
though often discordant with the various ecosystems, have been organised through a nested hierarchy of towns around some major city. A full
study of these diachronically changing patterns cannot yet be written.
Few of the detailed studies necessary for such a project have been published. Coverage is uneven, in both spatial and temporal terms. This book
attempts no more than an outline sketch. It aims to provide an historical
perspective on the human geography of the present.

Accordingly, considerable attention has been given in Chapters 5–8
to the political and military history which unfolded across the landscapes
of the region and left such profound marks on its townscapes. Political
and military events not only influenced the use of land, but were
themselves constrained by physical conditions, as well as current systems
of resource exploitation. The attempt to provide an historico-geographical
perspective on the present also led to a conscious decision to deal at length
with the rise and expansion of Islam. Since the fourteenth or fifteenth
century AD Islam has been the dominant religious, cultural and social
system in the region. Other social systems have been influenced by it
and, to some extent, other religious communities have defined themselves
by Muslim norms. Christianity, Judaism and Zoroastrianism are given
less prominence on the grounds that, although of considerable importance at particular times and in particular places, they were long in retreat
in the region — at least until the beginnings of Zionist settlement in
Palestine — and their important legacies impinge less directly on the

present human geography of the whole Middle East. Despite these idiosyncracies, the main aim of the book is to trace the evolution of humanised, managed landscapes and, as far as currently possible, on a regional scale. The starting base is the emergence of food production in the region. The equally arbitrary terminal date is the change in long-established patterns of subsistence and land use which gathered pace during the nineteenth century AD, largely as a result of European commercial and political penetration. These changes produced patterns which are recognisably the direct ancestors of those found in the closing decades of the twentieth century.

The shift from food processing to food production around 8000 BC was the most fundamental change experienced by man in the region and the rest of the world. It is now irreversible. Chapter 3 examines the causes and immediate consequences of this revolution. The modes of living which subsequently evolved from it — and which persisted in the region until recently — are outlined in Chapter 4. Later chapters trace historico-geographical development through to the early nineteenth century AD, period by period. These are designated partly because they are regarded as important to an understanding of the historical geography of the region, though it is realised that political and cultural systems do not have the same periodicity as either other human systems or landscape change. The penultimate chapter is an attempt at stock-taking. It tries to establish what the Middle East was like — how far it had come — by the eve of the great changes brought about through vastly increased interaction with industrial, expansionist, secular Europe. The final chapter outlines the ways in which the geography of the region was affected and provides the forward links to the situation in the late twentieth century. A significant date in the transition was AD 1840.

By about 1840 western political involvement in the region was well advanced. The world-wide strategy necessitated by the Revolutionary and Napoleonic Wars (1792–1815) brought French and British rivalries — and armies — to the region. Russian expansion into the Caucasus and central Asia was at Persian and Turkish expense, but it also led to confrontation with Britain, then expanding its power in northern India, and whose officials were concerned about the possibility of externally inspired internal revolt (Yapp 1980). Britain imposed a maritime truce on the pirates — some would say privateers — of the Gulf in 1835, and followed it with a peace treaty in 1853. Aden was occupied in 1839. The 1850s saw British actions against the Persian coast to counter a Russian-inspired Persian threat to Herat which seemed at the time to jeopardise the security of India. European intervention, both diplomatic and military, forced

Muḥammad 'Ali (1805–49) to evacuate Syria in 1840 and secured him the hereditary viceroyalty of Egypt as compensation. Political intervention was instrumental in bringing about social and administrative reform, notably in the Ottoman Empire where the reform era (the *Tanzimat*, 'Reordering') was inaugurated with the *Hatt-ı Şerif* of Gülhane (1839), but also in Persia where less successful attempts were made to modernise and centralise. More influential on the landscapes and townscapes of the region was European commercial penetration.

Although trade with the Middle East had developed in antiquity, revived during Europe's Middle Age and expanded during the eighteenth century, the entry to a phase of European domination is marked by the Anglo–Turkish Commercial Convention of 1838 and the Anglo–Persian Commercial Treaty of 1841. These ended the old tariff barriers and began opening up Turkish and Persian domains to the factory-produced goods of Europe. Steam navigation came at about the same time, to be followed by the first carriage roads and railways. As the interior was opened up, so cultivation expanded and became more commercialised. The first steam-driven factories were opened. After about 1840, then, commercial contacts and the necessities of political survival increased the pace of change throughout the region. The decadent East, ridiculed by Kinglake and Morier (Kinglake 1844; Morier 1824) but romanticised as gorgeous and voluptuous by contemporary Orientalist painters (Jullian 1977), began to fade like the Arabian Nights themselves. The wide gulf between East and West, which appeared so fixed before 1840, began to shrink before the advance of the West (Spender 1911, xi). In the region itself, the space-time envelopes of the ordinary inhabitants both expanded and contracted. People and information travelled faster, so that places came closer together. As a consequence, however, individuals became more aware of a larger region lying beyond the physical and market horizons of their town or village, wider even than that discovered by those whose piety took them on the Great Pilgrimage to Mecca. The basic framework of mountains and seas remained the same, of course, creating friction in economic and political systems. The major features of climate similarly endured, though subject to short-run fluctuations which, through drought and deluge, still brought disaster. Physical conditions provide a matrix which cannot be ignored in studying the human geography of the Middle East, past or present. They are outlined in the next chapter.

# 2  THE MEASURE OF THE REGION

The treatment of physical conditions presents a major intellectual challenge in writing an historical geography, especially one attempting to cover 10,000 years. The natural environment cannot be assumed to have remained unchanged in every particular over such a time-span. Nor can man's perception and evaluation of it have remained constant (Glacken 1956). How should physical conditions be treated? It is clear that changes in the various elements of man's physical environment have taken place on different time-scales. Mountain-building and peneplanation take thousands of years to operate, so that even in the time-span covered by this survey man is likely to have seen few major changes on the regional scale, although from time to time the power of natural forces must have been apparent locally in volcanic eruptions, earthquakes and major landslips. Changes in climate and fluctuations in sea-level happen on a shorter time-scale but leave subsequent generations puzzling over abandoned activity sites inland and drowned harbour works at the coasts. Erosion and sedimentation are more instantaneous, perhaps the work of a single storm lasting no more than a few hours, but the way was prepared by generations of tillage. The consequences of this and skeletonal soils were long-term. Human activity itself brings about noticeable change in vegetational assemblages, some of which initiate erosion or water shortage, for example. Such consequences of simple, direct resource exploitation might even be indistinguishable from those resulting from purely natural processes. This complicates the task of interpretation.

One way of dealing with physical conditions in an historico-geographical study is to treat the major terrain units — mountains, plains and coastlands — as constants, exercising definite constraints on human activity at all periods. Their role may have been appreciated only gradually, as the horizons of local communities were expanded by incorporation into larger societies and as the technologies for interaction were developed. The dominant characteristics of local climates may be viewed in a similar way, as harsh facts to which life-styles and economies were adapted. Climatic change was comparatively slow. In consequence, human adaptation is likely to have proceeded in a series of adjustments and modifications, delayed by faulty perceptions, optimism and fatalism. But even the severe climatic regimes found over so much of the Middle East during the last 10,000 years may be considered as offering a range of possibilities

9

to the people living under them. Some of the options would be more constrained than others, and perhaps of limited duration. All responses, however, would have been affected by the varying structures of social organisation and political control which people themselves devised out of their value systems, their understanding of the world and their technical abilities.

In order to deal with physical conditions in the way proposed, the nearly 8 million $km^2$ of the Middle East have been divided into topographic sub-regions (Fig. 2.1). The framework is provided by the mountain ranges and the sea. Many of the units thus defined consist of plains or rolling country. Altogether, they form a mosaic of relatively stable pieces, whose long-term role in the historical geography of the region will be outlined in this chapter. Change in the more dynamic elements of the physical environment, especially in climate and vegetation, will be described in later chapters. Vegetational change, however, will also be treated as an indicator of the pattern and pace of human activity.

## The Mountains

The main mountain ranges diverge from a tangle of high, bleak lava plateaux, towering volcanic cones and steep-sided basins in what, for the lack of a more convenient and neutral regional name, may still be called Armenia. Two chains run roughly westwards, gripping the high Anatolian plateau between their open arms. On the north are the Pontic Mountains. Continuous mountain chains sweep westwards in a belt 30–40 km wide, divided by clearly defined valleys. The southern mountains are less continuous and consist principally of the Taurus and Anti-Taurus ranges, but in the far south-west include a high plateau and a number of shorter ranges. The Taurus Mountains vary in width between 50 and 100 km. In the west the four main ranges run roughly parallel with each other. They are generally separated by deep narrow valleys, though the innermost ranges are sufficiently far apart to accommodate the lakes of Beyşehir and Suğla. More lakes are found in small upland basins further west. The Anti-Taurus Mountains curve round eastwards from the Gulf of İskenderun and run south of Lake Van to merge with the so-called Kurdish Alps. The ranges are high, but discontinuous. They are separated by remote high valleys containing streams and rivers which drain to the Tigris and Euphrates. These rivers cut through the mountains in spectacular gorges. Held in the grip of the main westward-trending mountains, at the Aegean end of Anatolia, is a series of block-like mountains separated by broad,

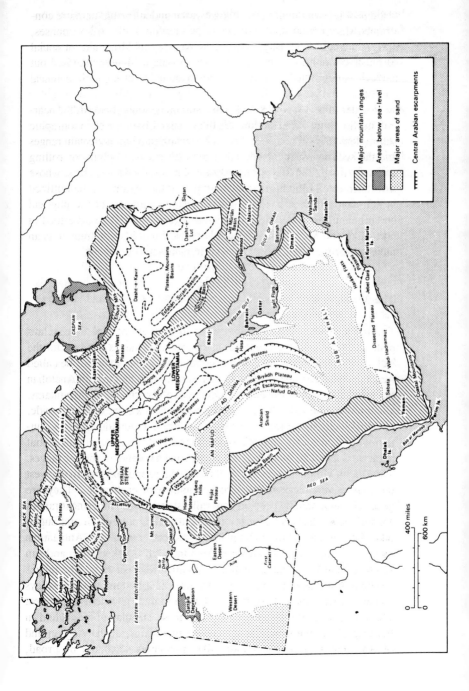

flat-floored valleys running roughly west-east and allowing maritime influences to penetrate deeply into the peninsula. At the other end of Anatolia, and diverging slightly from the Anti-Taurus Mountains, are the short, narrow ranges of the Amanus and Ansārīye Mountains, the line of which is continued into the Troödos Mountains of Cyprus.

A belt of immense, parallel mountain ranges of roughly similar height, and a maximum of 340 km across, stretches south-eastwards from Armenia for some 1,200 km towards the Straits of Hormuz before swinging east through Makran. The Zagros Mountains contain many sheltered basins linked transversely by deep, narrow gorges cut by perennial rivers. Across the Gulf, the Hajar range forms the core of Oman. Eastwards from the Zagros Mountains, and roughly parallel to them, are several smaller detached ranges. These separate the three great basins of inland drainage — the Dasht-e-Kavir, the Dasht-e-Lūt and Sīstan. In the far north of Persia, the single serated fold of the Elburz Mountains curves round the shore of the Caspian Sea and separates it from the high interior.

Spectacular mountains rise high from coastal plains on both sides of the Red Sea trough. On the eastern side they extend northwards towards the Anti-Taurus Mountains and curve south-eastwards through southern Arabia. In Yemen, they form a series of vast, steep-sided blocks, but along their southern edge runs the mysterious rift of the Wadi Hadramawt. In the far north, the mountains edge a variety of upland terrain types, including the black basalt plateau of Hauran and the lava wastes of el-Leja, as well as a series of basins paralleling the Red Sea and separated from each other by tracts of lava. On the west is the deep rift of the Jordan Valley, containing the Dead Sea and the Sea of Galilee. Further north is its continuation in the High Valley of Beq'a. In the triangle between the Mediterranean Sea and the Gulfs of Suez and 'Aqaba is Sinai — a remote and desolate plateau from which two peaks protrude.

Mountain climates are diverse, much depending upon elevation and exposure, as well as latitude. Above 30°N, the mountains are exposed to cyclones generated particularly over the Mediterranean Sea during the winter months, and orographic effects ensure that the amount of precipitation is high.[1] Totals above 600 mm are normal but rise to more than 1,500 mm over parts of the Pontic Mountains, the Taurus, Anti-Taurus and the western part of the Elburz Mountains. The eastern section of the Pontic Mountains and the western end of the Elburz Mountains receive some precipitation even in summer. Much of the winter precipitation everywhere in the mountains north of 30° falls as snow. In Armenia it lies from 80 to more than 120 days every year. It may frequently last just as long over the Elburz and Zagros Mountains. Lebanon, Hermon and Ulu Dağ (Mt

Olympus), amongst others, were celebrated in the past for their long-lasting snow-caps.

Cyclones penetrate less frequently into the areas lying below 30°N, with the consequence that mean annual precipitation over the Red Sea mountains is less than 100 mm. Over the ranges of southern Iran and Oman it falls progressively with distance east, from nearly 400 mm in a year to 100 mm. Totals attain 400–600 mm again over Yemen, but, while mist is frequent, most of the precipitation is brought in a few intense storms generated by tropical maritime air rising over the mountains. Elevation even here reduces temperatures. The consequence is that summers are more pleasant in the mountains than in the lowlands or along the coast.

Over much of Armenia and large parts of the Pontic, Taurus and Elburz Mountains, mean daily temperatures are lower than 20°C. Elsewhere they lie between 20° and 30°C. Mean daily temperatures in the northern mountains are below freezing in January, except near the coasts, and over Armenia and the high ridges of the Taurus fall as low as −10°C, with some days, of course, lower even than that. Further south, over the Red Sea mountains and southern Iran, mean daily temperatures in January lie between 10° and 20°C, while in Yemen and the higher parts of Oman they fall further still, but remain above freezing-point. The mean annual temperature range in these areas is less extreme than over Armenia, where it exceeds 25°C.

Elevation and exposure not only condition the detail of mountain climate; they also produce distinctive vegetation assemblages. Although much of the 'natural' vegetation has now vanished, enough survives in the mountains for its broad characteristics under the present climatic regime to be reconstructed.[2] Above about 3,000 m Alpine vegetation consisting of dry-top lawns containing such species as *Astragalus, Onobrychis* and *Gramineae* is characteristic. Between 2,500 and 3,000 m open coniferous forests of spruce, pine[3] and fir are found, except in the Zagros Mountains and the Red Sea mountains where juniper-steppe formations take their place (Zeist and Bottema 1977; Vesey-Fitzgerald 1955 and 1957a). Greater diversity exists at elevations below 2,500 m. From about 800 m upwards a rather open mixed woodland of deciduous and coniferous species may once have existed, with pine[4] and oak[5] generally dominant, but with pure stands of cedar, fir and juniper[6] important in particular localities. Oak forest, dominated by *Quercus brantii*,[7] but generally rich in species, may once have characterised the western Zagros Mountains, while beech and beech-fir[8] may have clothed large parts of the Pontus (Beug 1967). An evergreen forest was found around the eastern

Mediterranean, in a zone between sea-level and 700–800 m. It was probably dominated by oaks[9] and pine,[10] but also contained carobs, olives and pistachios, as well as an undergrowth of vines and lentisk. Below 700–800 m in the Zagros Mountains a mixed pistachio-almond forest seems probable (Zeist and Bottema 1977). On the south-facing slopes of the Jebel Qara in south Arabia, which are bathed in mist and rain brought by the monsoon in summer, an almost jungular forest existed. This is the home of *Boswellia sacra* and *Commiphora myrrha*, the trees which produce incense and myrrh respectively (Groom 1981, 96–120).

Such richly diverse resources have been readily exploited, in many cases so successfully that the original condition of the vegetation is difficult to reconstruct. Mountains, however, have not been just a source of gums, timber or fuel. They have had other values over the centuries. In general, mountains divide region from region and hamper communications between them. Indeed, the Hijāz region of western Arabia takes its current name, 'The Barrier', from precisely this characteristic. Everywhere, however, many mountain valleys which appear to offer the possibility of through routes turn out to be culs-de-sac. Others, like the deep and gloomy gorges on the Bitan and Greater Zab Rivers in the Zagros Mountains, cannot be negotiated for their entire length. The consequence is that through routes are confined to a few passes. These have had considerable strategic importance.

Mountain areas are remote and isolated. Severe climates and short growing-seasons make human life harder than in the plains, and render it more precarious. The frequency of earthquakes and avalanches makes certain localities extremely hazardous. Understandably, the mountains of the Middle East have been sources of labour for the more fortunate neighbouring areas. They have had a number of other positive values as well. As Braudel (1972) observed of the mountains bordering the Mediterranean, the waves of high civilisation have often passed them by. Their isolated valleys have frequently sheltered heterodox religious sects, like the Kiribas and the Nestorian Christians found today in parts of the Anti-Taurus Mountains. Languages, such as those of the Semitic group spoken by the Mahra and Shahra tribes in south Arabia, have been preserved. Relatively independent principalities have been sheltered from time to time, as in Mt Lebanon and the mountain fastnesses of Armenia. Economic and social diversity have been supported from the favourable juxtaposition of extensive summer grazings, aromatic shrubs and fine timber on the slopes, with cultivable land in upland basins and valleys. A vertical succession of cropping was sometimes possible. The effect of trapping rain and snow is to turn the water-surplus mountains into reservoirs for the thirsty plains, to which the vital water is transferred by rivers, surface canals and *qanāts* (sub-surface canals). Mountain snow was more directly appreciated in the sweltering cities of the plains in the days before refrigeration. Finally, the mountains have been major sources of raw materials for other areas — obsidian from the gigantic flows of volcanic glass on Memrut Dağ in eastern Anatolia, copper from near Anārak in central Persia, and building stone from practically every individual mountain.

## The Plains

Separated by the mountain ranges, and to some extent caught up within them, are plains of varying size, height and form. Between the Taurus and the Pontic Mountains lies the so-called Anatolian plateau at an average height of about 1,000 m. It is really a series of enclosed lake basins, many of them now dried out except seasonally, when they receive the drainage of their surrounding hills and the more distant mountains. They are separated by rolling hills, except in the south-east, where the general monotony is broken by some groups of largely volcanic mountains. Mean daily temperatures in July exceed 20°C, but mean daily temperatures in January are below freezing. Precipitation is less than 400 mm in most areas, some of it falling as snow which may lie for more than ten days in the year. The total amount of precipitation varies from year to year. *Artemisia* (sage-bush) steppe may have been typical of the 'natural' vegetation, but some thin woodland may have existed in higher damper areas. Somewhat similar conditions are found in central Persia (900–1,500 m), although mean daily temperatures in January exceed 0°C and the daily mean in July is above 30°C. Annual precipitation totals less than 300 mm and is frequently below 100 mm, with an inter-annual variability of 50 per cent (Brichambaut and Wallén 1963). Snow may cover the ground from 80 to 100 days in most years. The most striking features of the terrain are the Dasht-e-Kavir and the Dasht-e-Lūt. The former lies at about 1,000 m above sea-level, and consists of former lake beds, encrusted with salt-covered mud, and great spreads of gravel washed from the neighbouring mountains. The Dasht-e-Lūt is a lower but similar tract of country, except that there the salt mud has been eroded into fantastic shapes recalling the devastated 'City of Lot'. Halophytic vegetation, sparse in trees, was probably characteristic before large-scale human interference. Further to the south-east, a freshwater lake still survives in the basin of Sīstan, though its extent varies annually. At the opposite

end of Persia, in the north-west and reaching into the angle between the
Elburz and the Zagros Mountains, is a monotonous rolling plateau at about
1,000 m. Cold and snowy in winter, mean daily temperatures in the area
during July range above 20°C, while annual precipitation is about 200–400
mm.

South and west of the Anti-Taurus and Zagros Mountains lies a high
(300–400 m) and stony lava plateau, crossed on the north-west by the
Euphrates and on the north-east by the Tigris. Annual precipitation is
400–600 mm and mean daily temperatures for July often exceed 30°C;
those for January are 5°C. The Mesopotamian Rise, as it has been called
in recent times (Raisz 1951), falls away in an irregular escarpment to,
on the one hand, the level limestone plains of northern Syria and, on
the other hand, Upper Mesopotamia or 'The Island' (Dhazīrat/Jezira) between the Tigris and the Euphrates. Rivers cut through the escarpment
in gorges. The steppe of northern Syria is a relatively dry (200–400 mm),
rolling plateau, broken by comparatively low ridges trending south-west
to north-east. They trap sufficient precipitation to water Palmyra (Tad-
mor) and a string of lesser-known oases below them. Although some
woodland may have existed previously in such relatively favoured areas,
*Artemisia* steppe was probably extensive. Pistachio and juniper trees may
have dotted the steppe, while acacia and *Genista* may have formed a shrub
layer, at least in some places. The steppes end against mountain chains
in the north-west and the basaltic masses of Hauran, el-Leja and Jebel
ed-Drūz in the south. In these relatively elevated areas, precipitation increases to more than 300 mm, but the inter-annual variability is greater
than in the north (30–40 per cent). In general, precipitation declines with
distance south. Wadis cutting down towards the Euphrates score the eastern
edge of the Syrian steppe; in the south they are so numerous as to give
this tract of country its current name of al-Wadiyan ('The Wadis'). Further towards the south-east the steppes merge first into the rocky plateau
of Hajura and then into the broad limestone shelf of al-Hasa, an oasis
separated from central Arabia by low, discontinous escarpments.

Upper Mesopotamia is almost a plateau, sloping gently southwards
from below the escarpment marking the edge of the Mesopotamian Rise,
but it ends in a line of cliffs marking the beginning of the vast combined
delta of the Tigris and Euphrates — Lower Mesopotamia. The plateau
merges into the Syrian steppes on the west, but on the east it extends
to the foothills of the Zagros Mountains. Mean daily temperatures in
January range between 5° and 10°C, while elevation mitigates those of
summer; in July temperatures vary between 20° and 30°C. Precipitation
rises from 600 to 1,000 mm, with an inter-annual variability of 30–35

per cent. Much of the 'natural' vegetation may have been woodland, possibly a mixture of pistachio and juniper, grading southwards into open steppe dominated by *Artemisia*. In this sub-region the Tigris and Euphrates are joined by copious tributaries from the east. The Greater and Lesser Zab flow from the Zagros into the Tigris. They are more powerful and unpredictable in their flows than the Belikh and Khābūr, which join the Euphrates from springs rising along the edge of the Mesopotamian Rise. Largely in consequence, the Tigris is the more copious and swifter-flowing stream, but its regime is much more uncertain and feared, especially in winter, than that of the Euphrates. The annual flood on the Tigris reaches its peak in March or April, a full month before it is attained on the Euphrates, and at this time its flow may be as much as nine times the size (3210 $m^3$/sec at Fatḥah 1931–36) (Beaumont 1981) of the mineral flows of September and October. The Euphrates rises a little in the early autumn, not so much from the onset of the winter rains over its catchment as from a slight reduction in the rate of evaporation when the heat abates. The effect of winter rains becomes noticeable from November onwards, but it is melt-water pouring down from the snowbound mountains of Armenia which brings about the main flood, whose 'foaming breakers' (Nābigha, quoted by Gibb 1963, 26) are apparent in Upper Mesopotamia in March and reach their peak (2,380 $m^3$/sec at Hit, 1932–66) (Beaumont 1981) in April and May at a level some nine times that of low water in September. Despite the volume of water available in the rivers, the upper valleys are too narrow and their flood plains too discontinuous to allow the large-scale development of flow irrigation. In addition, much of the land between the two rivers is far too remote and too high above the channels for much use to be made of lift irrigation before the availability of power pumps.

As it enters their combined delta, the Euphrates flows some 10 m above the level of the Tigris. There was thus considerable scope for the development of flow irrigation on the gentle slope between them. Irrigation was also helped by two further characteristics of the area. The finely textured alluvial surface of Lower Mesopotamia was, and still is, etched with the lines of abandoned channels, for the two rivers have shifted their courses continuously. It is also pitted by numerous depressions resulting from the lateral accretion of bed-load. The build-up of alluvium has been estimated at a minimum of 20 cm every 100 years (Mitchell 1958). In the far south of Lower Mesopotamia the two rivers finally mingle in a vast tract of marsh and swamp where they are joined by the Kārūn from the east. The mouth of the joint outlet, the Shatt al-'Arab, 'is very difficult to distinguish, as there are no sea-marks on its bar, which has more than fifteen leagues of sand banks' (*Travels of Abbé Carré*, III: 838). The whole sub-region receives 100–300 mm of precipitation in the year, almost entirely in the winter, when mean daily temperatures in January exceed 10°C. Summers are dry and hot, with mean daily temperatures in July exceeding 30°C.

Similar climatic conditions prevail in Arabia. The mean annual range of temperature is between 15 and 20°C, compared with 20–25°C in Mesopotamia. Mean daily temperatures in July exceed 30°C, but mean daily temperatures in January lie between 10 and 20°C. Mean annual precipitation, however, is less than 100 mm, while its inter-annual variability is 60 per cent. Permanent vegetation was almost certainly sparse in the past. It was probably confined to steppe species, including *Rhanterium eppaposum* and *Stipa tortilis* (Vesey-Fitzgerald 1957b), but even now almost the slightest shower of rain produces a short-lived carpet of annual and perennial plants. The terrain may be described as an immense but gentle slope running from the high mountains bordering the Red Sea to the salt flats and marine terraces along the Gulf. The surface is broken, however, by an immense granite dome averaging 100 m in height, faulted and spread with tracts of lava (the Arabian Shield), and then by a series of westward-facing escarpments arranged like huge strung bows. To the north and south of the escarpments lie vast stretches of sand, the Nafūd and the Rub' al-Khali. The Nafūd occupies an immense oval depression, '7 or 8 days" journey across (*Letters of Gertrude Bell*, 1927, 337). Dunes within it are heaped as high as 150–200 m and they are clearly visible from a distance, fringed with streamers of sand blowing from their crests when the wind rises. Many have a horseshoe shape, with the ends pointing north-westward in the direction of the prevailing winds which, generation after generation, have here piled the sand eroded from the Tubeiq Hills in the process of sculpting their curious mesa-like forms. Narrow belts of sand connect the Nafūd with the sand-filled depressions of the Rub' al-Khali. 'The Abode of Emptiness' itself is located between the mountains of southern Arabia, 'a bitter, desiccated land' (Thesiger 1959, prologue), and the escarpments of the north-centre, and impinges upon both the Indian Ocean and the shallow waters of the Gulf. Although broadly similar in character to the Nafūd, the scale of the Rub' al-Khali is altogether greater, with the largest, most fiercesome, red and tan-coloured dunes lying in the south and south-east.

By contrast, much of the Western Desert of Egypt, where precipitation is as low as over much of Arabia and lower than over the mountains of the Eastern Desert, is a low-lying plateau with an elevation of less than 200 m. Mean daily temperatures for July range upwards from 25°C to more than 30°C, while mean daily temperatures in January are above 10°C. The plateau merges with the immense sand sea of the south-central Sahara, but is itself characterised by tracts of bare rock and loose stone. Grey colours predominate in the north, where limestone is the base rock, but warmer tans and browns are characteristic where sandstone forms the surface in the south. The surface is dissected by shallow wadis draining to mud and salt flats. A few large, steep-sided depressions break the general monotony. Nile silt covers the floor of one of them, the Faiyūm, which is connected with the Valley.

The Nile itself emerges from the cataract zone south of Aswān into a sharply defined notch about 10 km wide.[11] On one side, it is bounded by the plateau of the Western Desert; on the other, it is separated from the Red Sea by the mountains of the Eastern Desert in a belt some 130 km wide. The steep-sided flood plain contrasts with those of the Tigris and Euphrates. Its cross-profile is convex, since it has been built up through the steady accumulation of silts or clays carried in suspension by flood water, rather than through the lateral accretion of bed-load within shifting channels, the process characteristic of the other two major rivers of the region. The average rate of accumulation has been estimated at some 10 cm every century (Butzer 1976). In broader sections of its valley, however, the Nile developed a secondary branch. Natural levees rise 1–3 m above the level of the flood plain, which is also crossed by the low sinuous humps of their predecessors. Together they enclosed a net of flood basins estimated to vary in size between 18 km$^2$ and 318 km$^2$ in their natural state.

Before the completion of the Aswān High Dam (1968), the striking feature of the Nile's regime was its majestic and predictable annual flood. Its arrival in Egypt was heralded in late June by the appearance of 'green water', coloured by the algae flushed from the Sudd by the rising White Nile. The water turned to a rich chocolate colour as the flow was enhanced by sediment-rich water entering the Blue Nile and the Atbara in Ethiopia. By the middle of August the Nile began to overtop the levees in southern Egypt and to spread relentlessly northwards to inundate the whole valley within a period of 4 to 6 weeks. Interestingly, the ancient Egyptian hieroglyph for the inundation season was a pool with lotus flowers (Finegan 1979, 182). In September the mean flow was 8,180 m$^3$/sec at Aswān, compared with 698 m$^3$/sec at low flow in May (Beaumont 1981). Both the flow and the height of the flood varied from year to year. In years of low flood some of the higher, more central basins would not be flooded at all, while in high-flow years even the highest levees might be briefly overtopped. In early October, the first basins in Upper Egypt would be dry again through a combination of falling head, natural flow back to the main channel, infiltration, sinking ground water and evaporation. Before the end of November all but the lowest-lying basins in the north would have been drained. While swamp plants like reeds and rushes were probably characteristic of such areas, a sub-tropical *galeria* woodland of tamarisk, aspen, oleander and acacia may be postulated for the levees.

The Nile flows from its notch across a delta now covering some 22,000 km$^2$, with a base along the Mediterranean Sea of some 220 km and a breadth of about 170 km. Its appearance is not unlike that of the back of a leaf. The major distributaries of the river and the natural levees form the veins. They are separated by alluvial flats and basins. Alluvium accumulated at 3 cm per century. Extensive areas of the Delta lay above flood level in the past. The flood was much lower in any case than in the Valley, since it was dispersed through the distributary system. The main distributaries form cuspate sub-deltas on the coast, where beach ridges and spits are characteristic, the latter cutting off a sequence of lagoons from the sea. The whole sub-region is still crowded with water birds during the migration season.

In retrospect, the Delta and the Valley of the Nile can be seen as offering exceptional opportunities for man, but some of these were shared, at least in kind, by other lowland areas of the Middle East, notably those crossed by the Tigris and Euphrates. The abundant aquatic life of rivers, lakes, swamps and marshes was important in early times. Grazing was available for wild and domesticated animals, while the potential for irrigation was considerable. Movement was always relatively easy along the great river valleys. It was probably even less difficult across the plains, except where tracts of bare rock and gravel, sand dunes and mud flats impeded advance. While goods could readily be transported and ideas exchanged in such environments, relative freedom of movement meant exposure to maurauding bands and invaders. Given the will and the steady exercise of authority, governments could control the plains more easily than the mountains, but when their efforts were relaxed they often became more insecure. Indeed, in diachronic perspective, the lowlands of the Middle East emerge as sub-regions of change. Not only are seasonal alterations as dramatic as in the mountains, from snow-field or verdant green to the parched yellows and brown of summer, but, on the time-scale of years, political shifts were frequently rapid. Similarly, the plains are sub-regions of physical hazard. Fierce winds blow across them, generally from the north. Bitterly cold in winter, in summer the winds are hot and dust-laden, while outbursts of khamsin-type in spring and autumn dry the skin and crack paper, and conditions are altogether unpleasant. The heat can be intense; where humidity is high, life is often far from comfortable. Dust devils dance in the distance on certain days. In some years, deep clouds of locusts wreath and flicker on their paths of devastation. Valleys and plains are frequently exposed to another kind of disaster, flood. Nowadays the principal causes are heavy rain and spring thaw, but a potent factor in the past was probably increased run-off occasioned by extensive forest clearance. Poor drainage encouraged salination, and the survival of pools of standing water provided breeding grounds for malarial mosquitoes.

## Coastlands and Seas

The sea penetrates deeply into the region, and along the coasts the alternation of mountains and plains creates a suite of distinctive environments. Along the southern shore of the Black Sea the coasts are steep, except where the Rivers Kızıl Irmak and Yeşil Irmak have broken through the mountains and built deltas into the sea. Communications with the interior are generally difficult, while the shore itself is exposed to violent storms. None the less, a number of small ports have existed since ancient times. Mean daily temperatures in January are above freezing, while those in July are 22°C. Largely as a consequence of the mountain backing, precipitation exceeds 600 mm and rises to more than 1,500 mm in the east. The Black Sea is linked to the Aegean and Mediterranean Seas through the drowned valley of the Bosporus with its powerful surface current, the Sea of Marmara and the narrows of the Dardanelles, swimmable by a Leander or a Byron. Penetration of the interior is not too difficult here, but the major east-west routeway is forced along the valley of the Sakarya River by more rugged, mountainous terrain lying north and south. Further south still, broad rift valleys allow the deep penetration of Mediterranean influences and contrast with the rugged mountain masses which separate them (pp. 10–12). The coast in this section of Anatolia abounds in sheltered inlets, but they have suffered progressive silting with material brought down by the major rivers. In the south-west corner of the peninsula, where the coast cuts across the grain of the relief, a series of bays and inlets has developed. Silting has proved less serious, but access to the interior is limited and penetration arduous.

On the southern shore of Anatolia, picturesque mountains frequently descend, shoulder after shoulder, directly into the sea, though a number of small bays offer shelter. High mountains, exposed to the cyclonic tracks, ensure that mean annual precipitation exceeds 1,000 mm. The ranges of the Taurus Mountains embrace two coastal plains. In the west is the wedge-shaped plain of Antalya, which rises inland in three great steps. Precipitation is still relatively high, while proximity to the sea moderates winter temperatures so that the January daily mean is 11°C. In July, however, temperatures rise to 28°C. More than 300 km to the east, the mountain ranges wrap around the extensive alluvial tracts of what the ancient Greeks called *Cilicia Pedias*, the deltas of the Seyhan and Ceyhan Rivers, territory separated from a higher inland basin on the Ceyhan River by a line of low hills. Mean daily temperatures here in January are below 10°C, but in July they exceed 20°C in the coastal sections and increase inland to around 30°C. Precipitation is between 600 and 1,000 mm a year. Although the coastline itself is predominantly smooth, small irregularities have existed and were exploited in the past as landing places and anchorages.

Good potential for port development existed along the Levant coast, especially north of Mt Carmel, where a number of small inlets and offshore islands offered shelter. In addition, there was the stimulus of relatively short, if sometimes arduous, crossings through wooded hills and mountains to the Euphrates and thence to Mesopotamia. The well-watered coastal plain is discontinuous but the coastal route is negotiable northwards from Egypt at least as far as Mt Carmel, where the plain of Esdraelon (Yizre'el) affords access to the Jordan Valley. From there the Yarmuk Valley leads out on to the steppe lying east of the rift. Only poor roadsteads exist in the south-east corner of the Mediterranean basin until the Nile Delta is reached. Here the lagoons and spits offered rather better opportunities for port development. Precipitation along the Levant coast declines progressively southwards. Mean daily temperatures increase in the same direction, though in winter the coastal strip contrasts sharply with the cold interior which is often snow-covered. In summer mean daily temperatures are highest in central areas, reaching 27°C in July.

The Mediterranean Sea is effectively tideless. Surface currents in its eastern basin move anti-clockwise under the prevailing north-westerly winds, which are especially strong and boisterous during the summer. The major current bifurcates either side of Rhodes, with a southerly arm turning south-westwards towards Crete and a northern arm running between the islands fringing the coast of Anatolia. In the neighbourhood of Chios and Samos this branch mingles with the current sweeping out of the Dardanelles. In general, then, the pattern of winds and currents is favourable to navigation in the eastern Mediterranean. The only major hazard is the distant lee shore of north Africa.

The Red Sea is some 2,200 km long and its entrance is constricted to a stormy strait ('The Gate of Lamentation'), some 26 km wide. Unlike the Mediterranean, the Red Sea is tidal and this increases the hazards to navigation. Coral reefs, especially luxuriant and numerous in the central section, parallel the shore and divide the sea lengthways into three principal channels connected by a number of transverse gaps. Access to the northerly Gulfs of Suez and 'Aqaba is impeded by more reefs and islands. In fact, so dangerous is the entire length of the Red Sea that down to recent times traditional craft preferred to sail only by day (Cortesão 1944; Villiers 1940, 6). Strong southerly winds are normal from October to April roughly as far into the sea as latitude 20°C, and gales are frequent (Naval Intelligence Division 1946). Mean January temperatures exceed 20°C. From May until September the predominant wind blows from the north-north-west. Summer temperatures are high, with the daily mean in July exceeding 30°C. In its northern section, the Red Sea is characterised by predominantly north-westerly winds throughout the year; they often reach gale force in summer. Along its entire length the Arabian shore of the Red Sea alternates between deep inlets, generally associated with gaps in the reefs, and discontinuous tracts of sandy plain which vary greatly in width. Access through the mountains to the interior is generally difficult, as it is on the Egyptian side.

Aden, 'the key not only of Arabia but of all the strait' (Cortesão 1944, 15), is the only really good harbour on the whole of the southern shore of Arabia, though inlets in the cliffed and mountainous coast afford a suite of roadsteads. Between November and March the north-east monsoon dominates the Arabian Sea, but yields to the south-west monsoon between May and September. Excellent shelter for ships is provided by deep inlets around Muscat and the Ra's Musandam peninsula in the eastern corner of south Arabia, but they are virtually shut off from the interior and, in the summer months, trap heat and humidity.

The Gulf is some 800 km long, its form a reversed version of the Arabic letter *raa'*. Like the Red Sea, its entrance (the Straits of Hormuz) is comparatively narrow (80 km) and it is congested with islands. Shores shelve gently into the shallow waters on the Arabian side. Sweet water comes to the surface in a number of places, while small indentations provide possible havens for ships, though the approaches are hazarded by shoals and reefs. A discontinuous and narrow coastal plain — largely waterless — runs along much of the Persian side. Salt marsh, mud volcanoes and sand dunes render this coast even more inhospitable, while the backing ranges of the Zagros Mountains restrict access to the interior. There are few good harbours inside the Straits of Hormuz compared with the situation outside. North-westerly winds predominate throughout the year but there is enough sea room to make navigation 'plain and easy', if rather 'boisterous and tedious' in the summer when the winds are particularly strong (G/29/25 1790, 284b). Perhaps this was why the Assyrians called it the 'fearful sea' (Lukenbill 1927, 45–6). Summer temperatures are high and the daily mean for July exceeds 30°C; the humidity is oppressive. Mean daily temperatures in January are greater than 10°C. Precipitation is generally below 100 mm a year on the Arabian side, but reaches 200–400 mm over the mountain-backed, north-eastern shore. Outside the Straits of Hormuz, in the Gulf of Oman, winds alternate in phase with the monsoon so that they are predominantly northerly in winter and south-westerly in summer. The particular set of the wind at any time has great bearing on the negotiability of the awkward S-bend of the strait, and delays were frequent in the days of sail (*Travels of Abbé Carré*, I, 117). Even so, with favourable winds the voyage from the head of the Gulf to Aden could take only 8 days (Leemans 1960, 5 n.4).

In the north of the region is a vast lake, the Caspian Sea. Its coasts of spits and lagoons are created by long-shore drift moving in an anti-clockwise direction and spreading material washed down from the Elburz Mountains by several short and fast-flowing rivers. The coast is gradually extending northwards. The resulting alluvial plains are drenched by an almost continuous rainy reason (600–1,500 mm per annum) and flooded by snow-melt from the mountains which virtually shut the area off from the Persian plateau. In winter the mountains also provide protection from the extreme cold of the interior and mean daily temperatures in January lie between 5 and 10°C. By contrast, mean daily temperatures in July are between 20 and 25°C. Under such near-tropical conditions, it is hardly surprising that the 'natural' vegetation seems to have been a mixture of swamp and jungle.

Off the coasts of the region are islands, more or less isolated worlds which have serviced the life of the adjacent mainland. The largest is Cyprus (9,251 km$^2$), a 'miniature continent' located in a strategic angle of the eastern Mediterranean only a day and a night's sail from Tyre (al-Mukaddasī in Cobham 1908, 5) and much the same from Anatolia. It is characterised by a variety of terrain, the luminosity of Greece and a mild climate. It possessed timber and was rich in copper. Rhodes (1,400 km$^2$) was once something of a rival: lying off the south-west angle of Asia Minor and at the entrance to the Aegean Sea, it too acted as a node in the communications of the eastern Mediterranean, even as an offshore commercial and banking centre for a time in antiquity (p. 136). The northward chain of islands fringing the western coast of Anatolia often served as a set of stepping-stones between Europe and Asia but, as small and precious jewels, they were always at risk. Both the Bahrain archipelago (6,000 km²), lying in an angle of the Gulf between the Qatar peninsula and mainland Arabia, and little Kharj, in the north-east corner of the Gulf, have shared something of the same role. In a similar way, location often imposed strategic significance upon Hormuz and Perim, lying at the mouth of the Gulf and the Red Sea respectively and at the ends of sequences of island steps within each strait.

Islanders frequently, but not always and everywhere, exploited the resources of the sea around their homes. Bahrain is the historic centre of pearling in the Gulf, largely because the oyster-beds lie in warm shallow water north and south of it. Pearls are also recovered from the Red Sea, especially near the Dahlak Islands. Sponges were once prolific near the tiny island of Arwad, off Tartūs in the eastern Mediterranean, but were especially abundant near Bodrum and Marmaris in south-west Anatolia. The best fishing grounds were found off the mainland, too — between Trabzon and Sinope on the Black Sea, in the Bosporus, off the entire west coast of Anatolia, in the Gulf, along the northern shores of the Gulf of Oman, and in the waters off southern Arabia. As well as providing the home ports for the fishermen, well-watered coastal plains offered potential for the development of cultivation and pastoralism. Reefs and shoals, cliffs and sudden squalls, however, were always dangerous to ships in coastal waters and added a macabre richness to the harvests of the sea (p. 80). Everywhere the sea-shores were exposed to assault by seafarers, driving wind and devastating flood.

## Size and Rhythm

Until comparatively recently, few of the region's inhabitants can have grasped either its scale or its shared experience. The land area is equivalent to about 82 per cent of that of the United States and is 30 times the size of the United Kingdom. The effect of size may be grasped from a consideration of the times taken to cross the region along different routes. For most of the period the fastest means of travel were the racing camel and the post horse. Late in 1799 Lord Elgin, then British ambassador to the Porte, estimated that letters could be carried between Istanbul, on the Bosporus, and Basra, near the head of the Gulf, in about 25 days via Aleppo (FO 78/24). On 2 December 1801 a *tatar* (messenger) arrived in Cairo from Istanbul in the same time (Wittman 1803, 387). In the middle of the ninth century, the former postal official Ibn Khurdādhbith estimated that the post could cross Arabia from Basra to Mecca in about the same time, while Nishapur in north-eastern Persia could be reached in 43 days from Baghdad. The journey from Cairo to Aleppo through Damascus took 22 days (Hitti 1964, 324).

Caravans and armies, of course, moved more slowly than couriers. John Sanderson took 40 days to travel from Istanbul to Aleppo with the suite of a provincial governor in 1595 (Foster 1949, 62–3). In the first half of the eighteenth century, large heavily-laden caravans covered the distance from Aleppo to Basra in 45 to 70 days, though smaller ones took about 25 days (Carruthers 1928, xxiv–xxxv). A through-time of 85–110 days between Istanbul and Basra compares with the 90 days taken by an embassy to follow the Royal Road of the Persians from Susa to Sardis (Herodotus V, 52–4) and the 84 days taken along a slightly different route by Xenophon's Ten Thousand on their march to central Mesopotamia. The subsequent retreat of the Greeks to the Black Sea was managed in around 81 days, even though it involved crossing some of the most rugged country in the Middle East when deep in snow (Xenophon). The French merchant Tavernier travelled with caravans which took 41 and 67 days to cover the route from Üsküdar (Scutari), on the Bosporus, to Tabrīz, in western Persia, by way of Amasya, Erzurum and Yerevan (Tavernier 1684, 3–15). In the early sixteenth century the journey from Hormuz, near the mouth of the Gulf, to Tabrīz was estimated to take 50 days by camel and that to the province of Gilan, on the southern shores of the Caspian Sea, more than two months (Cortesão 1944, 20, 23).

Sailing times allow the addition of another dimension to our perception of the region's size. A voyage from Alexandria to Constantinople took 18 days in the eleventh century (Goiten 1967, 326), compared with the 23 days taken by Sanderson on a galley in the opposite direction in the autumn of 1585 (Foster 1949, 39). A little earlier, Tomé Pires learnt that Jiddah was about 10 days' voyage up the Red Sea from Aden, 'with a (favourable) wind' (Cortesão 1944, 11). Bahrain was only 4 or 5 days' sail with a favourable wind from Hormuz (Cortesão 1944, 20), while in the spring of 1674 the voyage from Nakhīlu, 350 km to the west of Hormuz, to the head of the Gulf took 10 days (Fawcett and Burn 1947, 831–8). Finally, a ship carrying a party of British officers took 10 days to make the passage from İskenderum to Rhodes in June 1802 (Leake 1802).

Just as wind and weather usually determined actual travel times by land and sea, so the climate generally imposed distinct seasonal and diurnal rhythms on the lives of the people. Practically everywhere cold, snow and bad weather made winter virtually a closed season for all but the most determined traveller. Farmers retreated into the snug security of their houses in the snow-covered, frost-bound parts of the region, while nomads either sheltered with their herds in the milder plains and coastal districts after withdrawing from the mountain pastures or, in Arabia particularly, began to venture into grazing areas freshened by the autumn rains. Other forms of travel were generally reduced. Fewer caravans crossed the snow-covered mountains. In the Mediterranean sailing ships and galleys seldom ventured to sea between the end of October and the beginning of May. Everywhere the change of season was heralded as much by the arrival of the first major caravans and ships as by the appearance of migrant birds. The steady winds of summer not only drove the ships but were also harnessed in Sīstan and the Aegean Islands to power corn mills during what the West came to call the Middle Ages. Summer, when the days are 13 to 15 hours long depending on latitude, is the time for harvesting and threshing; winter and spring, when the rain has softened the baked soil and daylight lasts 9 to $11\frac{1}{2}$ hours, is the time to plough and sow. Seasonal flow in winter and spring worked mills on many streams. Mounting aridity forced the nomads back upon secure water supplies in Arabia and Egypt, but elsewhere sent them high into the mountains in search of grazing. By contrast, villagers from the mountains sought work in the plains during the few weeks before their own crops ripened and their meagre harvests were gathered in.

Heat and glare discourage work in the middle of the day, at least in summer. They put a premium on shade and coolness, making it worthwhile, in the days before air conditioning and refrigeration, to erect wind towers for cooling house interiors in southern and eastern districts and to transport snow over immense distances. People retreated to their houses to sleep away the afternoon, following the invariable early start to their day. Early morning and evening, 'when the plains appeared/ in silk brocade, the hills a veil of water' ('Abdallāh ibn al-Mu'tazz in Wightman and al-Udhari 1975, 81), are the most delicious parts of the day, savoured by those who have the leisure. Cool nights, with clear skies and ample light from moon and stars, made the dark hours a favoured time for land travel, especially in summer, despite the discomfort of comparatively low temperatures and broken sleep. Men must have contemplated the splendour of the heavens, and long before the formulation of astronomical systems, the regular movements of the stars were used to identify seasonal rhythms.

## Notes

1. Unless otherwise indicated, the climatic information is taken from Beaumont, Blake and Wagstaff (1976), 49–84.

2. The reconstruction of broad patterns of 'natural' vegetation is derived from Butzer (1970).

3. *Pinus nigra.*

4. *Pinus nigra* and *P. brutia.*

5. *Quercus cerris, Q. pubescum.*

6. *Cedrus libani, Abies cilicia, Juniperus axieba.*

7. *Quercus persica.*

8. *Fagus orientalis, Abies bornmnelleriana.*

9. *Quercus macrolepis, Q. ilex, Q. coccifera.*

10. *Pinus brutia.*

11. The topography of the Nile Valley is derived from Butzer (1976) and Issawi (1976).

# 3 DOMESTICATION AND THE BEGINNINGS OF FARMING

## Antecedents

Climate and vegetation were very different 40,000 years ago, towards the height of the Würm glaciation in the Alps. Large tracts of the Middle East appear to have been under steppe dominated by *Artemisia* and *Chenopodium* (goose-foot). Trees survived only in the mountains near the Mediterranean Sea and possibly around the southern shores of the Caspian (Zeist and Bottema 1977 and 1982). Although precipitation was low, there was enough moisture to nourish shallow lakes in the interiors of the Western Desert, Anatolia, Persia and Arabia and to activate the dendritic drainage system of the latter (Chapman 1971; Cohen and Erol 1969; Erol 1978; Holm 1960; McClure 1976; Roberts 1982a; Roberts *et al*. 1979). Temperatures appear to have been 5–7°C lower than today. Winters were particularly harsh. The snow-line, which today is above 3,000 m in the Zagros Mountains and north-eastern Anatolia, was 1,000 m lower in the western parts of the region and 800–900 m down in the Elburz Mountains (Butzer 1972, 296–7; Wright 1961). Strong winds may have been responsible for piling up some of the sand dunes along parts of the eastern coast of the Mediterranean. Sea-levels were much lower than at present. The Gulf did not exist, and the Red Sea had been reduced to a long narrow lake. The Black Sea was fresh water (Eisma 1978; Larsen and Evans 1978; Sarnthein 1972).

By this period *Homo sapiens sapiens* had emerged. He tended to live in single monogamous groups (Darlington 1969, 48–9). These used caves and rock shelters on a more than causal basis and in a migratory pattern that was probably seasonal in character and involved temporary camps. Isaac (1970, 26–7) has even suggested that notions of property can be inferred from the differential use of caves. If these extended to territories, with their resources of plants and animals, they may have been important in the transition to food production. There is some evidence that the predictability of the seasons was known and, as in southern Europe at the time, it would have been related to the movements of game (Solecki 1974). Man seems to have been dependent upon hunting a limited range of large animals (onager, aurochs, red deer, wild goat and gazelle), but the gathering of vegetable foods is likely to have been important to his

29

diet. Fire was in common use. Tools were made from long flakes of stone struck from a core. Bone awls and a particular type of scraper suggest that baskets were woven. A trend towards regional specialisation has been detected, and this may indicate that human groups were becoming less mobile than formerly (Redman 1978, 65).

The use of smaller and neater blades, often in composite wooden shafts, is evinced from across the entire region some 16,000 to 20,000 years ago. This suggests that the people of that time had developed greater mastery over their materials and become more efficient in their exploitation of resources. Hunting appears to have intensified and to have involved a wide variety of smaller animal species. More intensive use of small territories may be implied (Redman 1978, 67). The appearance at the site of Ein Aqev in the Negev about 17,000 to 18,000 years ago of the first basined grinding slabs indicates a shift in the preparation of food. In particular, it points to the grinding of nuts and possibly seeds (Kraybill 1977). Some 2,000 years later grinding slabs and hand stones were in frequent use on the edge of the Nile Valley, although hunting and fishing continued. On a few sites the grinders are accompanied by microliths, some of which have a silica sheen interpreted as indicating their use in 'sickles' for cutting grasses (Curwen and Hatt 1953, 4, 105). Despite the claimed recognition of barley pollen on one site, it is doubtful if such a winter-seeding plant grew in the Nile Valley at this early date. A variety of local grasses (*Aristida, Panicum turgidum, Eragrostis spp.* and *Cenchrus ssp.*) are much more likely candidates for harvesting (Clark 1971; Reed 1977).

Mortars and pestles are frequent on roughly contemporary sites in the rest of the Middle East. Although they may have been used for grinding the seeds of wild cereals, as many commentators have assumed, the pounding action typically associated with them may in fact suggest that their main purpose was the preparation of other foodstuffs, notably nuts and berries. Ethnographic studies reveal that in recent times acorn flour was made widely in the region and its production in the distant past is a distinct possibility (Bohrer 1972). Oak may have been a principal element in the woodland which seems to have been spreading over the western parts of the region in the period 18–12,000 BC as the climate ameliorated, though the advance of trees was delayed in the east and a cool steppe vegetation prevailed practically everywhere (Zeist and Bottema 1982). The region as a whole may still have experienced a precipitation regime different from that of today, though seasonal downpours are evinced (Vita-Finzi 1969a and b). Temperatures in northern areas appear to have been at least 5°C lower than today, while the snow-line was still comparatively low in the mountains.

Although man lived on open sites in Egypt and elsewhere, he continued to prefer caves in the more northerly parts of the region. Occupation was not generally long. There was, however, a particular concentration of sites in the coastal zone of Palestine, in what later became known as the Judaean Desert and the Negev. Sub-regional associations doubtless existed elsewhere but as yet they are less clear (Hours 1982). Some sort of contact across the region is shown by the presence on Levant sites of obsidian, which can only have come from the great glass flows in Armenia, and of shells which have been identified as originating in the Gulf, where marine transgression was in progress (Vita-Finzi 1978). Such long-distance contacts may have been important in the spread of innovations.

**Incipient Agriculture and Animal Husbandry** (Fig. 3.1)

Palynological evidence from a limited number of occupation sites in western Anatolia and the Zagros Mountains suggests that the spread of woodland was retarded in the period 10,000–6,000 BC. Since temperatures appear to have been rising towards their present levels, this must be attributed to 'greater dryness' than was found later. Precipitation levels lower than at present are not necessarily implied. Dryness could have resulted either from higher temperatures reducing precipitation effectiveness as far as vegetation is concerned or from a long period of drought during the summer months. Whatever the cause, a steppe-forest vegetation seems to have prevailed in many parts of the region until 4,000 years ago. At Köyceğiz in western Anatolia the present-day vegetation may have been established some 7–6,000 years ago. At Söğüt, however, oak and juniper appear to have been important around that date and it was not until some 3,000 years ago that pine (*Pinus nigra*) became dominant. A similar increase in pine and a corresponding decrease in cedar took place in the vicinity of Beyşehir, also in western Anatolia, around 5,850 years ago, while at Karamuk the dominance of pine and juniper appears to have alternated over a relatively short time-span (Zeist *et al.* 1975; Zeist and Bottema 1977 and 1982).

Conditions favourable to tree growth suggest that around 6,000 years ago the precipitation assumed its present characteristics, though minor fluctuations have occurred since. The shallow lakes of the Western Desert, central Anatolia, Persia and Arabia began to dry up (Chapman 1971; Cohen and Erol 1969; Erol 1978; Holm 1960; McClure 1976; Roberts 1982a and b; Roberts *et al.* 1979), while the snow-line retreated. The

Figure 3.1: Domestication

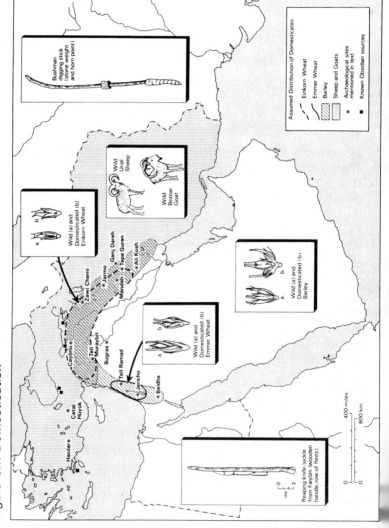

post-glacial rise in sea-level was rapid; the Gulf and the Red Sea began to assume their present forms. The Black Sea, however, was still an enclosed fresh lake, possibly connected to the Caspian Sea which seems to have risen by about 20 m at this time from the inflow of melt-water (Brice 1978b).

In the developing woodlands of Palestine and Syria groups of people with a shared culture now recognised as Natufian after the type site in Wadi en-Natuf, supported themselves by fishing and hunting (mainly gazelle), but some sort of plant food was obviously important to them. Grinding stones and mortars abound on every habitation site. Their importance is indicated in various ways. Some of the grindstones were fixed in the floors of huts or even cut into the rock in front of caves. Mortars became perforated with use and, in some cases, were then used to mark graves. Considerable effort must have gone into shaping them from such hard rocks as basalt. There are other indicators of the role of plant foods. The bone hafts of reaping-knives or 'sickles', grooved to take microliths, have been found, while some of the blades have the silica sheen suggestive of grass cutting. Carefully plastered pits indicate storage (Cauvin 1978; Hassan 1977). Most commentators have assumed that, taken together, the evidence points to the harvesting of wild cereals for human consumption. However, finds of actual grain come from later contexts, mainly from sites further east, and Bohrer (1972) has suggested an alternative explanation, as mentioned earlier. She points out that world-wide ethnographic evidence shows that in communities of gatherers, wild grasses were normally harvested for their seeds by scooping or beating movements. The cutting indicated by reaping-knives or 'sickles' stands out as exceptional. Ethnographic parallels indicate that cutting was normal only where animal fodder was required. Bohrer argued, therefore, that acorns could have been the staple in human diet and that wild grasses were initially cut to provide winter feed for livestock.

The dominant view, however, might be supported by pointing to the evidence for the deliberate locational preferences shown by Natufian sites. The distribution differs in important respects from that which immediately preceded it. Although there were settlements in the coastal plains and the Judaean Desert, as before, a distinct preference was now shown for the Judaean Hills, Galilee and the area of Mt Carmel. The claim has also been made that the sites are all located on basaltic soils or *terra rossa*, but frequently in places where a diversity of resources could have been exploited (Hassan 1977). Oak woodland probably covered these areas, but the locating factors might have been natural stands of wild cereals within it. If the seeds of wild cereals were indeed collected, there are two likely candidates. The first is two-row barley (*Hordeum spontaneum*), which is not only widespread in the dry-farming zone of the Middle East today, despite its sensitivity to extreme cold, but is also reported to form extensive natural stands in the Anti-Lebanon Mountains, the Jordan catchment, and the basaltic areas of the Hauran and the Jebel ed-Drūz. It is frequently found mixed with emmer wheat (*Triticum dicoccoides*). This is the most demanding of the wild cereals and is thought, on that account, to have occurred in restricted areas, such as that composed of eastern Galilee, the Golan Heights and the Hauran, but perhaps extending as far north as Mt Hermon. This limited habitat has suggested that emmer wheat might have been the prime attraction for the Natufian people (Harlan and Zohary 1966; Zohary 1969). Whatever the explanation, Natufian settlements were generally large (100–1,000 m$^2$), while their remains suggest a degree of permanency. For example, some of the mortars are too large to have been easily transportable and rebuilding phases are evident in the architecture (Cauvin 1978, 14–18; Redman 1978, 71–82).

Roughly contemporary sites have been found in the markedly different environment of the western foothills of the Zagros Mountains, near the Greater Zab River. They are mainly open sites, and stone foundations — as well as traces of rebuilding — point to occupation over a relatively long period. The presence of trough querns, grindstones, mortars and pestles shows that plant food was processed, while rare blades with silica sheen suggest that wild cereals may have been harvested. The hunting of wild goats and red deer was important (Redman 1978, 82–4). However, the large percentage of one type (80 per cent sheep) recovered from the upper layers of Zawi Chemi dated to c.9–8000 BC, together with their selective bias in age (50 per cent immature), has been 'interpreted as a statistical demonstration that the people were herding animals' (Redman 1977). The argument is that hunters would not discriminate, particularly over age, but kill whatever they could take and that this would probably mean a high percentage of adults. If the conclusions are correct, then the bones from Zawi Chemi are the earliest evidence for any form of animal domestication.

## Early Farming

The bones of sheep and goats showing distinct signs of morphological change indicative of domestication have been found on many sites across the region dated to about 7000 BC. By this date, emmer wheat, also displaying some of the morphological signs of domestication (*Triticum dicocoides*), was grown near a scatter of inhabited sites in an arc from southern Palestine (Beidha), through the Mesopotamian Rise (Cäyönü), to the foothills of the Zagros Mountains (Jarmo, Ali Kosh) and along a westward extension at least as far as the lake district in western Anatolia (Hacılar) (Fig. 3.1). Domesticated emmer-wheat was sometimes mixed with the other principal cereals of the region. At Jarmo and Beidha, for example, wild emmer was also being used, while at Hacılar it was wild and cultivated einkorn wheat (*Triticum boeoticum* Boiss. emend. Schiern; (*Triticum monococcum*). Elsewhere emmer was found in association with both wild and cultivated barley (e.g. at Tepe Guran, Tell Ramad, Jericho and Ali Kosh). Since the evidence for cereals is now unequivocal, the reaping-knives ('sickles') or sickle blades recovered from sites of the period were presumably used for harvesting and the querns and rubbing stones for turning the seeds into flour. The structures found at Hacılar and Jarmo, interpreted as ovens, may have been used to prevent germination by parching the grain prior to storage. The accumulation of deposits (8 m at Ganj Dareh, for example) suggests that settlements were becoming even more permanent. A change from building with wood and wattles to the use of mud is first known from Tepe Guran, but is evinced from many other sites of somewhat later date. In addition, rectangular architecture seems to have become increasingly preferred to the circular and oval forms common in earlier periods. At Jericho, in the Jordan Valley, sea shells indicate contact with the Gulf, while finds on that and other Levant sites show that obsidian was being drawn from Anatolia (Bökönyi 1976; Mellaart 1975, 28).

These communities were well advanced on the route which led to the fully-fledged farming villages of the fifth millennium BC and later. In the eighth and seventh millennia communities displayed much greater diversity, both in their methods of subsistence and also probably in their types of society (Stigler 1974). Thus hunting remained important. Pistachios, acorns, fruits and snails were collected. Legumes were apparently still gathered wild at Ali Kosh around 8000 BC, where they formed more than 94 per cent of the plant material recovered. Communities where cultivation and herding were manifestly well-established coexisted in the region with settlements which have yielded no evidence for either domesticated plants or animals but, none the less, had comparatively long histories of occupation (e.g. Buqras I, Tepe Asiab). Some communities may have continued with seasonal movements, perhaps taking domesticated animals with them in something like a pattern of transhumance (Oates 1973; Mellaart 1975, 70–90).

**Fully-developed Farming Villages**

During the seventh and sixth millennia BC two developments became apparent across the region, apart from Egypt and possibly central Arabia. Within those areas where precipitation is now sufficient to support the dry-farming of cereals, there was an increase in the number of sites yielding unequivocal evidence for domesticated plants and animals. Farming techniques may have been neither elaborate nor particularly secure, but the great majority of known communities now seem to have employed them. The techniques were even transferred to Cyprus, where there is no convincing evidence for Palaeolithic occupation. More remarkable was the second development. This was the spread of farming into areas lying outside the present limits for permanent dry-farming. Cultivation may have been possible in some places because a relatively minor fluctuation in precipitation allowed a temporary and localised expansion of cereals. In the case of sites lying well beyond the present limits of dry-farming, notably in Lower Mesopotamia and on the edge of the arid basins of interior Persia, cereal production implies the mastery of irrigation. The exploitation of seasonal floods would have been easy anywhere, while the construction of temporary dams cannot have been difficult, and both techniques may have been used in Syria–Palestine from an early date (Miller 1980). In the Mandali area, north of Baghdad, sites assigned to the Sawwānian culture (c.5600 BC and later) were seen to be arranged along low contours paralleling the neighbouring hills and to lie at right angles to the natural flow of water. During the fifth millennium there was a canal along this line, together with a number of smaller channels. Accordingly, the sixth-millennium pattern has been interpreted as associated with the use of channel irrigation (Oates 1973).

The main cereals cultivated across the region appear to have been those common in the early phases of farming already outlined — emmer and einkorn wheat, and two-row barley. They were joined by the hybrid six-row naked and hulled barley, as well as by bread wheat (*Triticum aestivium*), the latter the clear result of several generations of selective breeding (Renfrew 1973, 66–80). Large quantities of pea (*Pisum sativium*) and lentil (*Lens culinaris* Medik) have been secured from several sites. They are likely to have featured prominently in the diet, especially since they do not form large natural stands and some of the finds of lentils have been made in areas where the wild form does not apparently now grow (Zohary and Hopf 1973). With the domestication of legumes the basis of later cultivation patterns was laid, though fruit trees such as the vine, olive and fig were subsequently brought into local systems.

Grinding stones, querns and 'sickles' continued in use.

The domesticated plants were incorporated into farming systems with domesticated sheep, goats, cattle and pigs. While each species was probably domesticated in a different area and possibly at different times, all species were at home across the whole region, even in Cyprus where neither they nor red deer seem to have been indigenous. Combinations of species were normal, perhaps because these allowed for the supply of a variety of products. The sheep may have been kept principally for their wool, which would have become longer and heavier with selected breeding away from the wild sheep (*Ovis orientalis*), which has a reddish-buff coat above, with white below. Although conclusive evidence is more recent, goats may have been used to provide milk at this period. Cattle would have provided hides and horns, while the pig not only acted as a scavenger but was probably also an important source of meat. The emphasis in combining species varied from place to place, and all of the domesticates were not found together on a single site before the end of the seventh millennium (Bökönyi 1976). Once this step occurred, then the foundations of the Middle East's farming systems were virtually complete. The region's major beasts of burden, the domesticated ass, Bactrian camel and dromedary, were added in the fourth–third millennium BC, roughly at the same time as the horse (pp. 86–7).

Developments in settlement form and in technology are associated with the final transition to food production. Although oval and circular buildings continued to be erected, rectangular architecture became normal through the farming villages of the region. There was an associated change from free-standing buildings to a honeycomb structure in which walls were shared and groups of cells were separated by passageways and open courtyards (Flannery 1972). Pizé was the usual building material, though mudbrick made its first appearance during the period. Several settlements contained shrines or temples. Some of the figurines found are thought to represent a mother goddess. Most known settlements were large compared with their predecessors (p. 35). Thus, possibly the best-known early farming village, the site of Jarmo, covered an area of 1.4 ha and contained about 25 houses. Villages in central Mesopotamia were bigger still, 4–5 ha (Braidwood and Braidwood 1950; Oates 1973). But at least one larger settlement is known, Çatal Hüyük in Anatolia (12.8 ha). Indeed, the remains on this site, like those of Jericho at an earlier date (32 ha), have been described as belonging to towns or even cities. The principal reasons seem to include their relatively large areas of occupation, the degree of sophistication displayed in their culture and, in the case of Jericho, the possession of a stone wall and tower (Redman

1978, 78, 182–7). However, the functional differentiation of Jericho and Çatal Hüyük necessary for true urban status is far from clear. None the less, if archaeologists have produced two major claimants to urban status for the sixth millennium and earlier, geographers would probably conclude that there are likely to have been others. Major urbanisation, however, was to come later (pp. 97–105).

Pottery-making was a significant innovation attested from early farming villages. The earliest known evidence was recovered from Ganj Dareh, on the eastern slopes of the Zagros Mountains, and dated to before 7000 BC (Smith, P.E.L. 1975). The ware was very lightly fired and, since it probably would not have survived at all but for a massive conflagration, its discovery raises the question of whether pottery was used elsewhere, and at earlier dates, in contexts which have normally been described as aceramic. As well-fired pottery becomes abundant on settlement sites, so it shows steady technological and aesthetic advance. Techniques and decorative schemes, however, display considerable spatial and diachronic variation. These are evidence of human delight and inventiveness, as well as differences in skill. Once mastered, the techniques of pottery-making must have led to further important changes (Redman 1977). Pottery vessels allow food to be prepared in a greater variety of ways, anticipating the extensive culinary range of the region today. Experiments in metallurgy and glass-making would have become possible, preparing the way for significant technological advances. Copper was already being worked, and probably smelted, in the seventh millennium. Its use, together with the more widespread obsidian, and the limited provenance of ores, point to patterns of contact and hint at travel, even exploration. Finally, the discovery of spindle whorls on early farming sites and the disappearance of many types of leather-working tools indicate that spinning had been discovered and suggest that flax and wool were being made into clothes.

The region seemed set upon the path which, with hindsight, we know to have become the highway of future development. Yet at this very moment, there is evidence of recession from various sites, particularly in Palestine. Around 6000 BC many of the early farming villages appear to have been abandoned. There is an hiatus in the archaeological record lasting perhaps 1,500 years. It has been interpreted as the result either of total abandonment of the sub-region by its population in favour of districts further north or of a reversion to semi-nomadism, possibly involving pastoralism (Mellaart 1975, 68). The changes are not confined to Palestine. Buqras, a site on the Euphrates near the Khābūr, produced no grain in its lower levels but the people, who hunted wild cattle, bezoar (goat) and sheep, used 'sickles', querns and pounders. These disappeared from the later levels. In the third and final phase before the abandonment of the site, domesticated sheep and cattle may have been exploited. A similar change to nomadism immediately before desertion took place further south in Mesopotamia, at Ali Kosh (Mellaart 1970).

Evidence for the abandonment of farming shows clearly that food production was not a subsistence method which was bound to succeed. In the early stages it was obviously precarious. In the more southerly parts of the region, the transition was delayed until around 5000 BC. As long as swamps and shallow lakes survived in central Arabia, they provided support for groups of hunter-gatherers (Oates 1982). Gradually, though, increasing desiccation must have forced people into modifying their subsistence habits. Relatively affluent hunter-gatherer economies persisted in the Nile Valley until, in the fifth millennium, the plants domesticated further north were introduced and, quite suddenly, farming began (Butzer 1976, 4–7; Clark 1971). The reasons for the change are not clear. The resources on which the hunter-gatherers depended had not disappeared from the Nile Valley. It is unlikely that they had become seriously depleted, since various social controls would already have existed to maintain human populations below the maximum carrying capacity associated with particular exploitation techniques. The success of hunter-gatherer economies, the delay in the transition to food production in Egypt and central Arabia, and the evidence for recession in other sub-regions all raise the question of why farming developed at all.

## Explanation

Two points should be clear from the preceding sections. First, farming was not an inevitable development and, second, it took several millennia to emerge. Hunter-gatherer subsistence systems went through a series of transitions, largely without the people being conscious that a significant or permanent change had been made. Critical thresholds were passed in a period of 3–4,000 years, with the result that many communities in the Middle East became dependent upon producing their own cereals and herding domestic animals. It is not clear what the thresholds were or to what extent the process of change was completely reversible; the change to pastoralism in Mesopotamia may suggest that it was not. Nevertheless, for the series of transitions to end in food production, a number of preconditions were necessary. These were identified by Flannery (1969) and his scheme is largely followed here.

The first and most obvious prerequisite was the availability of the later domesticates. Whatever the experimentation with other plants and animals — and it seems to have been particularly rich in Egypt — the region's farming came to be based upon barley, wheat and legumes, sheep, goats, cattle and pigs. Even though the climate and vegetation were different from in the period when the first steps to domestication were taken, the domesticated animals were present in the fauna of the Late Palaeolithic of 40–20,000 years ago. Their reconstructed habitats were particularly wide some 10,000 years ago, but with a common area stretching from the northern coast of the eastern Mediterranean, across the Mesopotamian Rise, into the northern half of Persia and for some distance into Central Asia (Fig. 3.1). Wild cereals, on the other hand, spread from refuge areas with the post-glacial warming and the extension of oak-pistachio woodland. They became stabilised in the arc of the so-called 'Fertile Crescent': Palestine, the Mesopotamian Rise, the Anti-Taurus Mountains and the foothills of the Zagros Mountains (Fig. 3.1). The dispersal and germination habits of the wild cereals proved well-adapted to the present-day climatic regime as that began to stabilise some 10,000 years ago. The mature seeds readily break free from the brittle rachis soon after the end of the winter rains and are dispersed by the wind. Their shape allows them to become inserted into the drying ground sufficiently deeply to survive the long dry summers, while their large food storage capacity means that, like other grasses, they germinate and grow rapidly once the rains begin again. Since the region developed a low index of vegetational diversity, there was a relatively high probability of particular plants being found in pure stands — the cedars of Lebanon, for instance (Wright 1977b). Accordingly, wild cereals may have formed natural fields, as they do in parts of the region today (Harlan 1967).

The second prerequisite for domestication was a level of technical competence on the part of the people which would allow them to make a systematic and intensive use of resources. It is clear that they possessed this by 20,000 years ago in the form of different types of grinders. Reaping-knives and 'sickles', in use from about 14,000 BC, indicate that wild cereals were probably cut. This technique of harvesting, whether its purpose was to provide fodder for animals or seeds for human consumption (pp. 33), would gradually have assisted the recessive tendency in wild cereals for a tough spike axis to develop and thereby reduce their propensity to disperse almost at touch when mature. A low moisture content would probably also have been favoured and that would improve storage capability (Helbaek 1959). The argument, then, is that harvesting techniques modified the character of wild cereals and made them increasingly attractive as primary sources of food.

A third necessary pre-condition for domestication advanced by Flannery was the emergence of a 'broad spectrum' economy in which people exploited a wide variety of resources. It is argued that an intimate knowledge of the full range of plants and animals available and of their habitats would have been necessary for the experimentation which is presumed to have resulted in the selection of the domesticates. However, as Bohrer (1972) pointed out, modern hunter-gatherers may have an extensive knowledge of available resources but in normal times they depend upon a comparatively restricted number of particular items. It is possible that hunter-gatherers in the Middle East were in the same position and that gradually some of them came to depend upon wild cereals.

Long familiarity with the wild ancestors of domesticated animals has often been advanced as a prerequisite for their being tamed and then herded by man. Close observation and intimate associations would develop during hunting, it is suggested. Reed (1974), however, has argued against the dominant opinion. Although the habits and anatomy of wild animals would be well understood by hunters, a social bond leading to domestication is improbable where the basic aim of hunters was to kill their prey and that of the prey was to avoid death. Harris (1977) suggested that loose control over animals could possibly have been established by providing salt, for which cattle, and probably sheep and goats too, have a craving. Reed himself suggested that the keeping of pets was a probable path to domestication in communities which were already sedentary or largely so. The care of young animals would soon build a social bond between them and children, and ultimately between adults and animals. The question is not simple. Isaac (1970, 115) has drawn attention to 'the irrational springs of domestication'. In doing so, he revived the theory that the large-scale domestication of animals was the result of the veneration of large-horned specimens, combined with a wish to offer them in sacrifice to a fertility god (Isaac 1970, 103–15).

When seeking to explain the domestication of cereals, several persuasive commentators have developed stress models of change. In these, the operation of one or two factors is allowed to become of sufficient magnitude and duration to disturb the apparent equilibrium of hunter-gatherer economies. Climatic change is one such possibility, first advanced by Childe (1952), subsequently largely discredited, but recently revived (Lamb 1968; Wright 1977b). Its latest proponents see climatic change as facilitating the emergence of vegetational assemblages containing the eventual domesticates, and some use of these ideas has been made in previous pages. None the less, most scholars currently view population growth as the catalyst in a complex reaction which finally resulted in domestication. It is argued that population growth is an independent variable which must steadily create mounting pressure on resources and local carrying capacities. Devices which might have limited it are usually ignored, though Cauvin (1978, 139–42) has pressed the importance of culture change. Population pressure, it is suggested, would lead to an intensification of production by encouraging people to sow cereals in areas where they were already gathering them wild. An alternative response has also been envisaged. According to this version, some of the surplus population would move to areas away from the natural stands of cereals. In that case, subsistence would perhaps necessitate the transference of cereals into slightly different environments from those to which they were either already adapted or to which they could become adjusted without themselves undergoing drastic changes (Zubrow 1975, 114). The deliberate transference of cereals would complete the domestication process (Flannery 1969).

Unfortunately, various arguments can be mounted which undermine the security of the stress model with population growth as the controlling variable. Under conditions of stress, it is unlikely that people would have the foresight or the courage to hold back as much as a third of their harvest to provide the necessary seed for deliberate cultivation. It is just as probable that they would destroy the natural stands by careful seed selection and be forced to move on. Another argument against the primary role of population growth is that most hunter-gatherer communities in modern times developed social mechanisms which kept population numbers well within the supportive capacity of the environment. There is every reason to suppose that this was true in the remote past. The problem then becomes one of deciding how and why the constraints broke down.

A variant of the usual stress model was advanced by Bohrer (1972). It starts with the possibility that acorns were a major item in the diet (p. 30). The collecting of acorns is likely to have disrupted the balance of the local vegetation, leading to the gradual diminution of trees over several generations. Further disruption would result from the cutting of branches for fuel and possibly fodder, the latter on the argument that the dates currently available suggest that animal domestication took place before that of cereals. The reduction in the number of trees would not only remove the source of acorns and other nuts, but also encourage the spread of grasses, including the wild cereals. Wild cereals, however, may not have been collected at first principally for human food but, as may be indicated by the use of reaping-knives or 'sickles', for fodder. An emerging crisis in food supply might produce a change-over from dependence upon nuts to experimentation with cereals. A somewhat similar set of ideas has been advanced by Mellars (1976). On his argument, the burning of woodland — whether to drive game or to encourage the growth of certain types of plant — would have upset the ecological balance. This, in Mellars' view, would have forced either the development of conservation methods or the emergence of new systems of subsistence. These in turn may have been determined by burning. Burning would have been instrumental in concentrating herbivorous animals and, if the frequency of burning was controlled, in attracting certain species rather than others. The way may thus have been prepared for selective hunting and even domestication. By concentrating wild grasses in particular clearings, burning might also have encouraged cereal domestication. Settlements might have been attracted to such local concentrations of food. Relative security in food supply might have encouraged population growth, as well as the experimentation often considered necessary for domestication.

Clearly, the possibility of various routes to domestication cannot be ruled out (Harris 1977). These were not in competition for the prize of being the only way by which settled farming could emerge. Harris (1977) has emphasised the interconnectedness of the causal nexus and of the role of positive feedbacks in systems transitional to domestication. This underlines the importance of identifying the thresholds in the process of change. Redman (1977) believed that one of the most crucial was sedentarism. It was difficult to attain, but once reached by a proportion of the total population then other developments, including domestication, could follow. Synchroneity was not a characteristic of the transition: as the earlier sections of the chapter have revealed, different systems of food procurement coexisted in time and space. While independent development is possible, the transference of skill and knowledge is likely. The evidence for extensive contacts across the length and breadth of the region from an early date indicates that the isolation of the various physical subregions and individual human communities was relative. By perhaps 5000 BC systems of food production dependent upon domesticated animals and cereals were being used by many groups across the northern parts of the region. The relationships between man, crops and animals became so interdependent that they could be regarded, as Darlington (1969, 79) has suggested, as behaving like the single 'genetic system of a species, fused into one evolutionary unit or entity whose parts are mutually dependent and came to be mutually adapted'.

## Consequences of the Transition to Food Production

The creation of a mutually adapted evolutionary entity marks the end of what, in terms of its timing and its effects, constituted a revolution. The transition from food gathering to food production was accomplished in about 3–4,000 years, a very short span on the scale of some 5.5 million years of recognisable human existence. From the Middle East the revolutionary subsistence systems spread outwards into Europe and India, bringing fundamental changes. Although the nourishment provided by the new systems was probably no better than under hunter-gatherer arrangements (Flannery 1969), farming reduced the immediate uncertainties about food availability. To that extent, the chances of survival may have been increased, despite the probable developments within cereal-based communities of protein malnutrition (kwasiorkor) and ergot (an illness produced by flour contaminated by fungus) (Brothwell and Brothwell, 1969, 179–82). Nutrition as a whole actually improved with the emergence of bread wheat, the development of baking and the discovery of how beer could be brewed. The whole system, though, was precariously balanced. Dependence on a limited range of plants, and even on a small number of animals, increased the risk of starvation in years when precipitation was below the level required for successful dry-farming. Man's hunting and other activities (forest clearance, for example) increasingly ensured that alternative sources of food diminished.

Food production seems to have been associated with the relative permanency of settlements, though it was not necessary to sedentarism. More permanent settlements and less movement would require a concentration of exploitable resources and, at the same time, produce the labour necessary for their exploitation. It would also allow the population to grow by reducing the frequency of miscarriages and increasing the rate of conception (Harris 1977). Greater densities of population resulted. Flannery (1969) suggested that the carrying capacity of early dry-farming systems was 1–2 persons per $km^2$, compared with perhaps 0.1 persons per $km^2$ under hunter-gatherer systems. Carneiro and Hilse (1966) demonstrated that a rapid growth in the region's total population is unlikely in the period 8–5000 BC. They suggested that rates between 0.08 and 0.12 per cent a year were reasonable. On a base population estimated at 100,000 for 8000 BC, equivalent to 2,000 settlements of 50 inhabitants or 1,000 with 100 inhabitants, these growth rates could have produced a total regional population of between 495,000 and 1,000,000 by 6000 BC and 1,100,000 to 3,650,000 by 5000 BC. In 1981 the region's population exceeded 182,000,0000 (UN 1983a, Table 3). The average density across the region in 5000 BC would be between 0.13 and 0.46 people per km$^2$. There would have been a maximum of 73,000 settlements, each with 50 inhabitants, or 35,500 settlements each inhabited by 100 people. In the mid-1960s there were more than 36,000 villages in Turkey alone, most of them with fewer than 1,000 inhabitants (Dewdney 1971, 83).

Early farming systems probably did not reduce the amount of leisure for the people, but in time the amounts of work necessary for subsistence may gradually have increased. On the other hand, the bursts of activity necessary for sowing and harvesting may have left relatively longer periods free from necessary work than under gathering systems (Smith and Cuyler Young 1972), while the keeping of animals near the settlements would have reduced the range and arduousness of hunting trips. Although an increase in leisure, or in its time distribution, was not strictly necessary to the pursuit of crafts — as the production of stone implements indicates — some change in age-old systems might have assisted in further developments. The need to organise the care of crops and animals, along with the almost inevitable emergence of differential control over resources, may have begun to produce a stratified society such as is clearly evinced in the archaeological record of the fourth millennium.

Dependence upon cereal-growing would reduce the area from which the settled population could draw its food. Flannery (1969) observed that 65 per cent of the surface area of Iran was considered uninhabitable and marginal in 1956. He noted that this would have left 35 per cent in which hunter-gatherers could have lived during the Late Pleistocene, assuming no great change in climate. However, only 10 per cent of the surface of Iran was considered suitable for arable farming at that time. The calculation is only indicative, but if it is extended to cover the whole of the Middle East, we can estimate that farming economies could have exploited a maximum of 647,800 km$^2$, or 9.3 per cent of the region's surface area (FAO 1972, Table 1). This may be a reasonable approximation, even allowing for variations in the dry-farming area occasioned either by subsequent losses through erosion or even by periodic advances resulting from a run of wetter-than-average years in marginal locations. Indeed, the figure may be too high since it includes the major irrigated areas of the region in Lower Mesopotamia and the Nile Valley. Accepting the calculation, however, we can estimate the population density at 1.70–5.63 persons per km$^2$ of the cultivable area around 5000 BC.

The effects of human activity on the region's vegetation do not generally appear in pollen diagrams until after the period under discussion. In any case, on a regional scale they may have been relatively slight during the transition to settled farming, though locally they may have been highly significant (Willcox 1974). There was certainly no large-scale disruption of the vegetation (Roberts 1982b). The beginnings of agriculture were occurring at the same time as woodland was both expanding and also taking on its present-day characteristics in species composition. But, as the number of farming communities grew and settlements lasted for several hundred years, man must have had a growing impact on vegetation. An irregular pattern of fields would be one of the first features to develop, whether by encouraging the expansion of natural stands of cereals or by creating new patches for crop-growing. Digging sticks — long, straight, pointed and possibly weighted with a stone (Fig. 3.1) — were probably used in land preparation (Curwen and Hatt 1953, 63). Large trees and other obstructions are likely to have been left where they were. Although relatively large areas cannot be cultivated using digging sticks, quite sizeable tracts of ground would have to be opening up to support the estimated populations of the early farming villages. If we assume that the average consumption of cereal was about the same as in classical antiquity (200 kg per person) and that yields were similar (mean of 942 kg/ha for wheat and barley) (Jardé 1925; Jameson 1977–78), then the support of each person would have required an area of 0.2 ha. Thus, the estimated 150 members of possibly the most famous early agricultural village, at Jarmo, would have required a cultivated area of 31.8 ha. Allowing something like a third for seed, the total area involved in food production would be 42.6 ha. This is a sizeable amount of open space to create in woodland, even if it was dispersed in numerous small patches around the village. If we further assume that a slash-and-burn or swidden system was practised, the cultivated patches would shift from time to time as declining yields revealed falling soil fertility. Cauvin (1978, 79) has actually suggested something like this to explain the disappearance of cereals from the pollen record of Tell Muraybit. Gradual shifting of the cultivated land would mean that steadily larger and larger areas of woodland would have been affected. The collection of leaves and the cutting of branches for animal fodder would extend the disruption even further away from the focal settlement. Nut-gathering, like animal browsing, would curtail regeneration — which would have been slow in any case under Middle Eastern climates, as these stabilised within their present-day parameters.

Man himself made more demands on the woodland. The earliest known oval structures seem to have been built of wattle, but the use of pizé and mudbrick in rectangular structures argues for the use of poles to support the roof. Indeed, recourse to mudbrick might itself indicate a growing shortage of wood. Kindling and fuel were necessary for cooking fires, to heat bread ovens and to keep warm during the cold winters. Smaller trees and shrubs would supply much of the demand, which in any case could have been partly supplied by burning straw and grass, as in recent times. The pressure, however, was unremitting. It must have risen steadily as the population grew. The development of fired pottery would have raised the consumption of fuel by a large factor.

The sum effect of all man's activities would be to thin out the existing woodland around his settlements and to begin its final destruction. Some indication of the scale involved can be suggested by extending the Jarmo calculations to the whole region. If we assume that there were perhaps 35,500 villages of 100 inhabitants in the region around 5000 BC, as suggested earlier (pp. 44–5), each of these would require a minimum of about 42.4 ha of cereal-growing land. That would give a total of 1,505,200 ha for the entire region. The amount is equivalent to 2.3 per cent of the area cultivated in 1971. To obtain some idea of the sum effect of human and animal activity, that figure should be increased by, at a conservative estimate, half as much again. Even then, the extent of the disruption, though increasing generation by generation, was comparatively small.

None the less, the opening up and progressive clearance of woodland in certain areas would increase run-off. Floods would probably increase in height and severity. Soil erosion would be initiated. After a time-lag, sedimentation would increase lower down the valleys and in the plains, as streams entrenched themselves and soil transportation increased (Schumm 1977, 26–9). Terrace-building may have been one human response. Soils themselves would gradually have changed their characters as a result of erosion and direct exposure to heat and sunlight (Dimbleby 1977), as well as in response to cultivation. In time, many soils assumed the relict features typical over so much of the region today (Beaumont, Blake and Wagstaff 1976, 33–44). These changes, in turn, would have hampered the regeneration of woodland. The final consequence, then, of the farming systems developed by about 5000 BC was the beginning of the slow but progressive degradation of the soil and vegetation of the region from their 'natural' state. Under a climate characterised by a long dry season, this must be held responsible over a long period of time for the physiological aridity and physiognomical starkness which now leave such powerful impressions on visitors to the region.

# 4 MODES OF LIVING

## Introduction

Human groups have produced styles of life, or modes of living (*genres de vie*), through their instinctive concerns to survive from year to year and to provide for the next generation (Vidal de la Blache 1911). A close interdependence may be postulated between physical and social environments. This implies a continuous process of mutual adjustment within an interactive man-environment system. In consequence, the detailed configuration of any specific mode of living would change over time, even though some of its outward manifestations might remain stable for long periods.

Once the initial domestication of plants and animals had taken place, a wide variety of life-styles became possible within the diverse physical environments found in the Middle East. Over time, though, hunter-gatherer modes of living declined. On the one hand, forms of cultivation and herding reduced the habitat for game and simplified local ecosystems; only fishing survived into modern times as more than a leisuretime pursuit. On the other hand, the higher energy yields per unit of input made possible by the emergent modes of living must have gradually transformed the older ones by their conspicuous success (Kemp 1971). Still greater productivity can be secured from ecosystems by simplification of the range of species grown or reared, and through specialisation. Increasing complexity in social organisation is likely to have encouraged moves in this direction. The resulting ecosystems are inherently unstable, and at risk from natural hazard and human folly (Netting 1974; Rappaport 1971; Slofstra 1974). The degree of risk was increased in many parts of the Middle East by relatively high and unpredictable variations in annual precipitation. In circumstances where transport technology and costs prohibited large transfers of food over any distance (p. 135), it was necessary for modes of living to incorporate means of storing food and to evolve social mechanisms for surviving drought.

Although cultivation and herding produced precarious modes of living, they became consolidated in two ways. One was through the domestication of other species (e.g. camels) and further technological development (e.g. ploughs) which increased productivity still more. The other way to consolidation was forced by the lack of alternatives as

Figure 4.1: Scale of Possible Life-styles in the Middle East

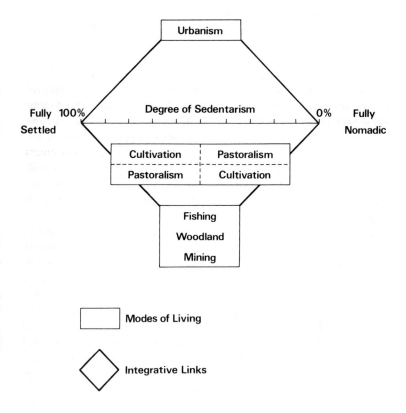

man-induced change in the physical environment actually tightened the constraints. Limitation of the opportunities finally produced for many parts of the Middle East a restricted continuum of possible life-styles which stretched between those of fully settled cultivators, perhaps specialising in growing a particular crop, and nomads whose range was extended by the endurance of the dromedary (Fig. 4.1). It is not entirely clear when in time the point of maximum spread was achieved, but fully nomadic dromedary-rearing does not seem to have been possible much before the end of the second millennium BC (pp. 86–7). Once the spectrum was established, groups of people were free to shift their mode of living within it. Such shifts might occur not only over several generations but even within the lifetime of one individual. The explanation is at least as likely to be related to the working of endogenous forces as to the effects of incursions of alien peoples. Economy and society would adapt themselves to local physical conditions. At the same time, human requirements would modify the most easily changed elements of the physical environment — vegetation, soil and water — and continue a dialogue of mutual adjustment.

Despite the existence of a range in modes of living, commentators have frequently characterised 'traditional' Middle Eastern society as a dichotomy between settled cultivators and pastoral nomads, mediated in some often ill-defined way by an urban style of living (Cohen 1977; English 1967 and 1973). As well as its extreme simplicity, this view has a number of other shortcomings. For example, urbanism fits awkwardly into a basically ecological model of society. In any case, it may itself differ from other modes of living only in the scale and degree of concentration in certain elements common to other forms of sedentary life (Ibn Khaldūn 1967, 43). Second, the very notion of a 'traditional' trilogy begs a number of questions. What span of time is meant by 'traditional'? When did the trilogy crystallise? How far is it possible to reconstruct past modes of living anyway? The concept of a continuum of life-styles is more realistic than that of a simple, 'traditional' trilogy. It allows for continual change over time as people adapt to circumstances — the advance and retreat of cultivation, the shifts into and out of nomadism, the decay and revival of towns. The following chapters will chart these developments. At the same time, the concept makes allowance for other forms of living which may also encompass elements of cultivation and nomadism but depend on other forms of specialisation.

The problem of reconstruction remains to be discussed here. Modes of living have spatial extent and geographical form. In addition, they have created the successive human geographies of the region. To appreciate the nature of their contribution and of their involvement with the physical environment, as well as to avoid repetition in later chapters, it is desirable to outline the salient characteristics of at least the main modes of living. The problem is that most of the relevant material comes from recent observers and modern studies. There is little alternative to using it. Devising a satisfactory way of doing so is more elusive. The method adopted in this chapter is to create a series of ideal types. There are dangers in this approach. They include over-reliance on historical markers which, perforce, assume a possible misjudged significance. Continuity is assumed, unless there is positive information to the contrary. This can be rationalised as follows. On the assumption that plant and animal requirements have remained much the same over time, then cultivation and herding, for example, can be seen to generate cycles of activity for the people dependent upon them. These are constrained by the packing problems created by finite time and limited space. There is a strong probability, then, that the basic elements — the building blocks for modes of living — will have stayed much the same for a very long time. They may have been arranged in different configurations in time and place. The specific influences will be outlined in the following chapters. The remainder of this chapter is concerned with indicating the nature of the basic blocks from which modes of living have been built for cultivators, pastoral nomads and townsmen, but attention is also paid to the activities of woodland and fishing communities, as well as to carrying and its parasite, brigandage.

**The Life of Cultivators**

The modes of living which developed around the deliberate cultivation of crops may be considered to be fundamental to all the others carried on in the region. Their viability is dependent upon the food production of the cultivators. Cultivation itself is of considerable antiquity. Its modern forms were already recognisable in the fifth millennium BC and associated even then with relatively long-lived villages (pp. 44–5). The siting of villages is a complex problem but water availability even in arid and semi-arid environments is only one of the decision-making elements, not a major determinant. Although too little excavation has taken place to be absolutely sure, Ayrout's (1963, 90) contention about Egypt is at least plausible diachronically for the whole region, namely that 'the same people living the same life have built the same kind of village'. Lowland villages today tend to be relatively compact agglomerations of houses, divided by narrow, meandering streets, while rough terrain and steep slopes have produced more fragmented forms in the mountains. Use of local materials and the climatic needs for different types of roof have produced a real variation in built form, but within recognisable areas the villages have tended to blend into the landscape. Indeed, they have often been so successful that from a distance, and in certain lights, they are almost invisible. Human presence is betrayed only by the delicate aroma of smoke or the faint lilt of chatter.

In the context of the present discussion, the life of the settled cultivators can be characterised, on the basis of criteria suggested by Thorner (1964), as a form of 'peasant economy'. More than half of its production is likely to derive from agriculture, which employs more than half of the active population. Towns have been important as centres of exchange from an early date and have been associated with villages in a variety of territorial states (pp. 97–105). The typical unit of production is the household, and it may always have been so, though its size and composition may have changed through time. Not only is the household normally a biologically related group, but it is usually concerned with its own welfare as a unit. It reproduces itself. It organises its own subsistence. At the same time, the household has had to meet the obligations entailed by membership of a wider society. As a minimum, these have probably included some contribution to communal celebrations, the support of state and religious systems, the maintenance of various specialists, and forms of exchange. But the household's needs are related to its size and composition, while its ability to meet its own requirements as well as other demands is conditioned by the amount of labour which it can deploy, making due allowance for its age structure, as well as for sexual roles (Chayanov 1966; Shanin 1971; Wolf 1966). The biological cycle of the family's life would produce rises and falls in labour requirements and the household's ability to meet them. These probabilities would translate themselves into landscape effects by the propensity to acquire and cultivate land as numbers increased and matured, and a need to shed land and reduce the extent cultivated as labour declined. A balance between labour investment and a subjective distaste for hard manual work may be postulated as an important element in 'peasant economy'. Together with labour availability, it is likely to govern the household's response to outside stimuli, whether coercive or commercial in nature, and its willingness to modify or change its use of land. The ability to change, however, is further conditioned by the attitudes of the rest of the community, by the structure of the existing system which ensures its own replication, and by the delay imposed by the growth-cycle of crops already in the ground (Kolars 1974).

The precise patterns of activity which play such a large part in the modes of living of cultivators are constrained by the use of hand tools, a sexual division of labour, the growing requirements of crops, and the particular mix of crops grown at any one time. Soil texture and slope angle obviously determine where crops can be grown. Distance from the village and labour input — high for vegetables but relatively low for cereals — are influential and give rise to broad zones of land use (Fig. 4.2). The growth-cycle of crops gives a seasonal rhythm to farm work and shapes the changing appearance of the humanised countryside. It was early seen to correlate with the rising and setting of particular stars and constellations so that, within limits set ultimately by inter-annual variability in

Figure 4.2: Zones of Land Use around Villages in Central Turkey

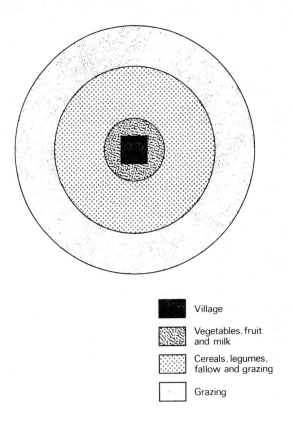

| | Village |
| | Vegetables, fruit and milk |
| | Cereals, legumes, fallow and grazing |
| | Grazing |

Source: Mitchell (1971).

weather, the yearly round was predictable and thus capable of being managed. Latitude and elevation obviously fix the beginning and end of the growing season, as well as its duration. None the less, the broad patterns of seasonal activity can be illustrated by one example, Judaea (Turkowski 1969; Wilson 1906). This has been deliberately chosen from a range of possibilities because of its mid-latitude position in the region and its moderate elevation, while the antiquity of a mode of living formed around the cultivation of wheat, barley and vegetables is attested by the calendar ditty on the Gezer Tablet of the late tenth century BC (Albright 1963, 132, 200; Turkowski 1969, 21–33).

Land preparation requires the removal of vegetation by cutting and burning, then the lifting and gathering of stones. Field stones might be heaped to form rough boundaries to the cleared area or built up into the retaining walls of terraces. In Judaea, cleared land was dug over with spades, roots were cut out and further large stones removed. Ploughing followed. These processes immediately expose the soil to erosion, and begin its physical and chemical transformation.

Although spades are used in certain parts of the Middle East to prepare the land for sowing (Wulff 1966, 261–2), the plough is used in Judaea and is almost universal in the region. A variety of simple ard-type ploughs pulled by a pair of animals has been observed in the region since at least medieval times (Fig. 4.3) (Haudricourt and Delamarre 1955), and a diversity of evidence confirms their antiquity (p. 84). Convincing reasons for survival have yet to be advanced. They may include the negative ones of there being no need to cut a furrow, either to turn a grass sod or to aid drainage, and the necessity of avoiding the inception of gully erosion in the Middle East's high-energy geomorphic environments. In addition, soil temperature and moisture conditions may make a relatively deep seedbed unnecessary. Finally, farmers may have lacked the capital to acquire a heavy mouldboard plough, which demands a team rather than just a pair of draught animals. At least in some districts and at certain times, even the ard was made by specialists rather than by the farmer himself.

Ploughing cannot readily begin until the baked ground has been softened by the first rains. Once that has happened, the farmer makes use of every dry day. In the Judaean Hills he is ploughing through November and December. Cereals may be sown then, but a second ploughing is often undertaken in January. Large clods of soil may be broken up with spades or large mallets, but in some parts of the Middle East simple wooden harrows are used. The timing of sowing is critical. If it is too early, frost may damage the young shoots; but if it is too late, drought will have adverse effects. Cereals, and later lentils, are usually sown broadcast, a skilful job since a successful harvest depends upon an evenly spaced crop. The seed is covered by a third ploughing. Beans, which are sown a little later, are usually rowed. The tedious job of weeding takes place during April in Judaea and, unlike ploughing and sowing, is usually undertaken by women. Meanwhile, the fallow may be receiving the first of several ploughings. Fallowing is as much a reflection of extensive land use as it is a response to concern over soil moisture conservation and fertility maintenance.

Lentils ripen in Judaea about the middle of May, followed by beans and lupins. Barley matures at the end of May and wheat towards the end of June. Legumes and pulses are harvested by pulling, generally women's

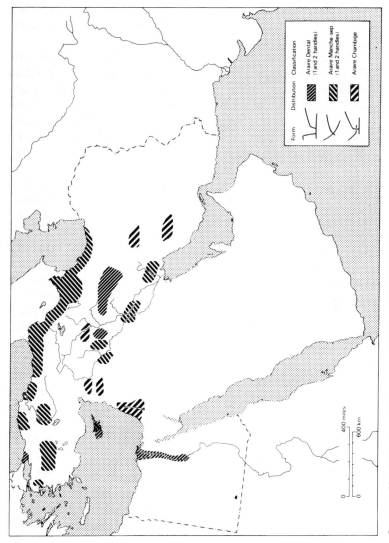

Figure 4.3: Partial Distribution of Ard-type Ploughs

Sources: Haudricourt and Delamarre (1955) and Wulff (1966).

Figure 4.4: Threshing Implements

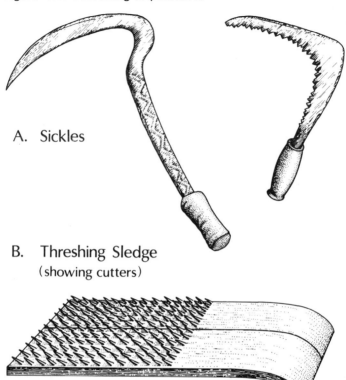

A.  Sickles

B.  Threshing Sledge
    (showing cutters)

C.  Threshing Wheel or Wain
    (after Wulff)

work. Thin stands of barley may be harvested in the same way, but wheat and barley are generally cut with sickles by a mixed team of men and women working steadily across the field (Fig. 4.4). Abundance of labour and a relatively long harvest season might explain the historical survival of the sickle until its virtual replacement late this century by the combine. The cut stalks are bound into sheaves and left on the ground until the end of the day, when they are stacked. When reaping is finished, the sheaves are carried by animals to the threshing floor.

Threshing floors are normally circular and, in the past, may often have been communal property. Generally, they are sited on an exposed hillside away from houses and vegetable plots so that winnowing is facilitated and the fine chaff dust does not damage eyes, spoil cooking or ruin vegetables (Serjeant 1974, quoting a fourteenth-century manual). Various methods of threshing coexist in the region, and may have done so for at least several centuries. Legumes and pulses are usually beaten with sticks. Cereals can be trodden by a team of animals driven round and round the threshing floor in a line, a technique found in modern Judaea but also depicted in ancient Egyptian paintings. In some areas of the Middle East a threshing sledge is used (Fig. 4.4). It consists of two or three heavy boards, curved upwards slightly at the front, and shod with either iron or chipped-stone cutters. A frame-like device fitted with wooden beaters or iron roller-cutters (Fig. 4.4) has been described from various parts of Iran (Wulff 1966, 273–5). In Judaea, threshing occupies from June until August. Once the straw has been broken and cut, grain is separated from chaff by constantly throwing the mixture into the air with long-handled wooden forks and spades so that the wind separates the heavier grain from the lighter chaff. Further separation requires a process of sieving, using different gauges of mesh. It was at this point in the process, when the grain can be measured, that share-cropping contracts and tax obligations were normally fulfilled. The bagged grain is finally carried to the farmer's home. The straw and chaff are saved: straw is valuable for feed and bedding, while chaff can be used to fire domestic ovens and to bind mud bricks and seed cakes.[1]

Further processing is carried out in the home by the women. Grain, legumes and pulses are carefully sieved and patiently picked over to remove impurities before use. Although water mills and windmills have long existed in the region (p. 27), hand querns have remained in use in most households as a cheap, if labour-intensive and mournful, means of making flour. Bread is the staple food, though prepared in a great variety of ways. The most common types are unleavened and look like slightly thick pancakes. Ideally, they are baked daily, either by placing the cakes of dough on a metal plate heating over an open fire, often outside the house or tent, or by putting them on the smoothed interior of a preheated, dome-shaped oven. Although best eaten fresh, this type of bread can be reconstituted for some time after initial baking with water and skill. Fuel is harvest residues or else sticks and dried vegetation carefully gathered by the women from the scrub-like vegetation which commonly covers uncultivated land. This relentless activity has been a major force in creating the bare, open landscapes familiar around and between many modern villages. Cakes of cattle dung, which give a slow and lasting, if smoky, fire, are now burnt in many areas (Fenton 1972). Bread is supplemented by fresh and dried vegetables prepared in various ways, and by fruit. Dairy products might be consumed as well. Meat is a rare luxury, associated with sacrifice, ceremonies and hospitality. September sees the ripening of grapes and other fruit. Picking is followed by the labour of drying and preserving. Grapes are trampled to prepare must for wine or, as with other fruits, for confectionery and non-alcoholic drinks. Olives are gathered slightly later, usually by shaking the tree or beating it all over with long rods. They are often preserved in brine, but pressing is essential to remove the valuable oil. Today this is done initially by using a vertical wheel, rolling round a concave platform, and then by squeezing the residue in a vertical press, but the antiquity of such methods is uncertain.

The cycle of activity outlined above is geared to the production of cereals, chiefly varieties of wheat and barley. These dominate land use, partly through their importance in human diet, but partly also because yields are relatively low and uncertain and a large area must be sown to guarantee subsistence. Yields in the system of 'dry-farming' or 'rain-fed farming' are totally dependent upon levels of precipitation, and survival requires levels adequate for plant growth. Empirical investigation has shown that the perennial growing of wheat and barley is possible only where annual precipitation exceeds 240 mm and the inter-annual variability is no more than 37 per cent (Fig. 4.5) (Perrin de Brichambaut and Wallén 1963; Wallén 1969). While cereals may fail in core areas during years of exceptional drought, they can be grown successfully beyond the mean dry-farming boundary in particularly wet years or runs of years. Perennial cultivation outside the limits specified necessitates irrigation.

Irrigation is essential to cultivation in Lower Mesopotamia and Egypt and its employment is, accordingly, of considerable age (pp. 90–7). It is attested at early dates from Muraybit in Syria and Jericho in Palestine (Miller 1980). It is vital in much of Arabia and Iran. Irrigation has also been used on a small scale in many dry-farming areas (Forbes 1965).

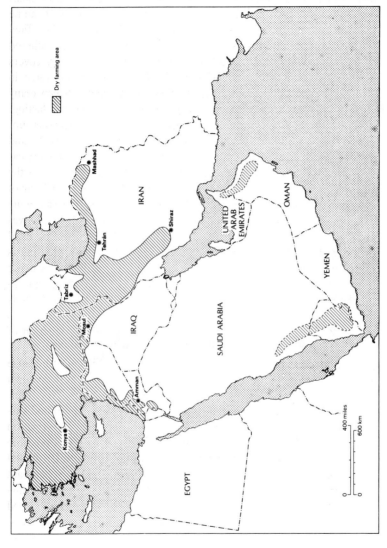

Figure 4.5: Theoretical Limits to Dry Farming in the Middle East

Although land use was often extensive in the past because water was available only in one season, the potential always existed for extending the growing season into the summer, increasing the range of crops grown and, by skilful application, generally raising productivity. The control and management of water, however, make irrigation a more labour-intensive form of production than dry-farming. The additional inputs have to be fitted into a similar yearly cycle of plant growth and possibly water availability. This requires careful scheduling, though not, as Fernea (1970, 130–2) has shown, the large-scale organisation and direction of Wittfogel's (1957) type of 'oriental despotism'. Rights to water were so critical in parts of the region that elaborate legal codes evolved to deal with the many problems which arose (Caponera 1954). Water loss is high — through evaporation, seepage and over-watering. Salination is a perennial risk because of the presence of soluble calcium and magnesium salts in the water and the concentrating effects of seasonal evaporation and capillary action (Beaumont, Blake and Wagstaff 1976, 45–6).

The methods of obtaining water may be divided into two sets, 'flow' and 'lift'. The simplest methods of flow irrigation used the seasonal floods of rivers like the Nile and the Euphrates, and initially involved little more than the elaboration of natural basins and distributaries (pp. 90–7). An interesting variant is the use of fresh water ponded up by the tide in the creeks of the Shatt al-ʿArab to irrigate date plantations (Naval Intelligence Division 1944, 440). The diversion of water from stream channels or springs required the construction of dams and weirs, as well as the maintenance of canals and distributaries. More elaborate still are the systems constructed to convey water on to flights of terraces, whether from springs, cisterns or, as in southern Arabia particularly, flash floods (Fig. 4.6B) (Ron 1969; Serjeant 1964). The use of gently sloping, underground gravity canals is extensive around the mountain fringes of Iran and southern Arabia, though it is found elsewhere in the region, too. Various terms are now employed to describe them, but the Arabic term qanāt ('lance', 'conduit') common in Iran will be used here (Beaumont 1968; Beckett 1953; English 1968; Noel 1944; Wilkinson 1974 and 1977; Wulff 1966, 249–54). Many qanāts are only a few metres long, but some range up to 50 km (Fig. 4.6A). Their presence is betrayed at the surface by lines of mound-fringed shafts spaced at 50–150 m intervals to provide access and ventilation. They are excavated by itinerant specialists (*muqannis*). Since they are slow and expensive to construct, qanāts were usually provided by the rich and powerful, though maintenance needs and inheritance practices have often produced multiple ownership. Bifurcating channels distribute the water from just below

Figure 4.6: Forms of Flow Irrigation

## A. Typical Qanāt
(after English 1968)

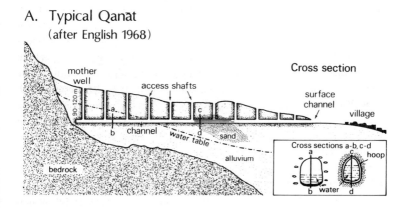

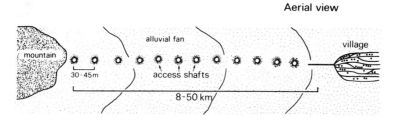

## B. Small-scale Flood Irrigation in South Arabia
(after Serjeant 1964)

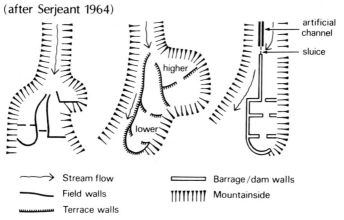

the point where the qanāt reaches the surface, but, as in other irrigation systems, the number of channels depends upon the volume of water made available and local customs concerning access to land and water. Land use becomes more extensive with distance down the slope.

'Lift' irrigation may have been a generally late development because of the technological requirements of all but the very simplest apparatus. These include the pulleys and windlasses necessary for lifting heavy leather bags of water (Derry and Williams 1970, 244; Wulff 1966, 256–8), as well as the weighted beam (*shādūf*) with its simple water container (Fig. 4.7A & B). Various kinds of wheels were subsequently devised (Forbes 1965). Some were geared in such a way as to be turned by animal power and they lifted water by a circle of pots (*sāqiya*). Others were more substantial wooden constructions (*tabūt*) and had their outer rims divided into compartments so that they could be turned by stream flow (Fig. 4.7C & D). Although lift irrigation is particularly associated in the popular imagination with wells, it was also used before the advent of pumps to take water from channels where water level was normally or seasonally below the surface of the fields. In both cases, the effort involved and the relatively low volume of water which could be raised in a working day tended to restrict the areas irrigated and to produce compact, but sharply delimited, forms. This contrasts with the linear and deltaic shapes of cultivated areas which result from flow irrigation.

## Pastoral Nomadism

The herding of animals is at least as old as cultivation (pp. 34–5) (Bacon 1954; Barth 1962 and 1968; Dyson-Hudson 1972; Johnson 1969; Krader 1959; Nelson 1973; Patai 1951; Planhol 1961–67; Stauffer 1965). Both activities were often carried on together, by the same people. Indeed, a minimal amount of animal husbandry was necessary for cultivators to supply draught animals once the plough was adopted. None the less, specialisation in nomadism has been characteristic of some groups in the Middle East, probably for millennia (Briant 1982; Krupper 1957 and 1959; Luke 1965). It was not associated of necessity with either areal and seasonal differentiation of grazing, or with nomadism. On the last point, Birks (1978) for example, has described a pastoral group from the fringes of the Rub' al-Khali in eastern Oman which moves less than 1 km in a year, uses the same water source throughout the year and is linked with a particular village.

Pastoral nomadism is essentially a subsistence strategy which permits

# Figure 4.7: Water-lifting Devices

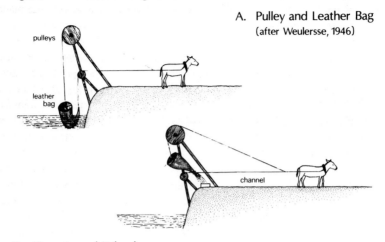

**A.  Pulley and Leather Bag**
(after Weulersse, 1946)

pulleys

leather
bag

channel

**B.  Flow-turned Wheel**
i) Left face (section)    ii) Method of construction    iii) Right profile (elevation)

**C.  Using Shādūf**
(from ancient Egyptian illustrations)

**D.  Animal-powered Wheel of Pots**

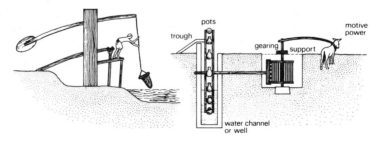

pots

trough

gearing    support

motive
power

water channel
or well

the keeping of greater numbers of livestock than the carrying capacity of a particular area would allow (Bates 1972). To do so, it exploits seasonal and areal variation in the availability of both pasture and water. Its essential discomfort, especially where long migrations and forced marches are involved; its association with a frugal diet and seasonal hunger; its 'deep and biting social discipline' (Lawrence 1962, 35); and its comparatively low yield in food per hectare compared with cultivation (Duckham and Masefield 1970, 10) all suggest that there is an alternative explanation to that of ecological adaptation often favoured by geographers.[2] The possibility is strengthened by the recognition that, over time, pastoral nomads have shown a propensity to settle and emphasise the cultivation aspect of their economies, while from time to time apparently settled communities have increased their involvement in animal husbandry and become more mobile. Moreover, pastoral nomadism has not been spatially discrete from cultivation.

The positive advantages of pastoral nomadism include the apparent ease with which capital can be accumulated in livestock numbers (Barth 1973; Swidler 1973), relative freedom from taxation because of the difficulties involved in assessment and collection, and the greater security which mobility has offered at certain times. Perhaps more important was a degree of choice which pastoral nomads could enjoy as to the form of polity under which they would live (Meeker 1979; Peters 1977). Genealogy is valued by many pastoral nomads. Their society is structured on the basis of real or fictive kinship, a device which facilitates co-operation and at the same time moderates competition and strife. Such an apparently loose-knit polity is a radical alternative to the centralised, impersonal state based on settled cultivation, though it was capable on occasion of creating territorial states (pp. 155–6). The way of life, involving mobility and familiarity with arms, allowed the almost instantaneous creation of a tough, highly mobile army which could be deployed in any power struggle.

Studies made since the late eighteenth century have shown that pastoral nomads in the Middle East, if not elsewhere in the world, have tended to live principally on cereals and/or dates. While it is clear that they can and do produce these themselves, specialisation means that nomads must acquire their staple foods through exchange. They offer a range of pastoral products (cheese, clarified butter, wool and hair) and the surplus male animals, as well as a variety of services (carrying and 'protection', for example). The precise mechanism of exchange may have varied over time through a range encompassing gifts, tribute and sales (Bates 1972; Stauffer 1965). Marketing in particular means that herd composition could be modified to meet demand; it was not simply determined by the grazing possibilities of local vegetation. Herd composition is critical in shaping the modes of living of pastoral nomads. The food and water requirements of particular animals, and their tolerance of heat and cold, shape both the frequency and the pattern of movement (Dyson-Hudson 1972). Productivity, especially in terms of milk and reproduction, controlled not only the size of the human group which could be supported but also the amount of labour required to manage a certain size and mix of herd.

Cattle are relatively slow-moving and cannot eat salty vegetation. Steers of 306 kg weight drink 32 l of water per head per day while feeding on herbage with 36–72 per cent moisture content (Wells 1970). It is not surprising, therefore, that cattle-rearing has virtually been confined to the marshes of Egypt and Lower Mesopotamia (Erman 1971, 436–41; Kees 1977, 29–30, 86–7; Wright 1969, 15), though the great marshes of southern Mesopotamia have also been asssociated with buffalo-raising in recent times (Salim 1962). The main exception to this broad generalisation today is located in the rolling meadows and jungular woodland on the southern face of the Jebel Qara in southern Arabia. Here the mists and rain of summer monsoons have allowed a form of cattle transhumance (Thesiger 1959, 27–8; Thomas 1938, 78–9).

The herding of sheep and goats is much more widespread (Cressey 1960, 171–3). Sheep, like cattle, are particular about their food. They must drink at least every 10 days if the vegetation is green and fresh, but every 2 days if it is desiccated. They cannot be ranged very far from water. Goats have a reputation for being omniverous, and they need less water than sheep. In addition, they are more agile and have both a higher reproduction rate and a longer lactation period. These advantages have often made them a favourite herd animal.

The dromedary was a comparatively late domesticate and its full potential as a transport animal was realised still more recently (pp. 86–7, 155). While it can live on a diet of parched grass and desiccated shrub, it cannot go without water for more than about 20 days and, in the extremely arid conditions of summer, may need to be watered every day (Thesiger 1959, 87). Dromedaries generally require more attention than sheep and goats. Reproduction rates are low. On the other hand, dromedaries travel faster than other herd animals and can range farther into arid areas.

It is reasonable to suppose that the support of a *tent* household at a satisfactory level of subsistence requires a herd size sufficient to produce a number of young and enough other products to allow for exchange (Swidler 1973). Opinion varies on the minimum number required, but ranges from 25 to 60 for sheep and goats and 10 to 35 for dromedaries.

Figure 4.8: Black Tent

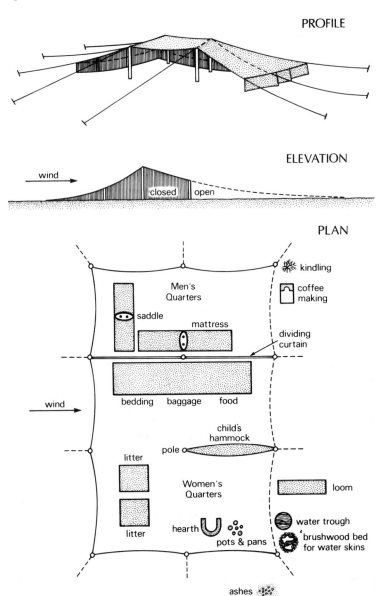

PROFILE

ELEVATION

wind

closed   open

PLAN

kindling

coffee making

Men's Quarters

saddle

mattress

dividing curtain

wind

bedding   baggage   food

child's hammock

pole

litter

Women's Quarters

loom

litter

hearth   pots & pans

water trough

brushwood bed for water skins

ashes

Figure 4.9: Patterns of Nomadic Migration

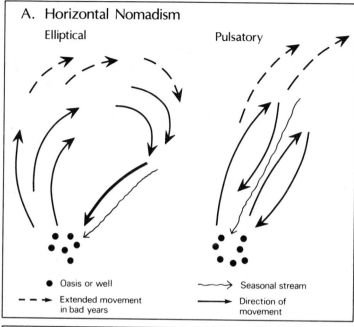

### A. Horizontal Nomadism

Elliptical

Pulsatory

● Oasis or well

⇢ Extended movement in bad years

⤳ Seasonal stream

→ Direction of movement

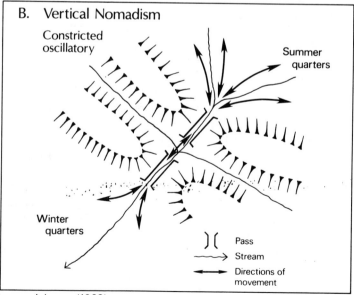

### B. Vertical Nomadism

Constricted oscillatory

Summer quarters

Winter quarters

)( Pass

⤳ Stream

⟷ Directions of movement

Source: Johnson (1969).

There seems a tendency to build up herds in years of good grazing as a way of accumulating capital against the inevitability of high losses during periods of extreme drought. In the past, thieving and raiding were often necessary and exciting diversions (Dyson-Hudson 1972; Sweet 1965), requiring the habitual carrying of weapons for defence and attack. Camps consist of five or more well-spaced black or brown tents (Fig. 4.8) facing down-wind, but spatially arranged along kinship lines, with here and there kraals of stone and thorn bush to protect the animals from predators and cold (Faegre 1979). Individual tents, and even the whole camp, shift location every few days for basically hygienic reasons, even when water and grazing are adequate. Their only traces are the blackened hearthstones whose power of evocation provided the standard nostalgic opening of the classical Arabic ode (Gibb 1963, 15–17). Relative isolation, as well as the group's vulnerability and its subsistence base, encouraged aggressive behaviour (Meeker 1979, 7).

The aggregate, seasonal movements of camping groups produce the configurations of mean migration types classified by Johnson (1969). *Horizontal nomadism* (Fig. 4.9A) takes place in level or rolling country. What Johnson calls 'elliptical' movements are governed by the seasonal expansion and contraction of the grazing range with respect to an area of relatively secure summer water supply. These movements are plotted in detail by small camping units drifting outwards from the base area at a leisurely pace, but then converging back rapidly along a much more concentrated route. Johnson's 'pulsatory' movements involve much greater concentration of the migrating groups, with advance and withdrawal taking place over much the same territory.

*Vertical nomadism* (Fig. 4.9B) arises from the exploitation of the altitudinal variations in seasonal grazing availability. As far as studies allow (Hütteroth 1973), it seems that in most of the region this type of nomadism conforms to Johnson's class of 'constricted oscillatory' movements. These involve related camping groups moving through mountain valleys and passes on their routes between relatively concentrated winter grazing in comparatively low-lying areas and the wider, more dispersed summer pastures in the mountains. Some groups have permanent houses on their upland grazings. Both vertical and horizontal nomadism take place in areas (*dirās*, 'tribal territories') where the groups concerned have established rights of access to water and grazing.

While Johnson's classification adequately describes the types of spatial patterns which nomadic movements make in particular areas, it assumes that they are always determined by physical conditions and that they would be replicated by whatever pastoral nomadic group happened to be in the area at a particular time. No account is taken of probable economic and political influences, or of the fact that the historical record shows that pastoral nomadic groups have changed the location of their activities (pp. 166–7, 226–8). Some groups may have moved to save their animals from the effects of drought; others may have been forced out by militarily superior rivals. In both cases, their wanderings would have been shaped by the availability of water and grazing, as well as by the degree of resistence encountered from other pastoral nomads or central governments. Fighting would ensue, involving the destruction of villages and crops, before a new *modus vivendi* was worked out between competing groups.

Although the life of pastoral nomads is dominated by the needs of their animals, there is enough ethnographic evidence to show that it has also involved various subordinate but significant activities. Spinning and weaving necessarily derive from the production of wool and hair. Hunting with dogs, hawks, spears and bows was once possible on the open range and in the mountains. Also significant, at least in certain contexts, were the supplying of firewood to towns and villages, and the collection of honey, wax and dyestuffs (Bent 1891; Burckhardt 1822).

## Woodland Communities

Gathering was a particular feature of life in the woodlands of the region. The acorn cups of the valonia oak[3] were collected for use in tanning and the production of a black dye, while a parasite scale insect[4] was gathered from the kermes oak[5] to produce a red colour. Various trees were tapped to produce gums and resins. Among the most important products were pitch from the Phoenician juniper,[6] resin from the Aleppo pine[7] and turpentine from the Maritime pine,[8] but the most famous were incense and myrrh. Incense is obtained chiefly from the bush, *Boswellia sacra*, found almost exclusively on the slopes of Jebel Qara, while myrrh is obtained from the tree, *Commiphora myrrha*, which has a wider distribution in southern Arabia (p. 14). Trees are felled and sawn for timbers and planks, each type of wood having its own particular uses. Firewood is collected and charcoal is burnt (Cuinet 1890–94; Orgels 1963, 24).

All these activities could be combined with cultivation and nomadism. Recent travellers and commentators, however, have noticed a degree of specialisation which presupposes the existence of exchange systems. Specialist groups, like the Tahtaci of the Taurus Mountains in southwestern Asia Minor, follow a semi-nomadic form of existence, moving their small camps as their activities destroy or deplete the resources on which they depend (Bent 1891; Cohen 1967; Planhol 1950, 81–8; Planhol 1963). Regeneration is slow under Middle Eastern climates. Pressure would obviously vary according to demand. While specialist woodland communities are likely to have existed for many centuries, the sending of expeditions to Mt Lebanon from Egypt and southern Mesopotamia in ancient times (pp. 112–14) may hint at either a relatively late develop-ment or a degree of plundering.


## Fishing

Fishing, an ancient way of living, has also involved a degree of seasonal nomadism in recent times, both on the Gulf coast and among fishermen in nineteenth-century Galilee (Donaldson 1981; Masterman 1908; Sweet 1968). The behaviour of the fish shoals gave it this character. Again, while fishing has often been one of a range of subsistence activities, especially on rivers and lakes, specialist communities — the *ikthoufagi* ('fish-eaters') of the ancient travellers (for example, *The Periplus of the Erythraean Sea*, 31, 35, 39) — have long existed on coasts adjacent to especially prolific fishing grounds (p. 25). In recent times, their villages have been relatively isolated, and consisted of rather flimsy houses scattered behind the strand line. Part of the catch is consumed fresh by the fishermen themselves or sold directly to consumers and middlemen. A proportion is dried for trade inland or for export overseas. It is destined for human and animal food or for spreading on the land as manure. A large quantity of fish, however, was converted into oil in the past by the simple process of leav-ing the bodies to decay in pits on the sea-shore. The oil was used for lighting and for preserving the planking of ships (Ingrams 1966, 139–41). Many of the techniques used are shore-based, and therefore possibly older than the use of seine nets from various kinds of craft. Drag nets are com-mon, as are cast nets and lines. Spears are sometimes used. Tidal weirs and traps are found in some creeks and estuaries (Fig. 4.10) (Serjeant 1968; Sweet 1968).

Pearl fishing is found in water up to 20 fathoms deep at various points in the Red Sea and particularly along the southern shore of the Gulf (p. 25) (Belgrave 1934; Bowen 1951). It is another seasonal activity, condi-tioned in this case by the water temperature, which must exceed 30°C for continuous diving to be possible, but it also fitted in well with the laying-up and fitting-out of deep-sea ships (Villiers 1940, 296). In recent times, skin divers have worked from ships operating in groups of ten to twenty. The oysters are now usually opened on board, but the piles of

Figure 4.10: Fish Trap

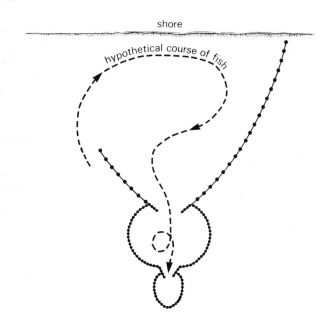

shells along parts of the Gulf coast suggest that the custom was once to open them on shore, a practice which could be related to a habit which survived until at least the Second World War, that of the smaller craft coming into the beaches for the men to sleep ashore (Villiers 1940, 299). Only about a third of the oysters retrieved actually contain pearls of reasonable size. Pearl fishing was thus a precarious activity. The life of the pearl fishers was harder even than that of other fishermen and sailors. Constant immersion reduced life expectancy, while the returns from about four months' work (June–September in the Gulf) were uncertain and yet had to meet family subsistence needs for the whole year. Debt was inevitable, and bondage was perhaps always a feature of the divers' life.

**Town Life**

Towns have been a feature of the region since at least the fifth millennium BC (pp. 97–105). They have contained a variety of life-styles. Many were found in other settled communities, but they were adopted by larger numbers of people in the towns, where the range and diversity of economic activity were greater than in any single village. These characteristics, together with larger total populations and higher residential densities, gave town life a special quality when compared with those modes of living found in the countryside. It was more animated, more variegated and involved a greater degree of social intercourse.

Although primary production was the concern of often quite sizeable groups,[9] a great many urban households were at least one remove from it. They were united by their dependence upon three particularly urban functions. The first was the control and mediation of political power in society (Murvar 1966). Throughout time, towns have been the preferred residence of the region's elites, of landlords and nomadic chiefs, of administrators and religious officials (Baer 1964, 177–203). Their power and wealth have been displayed in the size and magnificence of the buildings they commissioned. The rest of the urban community was, to an extent, dependent upon them, though the degree and nature of dependence probably varied from place to place and time to time. The withdrawal of an elite frequently initiated a downward spiral in political and economic importance, followed by substantial loss of population and the decay of the physical fabric. Similarly, the decision of a ruler to create a new administrative centre, and the consequent residence of the elite in it, brought growth and expansion (Ibn Khaldūn 1967, 263–6, 272–3, 289–91). The second basic function of towns has been the tranformation of commodities, often produced in the countryside, for further use and consumption. The third function is a related one. It is the co-ordination of exchange between the various modes of living found in the surrounding area. With the power and control exercised by the elite, both of these functions meant that the towns were, in a sense, parasitical on the countryside (Costello 1977, 37–8). In addition, they were able to structure the territory around them in perceptual and functional terms. The life of cultivators and nomads was dependent upon them. Cultivation, for example, expanded and contracted, while crops changed, according to urban demand (English 1966). In turn, urban life itself depended upon communications. On the one hand, each town was linked with its hinterland. On the other hand, each urban region interacted with others (Beckett 1966; Ehlers 1975; English 1966). The whole settlement structure might be given an hierarchical form by the integration necessary to political and commercial systems.

The three basic functions produced two interdependent life-styles which were widespread in towns and particularly characteristic of them. They are, first, that of the merchants and shopkeepers, and second, that of the craftsmen or artisans. Both of them, as well as the life-style of the elite, required servants and labourers. Between them, they may have helped to generate a significant informal economy which was carried on by the partially employed, hucksters, thieves and beggars (Lane 1842, 295–300).

A considerable diversity of manufacturing activity was normal (Wulff 1966), though the range expanded over time with technological development. Artefacts recovered by excavation at Ur suggest that at least five different crafts were practised there around 2800 BC, while contemporary texts add another three or four (Wright 1969, 40). Evliya Çelebi, the seventeenth-century Turkish travel writer, listed 136 separate trades in Cairo, while an incomplete list dating from 1801 names 74 (Raymond 1973, 204–5). A number of trades produced goods for everyday consumption (pottery, shoes, certain textiles, for example) and could have been located in both the town and the country. Many others produced goods which could only be afforded by the elite (e.g. seals, jewellery, elaborate textiles), and therefore required a concentration of patrons.

Detailed information from the time of the Arab caliphate onwards makes clear that extreme specialisation was characteristic of some crafts, notably according to type of fabric in textiles and particular colours in dyeing (Costello 1973; Lombard 1975, 181–6; Raymond 1973, 215–17). The units of production were usually small and totally unmechanised. Where work was not carried out entirely in the open air or, like spinning and some weaving, in the home, small cell-like workshops opening off the street were normal. Evliya Çelebi's figures point to an average workforce of a master and two or three journeymen in most workshops. Larger units were evinced in the historical record from time to time. Even then, the employment of more than about a dozen workers was exceptional and all production was still by hand (Raymond 1973, 219–23). Powered machinery was developed in the region comparatively late, and until the nineteenth century AD was virtually confined to grinding corn, olive pressing and cane crushing.

Craftsmen are often their own retailers and this is likely always to have been so, except where work was specially commissioned. Other forms of retailing in the Middle East today which are probably of some antiquity are the daily selling of perishable foodstuffs, especially bread in the towns, and small amounts of cooking oil and dry goods like spices, pulses and flour. As in other societies, the purchase of penny quantities is less a response to lack of storage space than a reflection of cash availability and consumer habits. The proliferation of shops selling an almost identical range of goods is striking today. It may be related to the kinship basis of buying and selling, and the need for credit (Potter 1955). Conceivably, it is an ancient pattern. As with workshops, retail outlets usually consist of one or two small rooms. Few assistants are employed: Evliya Çelebi's figures give a mean of 2.6 for seventeenth-century Cairo (Lane 1842, 291–3; Raymond 1973, 272).

The distinction between retailers and merchants is sometimes difficult to draw. In general, merchants are usually and chiefly engaged in wholesale trade, that is, in the bulking and dispersal of commodities, and act either on their own account or through intermediaries. A mode of life, diachronically similar, is revealed by cuneiform texts from Mesopotamia (Larsen 1976; Leemans 1960 and 1968), documents of the eleventh century AD from the Cairo Geniza, (Goiten 1967 and 1973), an Armenian merchant's commercial register (*roozlama*) of the late seventeenth century (Khachikian 1967) and the business correspondence of English factors resident in Aleppo during the early eighteenth century (Davis 1967). It has been characterised as 'peddling' (Steensgaard 1974, 21–30, 43–7). Trade was usually carried on by a partnership, in which one or more of the partners was responsible for the actual buying and selling away from home. This might involve constant travel or prolonged residence in a distant town, probably both. Means of transferring credit were early evolved. Good health was required, together with strong nerves, a sure feel for trends in the market, and patience to postpone buying and selling until the moment was right. Commodities were usually bought and sold on a small scale. Combined transactions were frequent, involving barter and cash. Substitution was easy and depended upon availability, demand and price. Specialisation was abnormal on the part of the merchant. For example, the Geniza documents mention about 200 separate items of wholesale trade, 40 of them of major importance (Goiten 1967, 209–10), while the Armenian, Hovhannes, son of Father David, handled 174 different types of goods in the decade for which we have evidence (Khachikian 1967). Trading was discontinuous and sporadic. This was partly a result of small consignments and low stocks, perhaps arising from a combination of under-capitalisation and the limited capacity of carrying systems. It was also a consequence of seasonality in cultivation, nomadism and gathering, as well as dependence upon the arrival of caravans and sailing ships.

The bulking of rural produce was almost certainly organised in a similar way. Bent (1891) reports that brokers visited the summer camps of pastoral nomads in south-eastern Asia Minor, buying cheeses, wool, hair and live animals from whoever would sell, and then retired to the towns when they had acquired as much as they could handle. Similarly, brokers toured the silk villages of Gilan, contracting with local farmers and returning a short time later to collect the cocoons (Lafont and Rabino 1910). The principals must subsequently have forwarded bulked consignments or themselves arranged for preliminary reeling and spinning. The webs of credit and clientship operating through these transactions extended urban influence and control deep into the countryside.

## Carriers and their Parasites

Wholesale trade required the conveyance of goods from one place to another. Indeed, urban life as a whole was 'linked to the arithmetic of distances, the average speed of travel along the roads, the normal length of voyages . . .' (Braudel 1972, 282) as well as with the carrying capacity of pack animals and ships. Land conveyance was organised in caravans, partly in order to carry largish consignments at any one time, partly for mutual co-operation and companionship on the journey, but basically for protection. Before the domestication of dromedaries and camels, the main pack animals were donkeys. One donkey caravan mentioned in a letter from second-millennium Mari was 50-strong (Leemans 1960, 134), but the average load per head was probably around 200 lb (90 kg). Although mules and horses were used after the second millennium, the classic pack animal became the camel (Bulliet 1975). Depending on breed and size,

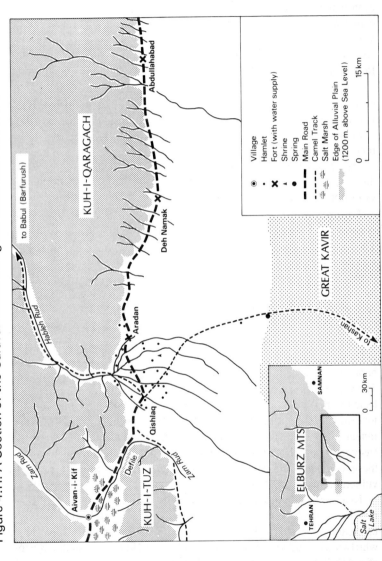

Figure 4.11: A Section of the Caravan Route along the Southern Foot of the Elburz Mountains

it could carry between 380 and 1,000 lb (171–450 kg), that is, the equivalent of the load of 3 to 5 horses (Tavernier 1684, 46, 50–1). The camel's endurance meant that more direct routes could be used than were possible even with donkeys. Fewer drivers were needed (Bulliet 1975, 22). The camel was faster than a donkey[10] and could cover 15–20 miles (24–32 km) comfortably in a day (Bulliet 1975, 24), though halts were made when convenient and according to water availability (Tavernier 1684, 45–54). Routes were plotted in detail according to the location of reliable water sources (Fig. 4.11). Night marches were common in summer to avoid the heat. Caravan travel was rendered uncomfortable by long wearisome hours in the saddle, the heat and cold, as well as by the slow monotony of the day's march, and constant anxiety over food, water and robbers. It was relatively expensive, partly because of the time factor but also because of the payments necessary as tolls and to secure escorts and safe conducts. There was no alternative to the caravan in winter when the seas were closed.

By the seventeenth century, and possibly long before, two main types of caravan could be recognised (Owen 1981, 53–4; Tavernier 1684, 45–6). The most important in terms of goods carried and regional connectivity was the type which arrived and departed at fixed times, to and from particular destinations. Great expectancy built up a few days in advance of the caravan's predicted passage through a village and its anticipated arrival at its destination. Long after the probable heyday of the caravans, there were two or three a year between Damascus and Baghdad in the late eighteenth century, four from Aleppo to Anatolia and two from Aleppo to Persia via Baghdad. The mean size was about 1,500 camels, but probably every seventh camel carried provisions and camping equipment. The second type of caravan set out only when enough travellers had collected to make a journey safe and economic. In the eighteenth century, for example, these caravans carried the trade between the eastern Mediterranean ports and the interior towns of Aleppo and Damascus. Travellers frequently complained of the delays and relative expense involved, and these comparatively small caravans seem to have been particularly vulnerable to attack.

River craft were used wherever possible. The current of the Nile carries boats steadily northwards, while the prevailing winds facilitate upstream voyages. However, continual shifts in the river bed, squalls and whirlwinds made navigation hazardous (Lane 1842, 302), and one letter of c.AD 1100 speaks of the 'great horrors' of the voyage (Goiten 1967, 297). Craft were made from papyrus and later timber. Downstream navigation was possible on the Tigris and Euphrates at all seasons, though the current was generally considered too strong on the Tigris for upstream movement between March and November. A variety of craft was used in Mesopotamia, but in the upper reaches of the river, rafts of lumber on inflated animal skins (*keleks*) were common from ancient times (Barnett 1958; Naval Intelligence Division 1943, 368, 466; Naval Intelligence Division 1944, 41, 114, 588, 561).

A variety of 'swiftly-gliding ships' (Qur'ān 51.2)[11] plied the seas from an early date, their type and size changing with technological development over time (Bass 1972; Landström 1969). For example, in the Gulf the use of loan words, usually from Portuguese, indicates the adoption of European techniques in ship-building (Johnstone and Muir 1964). The most notable was the replacement of the techniques of pegging and sewing the planking by nailing (Hornell 1942; Hourani 1963, 89–99). *Baghalas* ('she-mules', Figure 4.12) were the largest local sailing vessels in the region in recent times, their transom sterns betraying European influence. They are 100–140 ft (30–42 m) long, 20–28 ft (6–8 m) in beam and $11\frac{1}{2}$–18 ft (3–5 m) in depth. They have a capacity of 150–400 t (Hornell 1942). This compares with an average of 4–500 t for a European East Indiaman of around AD 1600 and 500–1,500 t for an English one of 1793 (Philips 1940, 80; Steensgaard 1974, 171). None the less, each baghala could carry at least as much as 300–800 camels, probably more, and at considerably less cost. The Geniza letters speak of up to 400 passengers on some ships, though most of them must have slept on the cargoes. Use of the lateen sail meant that baghalas and similarly rigged vessels could sail close to the wind, but they could not tack and needed large crews to wear ship (Hourani 1963, 109–10; Villiers 1940, 24). Navigation was always hazardous in Middle Eastern waters (pp. 21–4), but from at least the sixth century BC the risks were somewhat reduced by the use of sailing directions. From an early date, navigational instruments such as the *kemal* ('guiding line') and astrolabe made direct, long-distance voyages possible and minimised delay (Collinder 1954, 45–98; Hourani 1963, 106–9; Taylor and Richey 1962, 43–8). Even so, the importance of cabotage, as well as the 'peddling' trade of captain and crew, meant that coasting was frequent (Villiers 1940, App. 2; Villiers 1948).

Goods in transit are a temptation to the unscrupulous and desperate. Although various disadvantaged groups in the Middle East have become 'social bandits' in Hobsbawm's sense (Hobsbawm 1969, 13–14), others incorporated plundering into their normal mode of living.[12] This often happened where certain conditions were met (Semple 1916). One was the scarcity of more conventional resources. Another was ready access to a well-frequented route on which to prey. Natural hazards were

Figure 4.12: Types of Sailing Vessel (hulls only) Used
Recently in the Gulf, Indian Ocean and Red Sea

Baghala

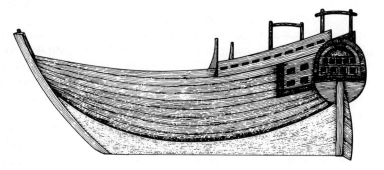

Būm

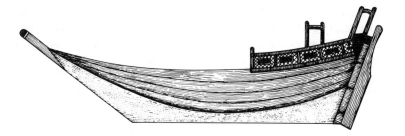

Sambūk

important, too. On land, these were primarily the rough and narrow passes through mountains like the Zagros and the mountains of Armenia. Caravans were forced to go slowly in single file and straggled over long distances, while cover for attackers was provided by rocks, steep slopes and bad weather. At sea, shoals, reefs and squalls could disable ships sufficiently to make them a ready prey for pirates. Creeks and bays, spits and dunes, provided hiding places. Isolation and difficult access made policing hard in such areas, while refuge was easy in caves, creeks and forests. Some of the booty secured by bandits and pirates was doubtless consumed directly. Most, however, must have been sold. Prisoners were usually held for ransom (Earle 1970, 73–93), like the youthful Caius Julius Caesar (c.102–44 BC) (Plutarch, *Caesar*, 1.4–2.4). Although the bandits were usually indigenous, from time to time the openness of the seas and the rich pickings of the region's trade tempted outsiders to try their luck. These included the Knights of Malta, who ranged the Aegean islands and the eastern Mediterranean almost annually from the 1530s down to 1798 in a crusade against the infidel (Earle 1970; Volney 1788, II, 87–8), as well as the notorious Captain Kidd, who was active off southern Arabia in the 1690s and was simply interested in loot (Cotton 1949, 142, 146–7; Serjeant 1963, 112–29). Their activities, however, required more than the light craft beloved of local pirates, whose operations were essentially hit-and-run affairs. They depended, in fact, upon a chain of technological development and transfer which was only just beginning when the earliest civilisations of the region began to flower.

## Notes

1. The addition of chaff prevents crushed olives and other oil-bearing seeds becoming too pasty for the subsequent crushing process. The 'cakes' are used as cattle feed (Wulff 1966, 397).

2. The ecological explanation goes back at least as far as Aristotle (384–322 BC) (Briant 1982, 26–7). Lattimore (1951, 331–4) suggested that nomadism was a controlled political response and not an environmentally induced reflex. See also Rowton (1973).

3. *Quercus macrolepis* Kotschy or *Q. aegilops* Boiss.

4. *Coccus ilicis* Planch.

5. *Quercus coccifera* L.

6. *Juniperus phoenicea*.

7. *Pinus halepensis* Mill.

8. *Pinus maritima* Lam. or *P. pinaster* Ait.

9. More than half of the population of Cairo in the mid-nineteenth century and 13 per cent of the labour force of Kirman in recent times (Costello 1977, 23).

10. Dickinson (1951, 409–19) gives the average speed of a laden camel as about 5 km/h,

though 8 km/h was not unusual. A riding camel at full speed could cover 23 km/h.

11. Translated by Dawood (1959, 116).

12. This is an ancient view, traceable at least to Aristotle (Barker 1948, 24; Briant 1982, 26–7).

# 5 THE EARLIEST CIVILISATIONS, c.5000–525 BC

The period surveyed in this chapter was critical in the historico-geographical evolution of the region. By its start the climate had settled into its modern forms, though the extreme aridity of Egypt, for example, may not have been attained until the third millennium (Beug 1967; Bottema 1975–7; Butzer 1972, 584; Horowitz 1971; Murray 1951; Zeist *et al*. 1975; Zeist and Bottema 1977). The sea had more or less attained its present level, providing a definite frame to parts of the region. Between about 5000 and 3500 BC village-based agriculture was consolidated throughout the Middle East, while its detailed repertoire was extended. The mysteries of metal-working were fully mastered over the next two or three millennia. On the alluvium of Lower Mesopotamia and Egypt the first literate civilisations emerged and began to record their affairs. 'Towns' may already have existed in at least some parts of the region, but they became both more numerous and more important socially and politically from the opening of the fourth millennium. Urbanisation seems to have been closely associated with the appearance of recognisable polities (Figure 5.1). Most of these early states were small and highly localised. They merged, collapsed and expanded in a kaleidoscope of political manoeuvring and warfare until by 525 BC Cambyses II, King of Persia (529–522 BC), united the whole of the Middle East for the first time into a single empire. Changes in subsistence patterns and technical ability were basic to the comparatively short-lived success of this enterprise and they furnish a convenient theme with which to begin the examination of 4,500 years of historico-geographical development.

## Subsistence and Technology

The life of the earliest agricultural villages in the region was based on growing a comparatively narrow range of cereals and pulses and herding cattle, sheep, goats and then pigs. It was perhaps more precarious than hunter-gatherer systems. Drought could decimate herds and destroy crops, while locusts swarmed with devastating effect; alternative sources of food were much reduced. Land preparation depended upon the use of digging sticks and hoes (Steensberg 1977). Power was restricted to human muscle and no draught or pack animals appear to have been used. Pottery

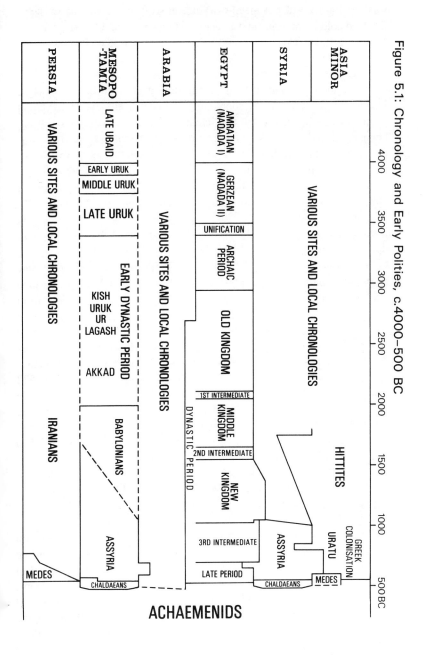

Figure 5.1: Chronology and Early Polities, c.4000–500 BC

was certainly fired, but metal-working, in so far as it already existed, involved cold hammering and annealing but not chemical transformation. While practically everything was to change in the period under consideration, the risk of failure in subsistence systems remained large; it may even have increased further.

Hand tools remained in use, but during the fourth millennium ideograms appeared on Mesopotamian cylinder seals and in Egyptian paintings which suggest that two-handled ploughs pulled by oxen were being employed. Early association with the two major riverine areas may indicate that the original development of the plough was related to irrigation, perhaps even its extension into the dry margins of naturally flooded areas. Elsewhere in the region, evidence for the use of ploughs appears rather later. It also shows the emergence of the single-handled implements of varying type which have remained characteristic of the region down to the present day (Haudricourt and Delamarre 1955, 65–78). Independent invention is possible, but uncertain. None the less, the effects of ploughing are clear. Land preparation was speeded up and larger areas were cultivated. The cost, however, was high. Large animals had to be supported specifically for the plough, necessitating the provision of winter fodder. The risk of erosion was increased.

Although fruit and nuts gathered from the wild were appreciated as much by the people of the earliest agricultural villages as by their more dependent ancestors, the available evidence suggests that deliberate planting and control came much later than the domestication of cereals and pulses. Olives appear to have been cultivated first in their 'natural' habitat around the eastern Mediterranean during the fourth millennium BC (Boardman 1977; Zohary and Spiegel-Roy 1975). The evidence for cultivated vines and — more sparse — for figs comes from the same areas, but a millennium or so later. All three appear to have become widespread in the rain-fed districts of the region during the third and second millennia BC (Renfrew 1975, 125-36; Zohary and Spiegel-Roy 1975). Vines and figs also did well in Egypt, but climatic conditions away from the extreme north coast and the Faiyūm were too stringent for the olive (Kees 1977, 81). The date palm appears to have been domesticated in Lower Mesopotamia, probably before 4000 BC. Commercial production remained a speciality of the original homeland, but cultivation of the date palm became widespread in the warmer southern parts of the region, including Egypt (Preussner 1920; Zohary and Spiegel-Roy 1975).

To be successful, all the species noted require vegetative propagation. This may explain their comparatively late domestication. On the other hand, so many varieties are present in the 'natural' vegetation assemblages of the region that, once the simple techniques were understood, adaptation and diffusion presented fewer difficulties than in the case of cereals. But if the amount of labour required in grafting and pollination was large, the eventual rewards were enormous. Similar opportunities were offered by flax. Although known in western Syria from perhaps as early as 6000 BC, it spread across the region during the next millennium and became associated especially with Egypt (Kees 1977, 77; Zeist 1977). While linen made comparatively light clothing, the new food crops increased human calorific and vitamin intake, perhaps improving life expectancy. Linseed and olive oil could be stored for future use and were burnt with wicks to give light. Wine supplemented beer in cheering the heart of man and enhanced his festivities (Psalm 103: 15; Stanislawski 1975). The concentration of sugar by drying dates, figs and grapes in the sun allowed preservation and storage, not only over winter but for longer periods of time. In all cases, the almost inevitable surpluses over immediate requirements must have sown the twin ideas of specialised cultivation and exchange or trade.

Meat was the prime aim of early animal domestication. As success was attained, so the importance of hunting for food declined, especially after the fifth millennium, though it remained a popular recreation with social elites (Bökönyi 1976). Milking may have been a subsequent development. At least, the earliest evidence is from the third millennium BC and comes from representations in a temple frieze at Ur in Lower Mesopotamia (c.2900 BC) and an Egyptian sarcophagus of the XIth Dynasty (c.2200 BC) (Brothwell and Brothwell 1969, 50–2). Fibre production is more certainly a secondary development since both goats and sheep possessed relatively short and coarse coats in the wild. Some mutation was required to increase the softer longer fibres, especially in sheep. Although it may have begun early (p. 7), the effects became increasingly clear in the representational art of the fourth-millennium Mesopotamia, while texts from the end of the next millennium frequently contain shearing lists (Bökönyi 1976). Large herds of sheep, goats and cattle are evinced in Mesopotamian and Egyptian texts, while great numbers of animals were offered in sacrifice: 50 rams for the daily sacrifices at a temple in Uruk, for example (Kees 1977, 89; Numbers 28:1–29:40; Pritchard 1955, 343–4).

The range of animals kept was increased during the same period. Zebu cattle, with their characteristic hump, appeared in Mesopotamian art by 3000 BC and became popular in Egypt from the middle of the second millennium (Zeuner 1963, 216–40). Buffaloes were known in Mesopotamia before 2500 BC but did not spread to other riverine areas until much later (Zeuner 1963, 249–50). Geese, at least partially domesticated, are evinced from third-millennium Egypt, but domesticated fowl do not appear anywhere for roughly another thousand years and are not clearly recognised elsewhere in the region until much later (Brothwell and Brothwell 1969, 55). Honey, 'probably the best natural source of energy available to man' (Brothwell and Brothwell 1969, 73–7), was consumed from earliest times, but the first evidence for domestication of bees is an Egyptian relief of mid-third-millennium date. It shows the extraction and packing of honey, as well as a set of hives constructed from clay pipes, piled in rows and not dissimilar from those in use today. Bee-keeping spread widely across the region over the next 1,500 years, though in Lower Mesopotamia date-syrup was more widely used than honey (Brothwell and Brothwell 1969, 73–7). Possibly even more significant than this development, because it increased human interaction, was the domestication of the transport animals: donkeys and camels.

The wild donkey had an extensive distribution, but before the end of the fourth millennium it appears to have been domesticated only in Egypt and was probably first used to carry burdens as a substitute for cattle (Isaac 1970, 86–9; Zeuner 1963, 374–7). Adoption elsewhere was comparatively slow, perhaps because of effective competition from cattle or onagers. The latter were certainly employed in Lower Mesopotamia, where they are shown drawing four-wheeled chariots. Onagers evidently went out of favour with the spread of horses and camels and, doubtless, of donkeys too. Once adopted, donkeys became the basis of both long-distance transport and pastoral nomadism. Although they can carry less, they possess definite advantages over cattle. Donkeys can travel more quickly. They require less frequent watering and eat a wider variety of plants. Thus, donkeys were not only more economical to keep than cattle, but they also allowed relatively large amounts of goods to be carried across some of the more arid parts of the region. Long-distance trade was facilitated, while nomads could range further than before; in particular, they could take advantage of the seasonal availability of grazing beyond the normal limits of the annual rain-fed zone. Range and carrying capacity, however, were increased with the domestication of the Bactrian camel (*Camelus bactrianus*) and the dromedary (*Camelus dromedarius*).

The Bactrian camel is differentiated by its double hump and shaggy coat. Little is known of its early history, but domestication seems to have taken place in central Asia during the third millennium (Isaac 1970, 89–92; Zeuner 1963, 359–62), perhaps initially for pulling ploughs and waggons (Bulliet 1975, 153–6). References to its presence in the Middle East became fairly common from about 1125 BC but, because of its greater tolerance of cold than the dromedary, the pure Bactrian camel was used in the northern parts of the region, especially in the interiors of Persia and Asia Minor (Mikesell 1955, 231–45). Hybrids with dromedaries became favoured in these areas for their fighting abilities, as well as for their strength and stamina.

The early history of the dromedary, or single-humped camel, is equally obscure. It was part of the natural fauna of Arabia (Zarins 1978), and both the ethnographic and the literary evidence suggests that domestication took place in the south, possibly before 2500 BC. Milk may have been the initial objective, but the development of crude, pad-like saddles allowed the dromedary to be ridden and to carry loads (Bulliet 1975, 34–78) and it began to be employed by nomads and merchants. During the second millennium it became known as a beast of burden in the borderlands of Arabia, especially on the west and probably in association with the trade in myrrh, incense and spices from the south (Bulliet 1975, 57–65; Zeuner 1963, 341–4). Settled communities seem to have resisted adopting the animal themselves for several centuries. Zeuner (1963, 364) attributed this to its low reproduction rate, its notorious bad temper and the consequent difficulty of training it, its susceptibility to disease outside the more arid parts of the region, the need for extensive pasturage (which was not available in the settled areas) and, finally, to its unpleasant smell. The list should be increased to include the availability of a viable alternative in the donkey and the imperfections of the saddle, which at first made the dromedary relatively inefficient for transport and riding. None the less, the expansion of Assyrian power into Syria and northern Arabia brought contacts with tribes which were not only breeding dromedaries but also apparently employing them in a military capacity, perhaps as mounts for archers. Appreciation of their possibilities led Ashurbanipal (668–626 BC) and then the Persian kings, Cambyses II (529–522 BC) and Xerxes I (485–465 BC), to make some use of them as transports and light cavalry during their campaigns (Zeuner 1963, 342–53). However, the great breakthrough, which led to the widespread use of dromedaries and hybrids in the region, came just after the period now under review. It was the invention of the rigid, north Arabian saddle (Bulliet 1975, 87–104).

The most important technological advance between 5000 and 500 BC was probably the mastery of industrial technology. Copper and gold were already being worked by the beginning of the period, a development perhaps stimulated originally by the colour of the metals, the availability of ore in native form and relatively easy access (Jesus 1980; Wertime 1973). An early origin may be postulated for smelting, since lead —

evidence for which occurs around 6400 BC at Çatal Hüyük in central Anatolia (Mellaart 1967, 21) — can only be produced by this means. However, the techniques do not appear to have become common until the fifth and fourth millennia BC, when they were applied particularly to copper. Several scholars have argued that they spread across the region along pre-existing trade routes, especially those leading to the obsidian sources of Anatolia (Jesus 1980; Wertime 1973), a sub-region which is also rich in non-ferrous metals, but the notion of independent discovery cannot be ruled out. Copper-smelting, of course, implies an ability to achieve temperatures of at least 800–900°C in the furnace. The use of charcoal as fuel is indicated because of the higher temperatures required to reduce the non-oxide ores used from the beginning of the third millennium (Jesus 1980, 37–9). Once developed, smelting and the greater use of the earlier-discovered techniques of casting must have eased production, increased the general use of metals and stimulated mining. These developments, in turn, may have encouraged the concentration of metallurgy on and around the major ore deposits, leading to a complex pattern of trade, probably embracing the whole region (Jesus 1980, 148–9). The use of arsenic bronzes would have emerged naturally when the mixed sulphide ores of Anatolia, Persia and Palestine came into general use. The development of tin bronze, however, remains something of a mystery. There is no doubt that it replaced arsenic bronze during the third millennium BC (Wertime 1973), but where did the tin come from? Tablets found at Kültepe (ancient Kanish) in Anatolia reveal the transport of tin to that point from Asshur in Upper Mesopotamia around 1900 BC (Wertime 1973), while a text from Mari on the Euphrates indicates that about 1780/1760 or 1715/1695 BC, depending on the chronology accepted, tin was being distributed westwards and northwards through that town. In both cases, the sources may have lain in north-western Persia (Leemans 1968; Malamat 1971). The authority of Strabo and the recent discovery of ore near Mashhad point to interior Persia as another possible source area (Strabo 15.2.10; Wertime 1973). Several possible sources are now known from Anatolia (Jesus 1980, 55–6) but, like those in the southern parts of the Eastern Desert of Egypt, they may not have been worked in the second millennium BC (Waldbaum 1978, 65). Wherever the tin came from, the employment of bronze allowed the production of hard and durable tools, including many of the types still used today (Derry and Williams 1970, 118), thereby increasing the manipulation of other materials. It also led to the production of swords.

Iron objects were made in the region from at least the third millennium (Waldbaum 1978, 17–23). Although magnetite was a probable early source, iron could just as well have been formed as a by-product of mixed-ore smelting (Wertime 1973). Initially a rare commodity, worth 60 times more than copper and twice the value of silver according to documents from Ugarit, iron seems to have become more common during the twelfth century BC (Waldbaum 1978, 17). None the less, Waldbaum's study of the Bronze Age/Iron Age transition in the eastern Mediterranean region shows that there were still many more bronze objects on archaeological sites in Anatolia, Syria, Palestine and Egypt yielding iron objects (about 88 per cent of the combined total of bronze and iron objects). During the eleventh century, the number of iron objects increased to 23.5 per cent of the combined total of iron and bronze objects. In the next century, they reached 31.4 per cent; iron was predominant for tool-making (63.7 per cent of the combined total) but bronze was still preferred for weapons and armour (53.3 per cent of the combined total).[1]

This relatively slow transition to the predominance of iron, equally apparent outside the Middle East (Pleiner and Bjorkman 1974), has been explained as the result either of a Hittite monopoly over the production of a 'secret weapon' or of an early mastery of iron-making techniques in Anatolia, as claimed by several ancient authors (Wertime 1973). A more convincing argument has been advanced by Waldbaum (1978, 67–73). She points out that the superior properties of iron in hardness, availability and cheapness were by no means self-evident in the second millennium, or indeed later. Greater hardness, in particular, is only achieved by adding carbon (which might have been accidental when heating with charcoal) and by quenching and tempering (both techniques inimical to the production of good-quality bronze). The application of these techniques seems to have been inconsistent and accidental until much later than the second millennium, and their importance was not fully appreciated until very recent times. In addition, the production of iron from the ore required considerable labour before the inventions of the tilt hammer and the blast furnace in medieval Europe, notably in repeated hammering at red heat to remove slag from the bloom (Derry and Williams 1970, 121; Healy 1977, 182–3). By comparison, then, bronze is not only easier to smelt and cast but, when hammered cold, is actually superior even to air-cooled steels (Waldbaum 1978, 68). Waldbaum goes on to argue that iron came into common use only when a shortage developed in the supply of tin. Similar chronologies in the increase of iron usage suggest that the crisis must have affected the whole region. From about 1200 to 800 BC one disaster after another seems to have struck the excavated sites of the region and considerable cultural changes took place. Waldbaum

(1978, 72) suggests that 'it is tempting to see in these events and their aftermath a situation in which the main supply link to the eastern source of tin was cut and established trading systems completely disrupted'. There was little choice but to make greater use of a second-rate, but available, material — iron. Easier availability and cheapness may then have given iron a competitive edge difficult to overcome, even though the brief revival of bronze in Greece after 900 BC (Waldbaum 1978, 71) may have affected the Middle East as well.

## Environment

Domestication was extended and metallurgy mastered within a physical environment which was itself dynamic. All parts of the region must have been affected as, on the one hand, woodland was cleared for cultivation and timber was cut for fuel and construction and, on the other hand, precipitation fluctuations modified both vegetational assemblages and landforms. Human interference on a massive scale during the second and first millennia BC is apparent in pollen diagrams constructed for western Anatolia. Not only are woodland clearance and cereal-growing apparent, but the introduction of fruit trees is obvious, and it has been suggested that the selective cutting of pine encouraged the greater relative importance of oak (Zeist *et al.* 1975). Development of the qanāt by the end of the second millennium allowed settlements to spread along the foot of mountains in the interior of Persia (Goblot 1963). Other techniques of water management facilitated the penetration of extremely arid areas like the Negev, with significant positive and negative effects in the catchments of seasonal streams (Stager 1976). Change in ecological systems, however, is at present best documented for Egypt and Lower Mesopotamia, especially the former.

### Egypt (Fig. 5.1)

During the fifth and fourth millennia the vegetation on perhaps half of the Nile flood plain had been disturbed by hunter-gatherer activities and herding. Cultivation was probably of secondary importance.[2] Over the 600 years of the Old Kingdom (c.2950–2350 BC) nomadism declined. Large game was almost totally eliminated, probably as a result of the destruction of natural habitats through the extension of cultivation. Cereals became the staple foodstuffs, with barley predominant (Kees 1977, 74). This forced greater dependence upon the annual flood and managing it within the framework of natural flood basins by raising levees, subdividing basins and providing water channels (Fig. 5.2). The natural deposition

Figure 5.2: Linear Basin Irrigation in Upper Egypt, c.AD 1850

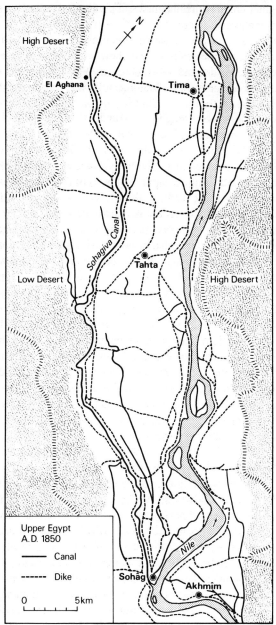

Note: This form of irrigation was more widespread in Egypt in earlier centuries.
Source: Butzer (1976).

of silt meant minimal preparation of the land for sowing and, under a regime of winter cropping, maintained soil fertility. Salination was largely prevented by natural drainage.

The level of the flood was crucial to successful farming. Variations from year to year were expected, but scholars have recognised the existence of secular trends (Bell 1970, 1971 and 1975). The mean level of the flood fell during the third millennium; this was primarily responsible for giving the present flood plain its overall form. A series of extremely low floods between roughly 2250 and 1950 BC brought drought, famine and, it has been argued, the social and political crisis associated with the so-called First Intermediate Period (Bell 1971). Several extremely high floods are documented from the period 1840–1770 BC (Bell 1975) and repeatedly from the ninth, eighth and seventh centuries BC (Butzer 1976, 29). As in more recent times, these probably brought short-term disaster: 'all the temples of Thebes were like marshes' in 877 BC (Breasted 1927, IV, 743). People and animals drowned, villages were washed away and, if sowing was delayed, poor harvests must have occurred in the season immediately after the high flood. Recovery, however, would have been comparatively rapid.

Such variations in flood level meant repeated changes in the location of meanders, their amplitude and wavelength. They forced an eastward shift of the general axis of the river in the 400-km section between Akhmīm and Cairo during the period surveyed in this chapter, thereby literally eroding the economic base of east-bank settlements. Modifications to the levees and flood basins must have been almost continuous and would have required constant adjustment to property boundaries and local irrigation systems (Butzer 1976, 33–6), thereby providing a practical input to the development of geometry. The necessary adjustments were successfully made, on the whole.

Cultivation expanded, perhaps first in the narrower sections of the valley, where comparatively small sections of the flood plain made irrigation more manageable. Development of the shādūf is confirmed for around 1346–1334 BC. Although used primarily on the small scale of garden plots, it may have increased the cultivated area overall by perhaps 10–15 per cent (Butzer 1976, 46, 82). During the first millennium emmer wheat replaced barley as the principal cereal (Kees 1977, 74). Despite phases of decline, population increased from perhaps 0.25 million in the middle of the fourth millennium (a mean density of 30 per km$^2$ of cultivated land) to around 2 million by 500 BC (a mean density of 210 per km$^2$ of cultivated land). Its distribution, however, was uneven. In the first half of the second millennium, for example, the greatest concentrations seem to have been in the area of Memphis and between Aswān and Qift, while the intervening tracts were relatively thinly settled (Butzer 1976, 57–92).

While expansion of cultivated land, settlements and population was taking place in the Nile Valley, similar changes were in progress in the Delta. High sea-levels between 3000 and 1200 BC did not extend the coastal lagoons appreciably and there is little evidence for back flooding, probably because of the offset effects resulting from increased alluviation. None the less, during the period under review, the Delta remained very much a watery environment, characterised by the lagoons, marshes and papyrus thickets which made it a paradise for royal hunts (Kees 1977, 32–4). Nomadism was well adapted to the environment and from early times remained a major economic activity. Land availability, however, meant that the Delta was ripe for colonisation by cultivators (Kees 1977, 33–4). The northernmost third may have remained almost deserted during the third millennium but the population of the drier south and centre may have doubled, while the whole region might have supported 750,000 people by 1800 BC (a density of about 75 per km$^2$ of cultivated land). Population may have doubled again during the second millennium, to reach 1.2 million around 1250 BC (a density of about 90 per km$^2$ of cultivated land), but with special attention given to development of the eastern frontier for strategic reasons. The Faiyūm was another area of colonisation, particularly around 2000 BC, when up to 450 km$^2$ may have been cultivated as a result of a fall in the level of its lake and the development of a canal system radiating from an arm of the Nile now known as the Baḥr Yūsef, a system atypical of ancient Egypt (Butzer 1976, 92–3). The population was perhaps 61,000 around 1800 BC (a mean density of 135 per km$^2$ of cultivated land). This would represent about 3 per cent of the total population of some 2 million at the time. By 500 BC the total for the whole of Egypt, including the desert oases, was probably about 3.5 million.

*Lower Mesopotamia* (Figs. 5.1 and 5.3)

Developments in the relationships between man and land through the system of irrigation were very different in Lower Mesopotamia from those in Egypt. First of all, there are two major rivers in the sub-region — four if the Diyālā and the Kārūn are included. Second, floods are less predictable than on the Nile and arrive at a time when they can destroy, rather than benefit, crops. Third, since the Tigris flows in an incised channel for much of its course through Lower Mesopotamia, the major source of irrigation water has always been the Euphrates. Accordingly, some of

the characteristics of this river are important in changing the relationships of man and land. The Euphrates runs above the level of the plain,
thus making the risk of inundation particularly acute when levees are
breached or overtopped. The notion of a sudden great flood was not only
the basis of an episode in the popular Gilgamesh epic (and also possibly
in Genesis) (Genesis 6:5–9:17; Sandars 1972, 108–13), but was also used
as a major divide in the so-called Sumerian King Lists (Finegan 1979,
23–6). Dry land gradually re-emerging after a flood episode constantly
reminded the Sumerians, in the southern parts of the plain, of the beginning of creation when inhabitable land appeared from the primeval waters
(Beek 1962, 12). In the contemporary world, as opposed to that of myth
and legend, floods left seasonal and permanent swamps as marked features
of Lower Mesopotamian landscapes but offering a variety of sources of
food.

The 'Sippar River', as the Euphrates was popularly called during much
of the first millennium (Adams and Nissen 1972, 42–4), is a braided stream
with originally as many as five major channels diverging across the plains
from the vicinity of Sippar (Adams 1958) (Fig. 5.3). These were subject
to continuous silting and frequent shifts of location. The possible depth
of sediments accumulated over the centuries remains a problem, despite
the efforts of Nützel (1978) to retrodict the level of land surfaces at successive dates. This is partly because of the effectiveness of wind erosion, at least in some areas of the sub-region (Adams and Nissen 1972,
6–7), and partly because of continued controversy over the rate, timing
and extent of subsidence in the geosyncline which underlies the whole
of southern Mesopotamia. The question of subsidence also has a bearing upon the likely position of the sea in relation to early settlements
such as Ur. On the basis of their interpretation of Classical writings, early
researchers maintained that the head of the Gulf was much further north
than in recent times and that the combined delta of the Tigris, Euphrates
and Kārūn must have steadily advanced southwards, perhaps by as much
as 150–200 km since about 3000 BC. Some of the recent analyses of
sediments have tended to confirm this view, at least in outline. Other
researchers, however, have been impressed by the variety of physical
evidence for regional subsidence. They have argued for the general stabili-
ty of the coastline on the basis of further sedimentary analysis, the
apparent long-term survival of the southern marshes and a re-examination
of the literary evidence. In their view, a delicate balance has been maintained between subsidence and aggradation. In fact, both sets of protagonists are right. Interpretation is complicated by: the blurring of
marine and continental facies in an estuarine environment; variations in

Figure 5.3: Irrigated Plain of Lower Mesopotamia: Ancient
River Courses and Early Sites

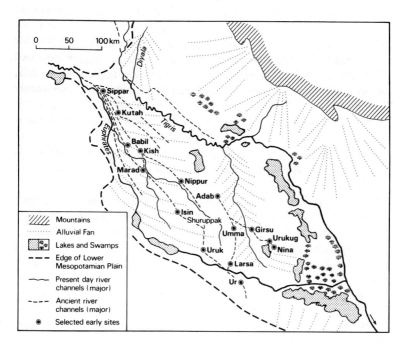

Sources: Gibson (1974) and Vaumas (1965).

world-wide sea-level; diachronic differences in the amount of sediment
carried by the rivers; and local, as opposed to regional, seismic activity
(Lees and Falcon 1952; Sarnthein 1972; Larsen 1975; Hansman 1978; Mcfadyen and Vita-Finzi 1978; Larsen and Evans 1978). Almost certainly, the
boundary of land and sea changed several times during the four and a
half millennia under review and in all probability river estuaries combined, divided and combined again as channels changed their positions.

Within this changing and uncertain environment, agriculture was
dependent upon obtaining irrigation water since precipitation is too slight
and unreliable. Water was obtained in two ways. The first was through
the uncontrolled, seasonal flooding of natural depressions left by the processes of silting and channel change. The second method was more dependable. It involved cutting short lengths of canal to control the flow of
water and distribute it to the fields when the need was greatest. The earliest

canals formed 'multiple-stranded, diverging and intersecting chains' leading off the main river and made considerable use of natural channels (Adams and Nissen 1972, 4), both seasonal and abandoned. Siltation was continuous and clearance an annual, though not necessarily arduous, task. The two systems, however, were essentially localised and small-scale.

A branching network of channels was equally characteristic of the lower Diyālā Valley and remained essentially unchanged until well into the first millennium. Further south, considerable modifications took place in the hydraulic system, certainly around Uruk and possibly elsewhere. The number of river channels in this area was progressively reduced during the third millennium, but the surviving channels remained in use until after about 1740 BC. Their survival probably indicates successful human intervention. This is also suggested by 'a quite unprecedented linearity' in two of three principal channels which was apparent around 2500 BC (Adams and Nissen 1972, 12). Much of the total system of natural, managed and artificial channels lost its interconnecting, looped character and became markedly dendritic in form (Adams and Nissen 1972, 38). A further major change took place after about 1700 BC. The principal channels of earlier times were abandoned and the whole of the Euphrates system between Ur and Kish shifted bodily towards the north-west. In the far south this coincided with a decline in state authority which may have reduced the capacity to deal with such drastic change and thus led to the widespread abandonment of land (Adams and Nissen 1972, 39–41).

Subsequent recovery was associated with the re-establishment of political authority and the re-creation of a dendritic system of channels by imperial planners (Adams and Nissen 1972, 55–7). While the extent and planning of the system was perhaps new, large-scale irrigation works were not. The earliest known major canal was constructed in the middle of the third millennium to lead water from the Tigris to a disputed area between Girsu and Umma which had previously been supplied from a branch of the Euphrates (Gadd 1971). By c.1700 BC the canal was sufficiently large to be called 'The Tigris'. Additional references to diversionary dams and large-scale canals appear in the records during the late third and early second millennium. While making more water available and probably extending the cultivated areas, large-scale irrigation schemes appear to have produced less happy results in the southern parts of Lower Mesopotamia, where gradients are negligible. Temple records seem to indicate, first, the emergence of salination and then its spread to more and more land. This can only be attributed to a rise in the water table resulting from over-irrigation and the neglect of drainage. Creeping salination, in turn, has been offered as an explanation for another change which emerges from the records. This is an alteration in the proportions of wheat and barley harvested. The two cereals were of equal importance around 3500 BC. By 2400 BC barley, which is more tolerant of salt conditions, had increased to 84 per cent of the recorded harvest and by 2100 BC it had reached more than 98 per cent. About 1700 BC no wheat at all was apparently harvested. The increasing proportion of barley, however, was accompanied by a continuous decline in yields. Farmers were evidently aware of the problems. An agricultural manual dating from around 2100 BC indicates that salination could be combated by growing only one crop in the year, alternating with fallow and preventing over-irrigation (Beek 1962, 16; Jacobsen and Adams 1958). Fortunately, conditions did not deteriorate so badly in the more northerly districts of Lower Mesopotamia. Indeed, the difference has been proposed as part of the explanation for the decisive shift in political power away from the southern cities of Sumer in the eighteenth century BC (Adams and Nissen 1972).

## Urbanisation and Early Towns

Modification of the water courses in the vicinity of ancient Uruk was accompanied by a series of significant changes in settlement patterns (Adams and Nissen 1972). While settlements of 1–2 ha proliferated along the water courses during the fifth millennium, there was already some differentiation in size. A few settlements covered as much as 5 ha. These have been identified as 'towns', while Uruk (with an area of over 10 ha) may already have been a regional centre — a 'city'. Settlement numbers increased appreciably during the early and middle parts of the fourth millennium, but the overall pattern took on a more concentrated look. Over the next 4–500 years more dramatic changes occurred. Many of the smaller settlements were abandoned and the entire pattern became closely related to a small number of major distributaries. Average settlement size increased to 6–10 ha and hierarchical differentiation became even clearer. Uruk probably attained its maximum size of 400 ha and perhaps 40,000 people before 3000 BC, when it was given its celebrated defensive wall. Thereafter its dominance was challenged by the emergence of a number of other settlements covering more than 50 ha. Concentration of population continued so that by about 2500 BC the great majority of the people in the territory of Uruk lived within walled centres of 'unquestionably urban proportions' (Adams and Nissen 1972, 19).

Experience varied elsewhere in Lower Mesopotamia, despite the apparent similarities in the physical environment (Adams 1972; Adams and Nissen 1972, 87–90). Around Nippur, urbanisation seems to have developed in much the same way as in the vicinity of Uruk, though the desertion of outlying settlements may have begun earlier and been associated with the abandonment of an important river channel. Urbanisation came later in the districts of Ur and Eridu and it was not accompanied by significant reductions in the number of agricultural settlements. By contrast, clusters of relatively large 'towns' appeared quite suddenly around Umma towards the end of the fourth millennium, but within the next 500 years or so their populations had been absorbed into the single regional centre. Rather different again was the experience of the deltaic fan of the Diyālā River, an area characterised by a system of bifurcating river channels which remained remarkably stable from around 4000 BC until the middle of the first millennium (Adams 1965, 33–45). Settlement numbers here continued to increase, though small groups of 'towns' and villages emerged, consolidated and, like the riverine distributaries, remained stable for a long period. They were not greatly affected by the comparatively late appearance of large 'towns' in the area.

The only other part of the Middle East where settlement pattern development leading to the emergence of large and dominant towns has been traced is the upper plains of Khuzestan, some 200 km north from the head of the Gulf (Adams 1962; Wright and Johnson 1975; Wright, Neely, Johnson and Speth 1975, 129–47). The area now has sufficient precipitation for dry-farming and this was probably true in ancient times. It was crossed by the bifurcating channels of the Rivers Kārūn, Dez and Karkheh, which offered possibilities for irrigation. Here the process of urbanisation was marked by an initial increase in settlement numbers and by colonisation of those districts which lay just outside the limits for successful dry-farming. A decrease in overall numbers followed during the late fourth and the third millennia and was accompanied by increasing differentiation of settlements by size. Susa (25 ha) became dominant.

Archaeological reconnaissance has been less comprehensive in other parts of the region. Consequently, while individual 'towns' have often been recognised, the circumstances of their development within a general history of settlement are frequently obscure. 'Towns' existed, for example, in the upland plains and valleys of south-western Persia and in Sīstan during the first half of the third millennium, and the formative stages must have taken place much earlier (Lamberg-Karlovsky 1973). Urbanisation in the interior of Persia, however, may have been retarded for perhaps as much as a millennium by the sparse and scattered distribution of population which seems to have been characteristic of these areas

(Ghirshman 1965, 42). By about 2300 BC a large walled settlement existed on Bahrain and other 'towns' may be postulated for the lower Gulf, though they may not have appeared in south-western Arabia until the latter part of the second millennium (Doe 1972; Globb and Bibby 1960). Some 2,000 years earlier 'important city-state sites' could be distinguished from 'royal fortresses' over much of Asia Minor and in some of the adjacent islands (Mellaart 1971). The early emergence of 'towns' is also evinced on the coast of Syria and in various districts of Upper Mesopotamia (Drower 1971). Urbanisation in Cyprus, however, was not apparent until the second millennium BC, when it seems to have taken place against a background of increase in settlement numbers and noticeable shifts in distributions (Catling 1973). In Palestine an increase in settlements occurred during the fourth millennium. It appears to have been followed by some concentration of population and the emergence of the forerunners of the 'fortified cities with high walls, gates and bars' which became widespread in the sub-region during the second millennium (Deuteronomy 3:5; Vaux 1971). A similar development has been suggested for Upper Egypt in the fourth millennium. Contrary to opinion based largely on the textual material, there is evidence, which is quite compatible with that from Lower Mesopotamia, for the existence by the middle of the third millennium of walled towns of varied size and character (Kemp 1977).

Having pointed to the early existence of towns across the region, we should pause to clarify various issues raised by the previous discussion. The terms 'towns' and 'urbanisation' have been used freely, as in the literature drawn upon, but without having been defined. An explanation for the emergence of 'towns' is needed, while it is important to attempt a characterisation of their layout and functions.

Most archaeological writers, but particularly those dealing with Mesopotamia, have assumed that settlement status is broadly shown by two indicators. These are, first, the size of the inhabited area, as given by the extent of surface scatters of pottery or the area of *tells/hüyüks* (mounds created by the decay and rebuilding of structures in mudbrick on the same sites) and, second, the presence or absence of monumental architecture.[3] 'Towns' are taken to be indicated by relatively large inhabited areas, especially when distinct breaks can be shown to occur in a rank-size distribution of all settlement sites from a particular area. Thus, in his study of the Diyālā fan settlements, Johnson (1972) worked with the assumption that areas of more than 15 ha indicated 'large towns' in the late fourth millennium, while areas between 9 and 15 ha suggested 'small towns'. The identification is more convincing where monumental public architecture is evinced. Temples and defensive walls, for example, point to some of the other characteristics of the 'town', namely, an association with a complex stratified society (also revealed by the discovery of specialist craft goods and palaces, commemorative statues, seal-stones and writing) and the focusing of control. Neither size nor large structures, however, is sufficient by itself to distinguish 'towns' from other settlements in terms of functions (administrative, commercial, industrial and religious), the aspect which geographers often take as characteristically 'urban' when above-average concentrations are apparent in settlements. Unfortunately, it is not always clear for sub-regions outside Mesopotamia and perhaps Upper Egypt that anything other than simple definitions of 'town' have been employed in the literature. 'Urbanisation' may have been used in a loose sense. What often seems to be meant is an apparent change in settlement pattern over time, from one of a relative dispersal of small sites to one of greater concentration and fewer but larger sites. Such changes are important, but the interpretation of them as 'urbanisation' may be misleading unless accompanied by other indicators of a shift towards urbanisation. That 'difference in kind rather than of degree between town and village' mentioned by Vidal de la Blache (1926, 471) should always be sought, though it is clearly significant that neither the Sumerian nor the Akkadian language distinguished 'town' from any other permanent settlement (Oppenheim 1970, 115). Failure to be explicit in the use of terms means that, sadly, a degree of scepticism is necessary when reflecting on the widespread distribution of 'towns' claimed for the second millennium, and probably earlier, or when considering the spatial intensity of the urbanisation process preceding it. The pattern may have varied quite considerably in time and space.

Ambiguities of definition, accompanied by uncertainties over chronology and the known differences in the experience of quite small adjacent regions in other aspects of life, together compound the difficulties of explaining the emergence of 'towns' which undoubtedly took place in the region. In addition, as Mumford (1966, 70) rightly observed, 'by the time the city comes plainly into view it is already old'. Yet the problem must be faced, chiefly because towns and urban life were such important elements in the subsequent historico-geographical evolution of the Middle East.

It seems clear that, whatever early aberrations there may have been in the sense of counterparts to 'urban' Jericho and Çatal Hüyük (pp. 37–8), the major phases of urbanisation followed a chain of significant developments; it may be claiming too much to say that urbanisation was consequent upon them. The developments in mind include both the spread of agricultural settlements across much of the region and a general increase in numbers. Mounting population density in areas of circumscribed resources and limited technology may have promoted a search for increased productivity as an alternative to colonisation. As Young has argued for Lower Mesopotamia, one solution to the problem would be to concentrate labour and provide more direction to its deployment (Allan 1972; Young 1972). A simple desire for rationalisation would have had a similar effect. Existing leaders, such as chiefs and priests, could have provided the necessary organisation. Evidence for destruction and warfare from a variety of sites suggests that security may have been an important driving force in many areas, while coercion cannot be ruled out. Differential access to land and water would promote social differences and create a 'surplus' of agricultural products. Older explanations for the emergence of 'towns' saw these developments as critical. Inequality could have created a demand for specialist products and led to both the careful regulation of land and the precise recording of its products, producing writing, geometry and maps as necessary tools. The existence of a 'surplus' allowed the support not only of an elite increasingly concerned with organisation and direction, construction projects and war, but also of elaborate social and religious institutions, as well as numbers of specialists (soldiers, scribes, priests and craftsmen). Aggregation of population is assumed to have resulted from both processes. Trade has been assigned an important role in recent studies (Blouet 1972). Some have stressed the importance of control over the supply of valuable raw materials used in distant lands (Kohl 1978). Others have emphasised the standardisation and mass production of pottery, the widespread distribution of which is archaeologically attested (Renfrew 1975; Wright and Johnson 1975). Both schools of thought are compatible. They assume that relatively large stratified populations could be brought together by communal needs and would be supported on the proceeds of industry and commerce. It is not a large step to point out that similar effects could result from the bulking and redistributive activities shown by texts from such widely separated 'towns' as Mari on the middle Euphrates (before 1700 BC), Kanish in Asia Minor (c.1900 BC) and Tilmun in the Gulf (third–second millennium) (Gadd 1971; Larsen 1976; Leemans 1960; Parrot 1938).

Contemporaries in Lower Mesopotamia during the second and first millennia BC do not appear to have regarded the 'town', in which they delighted, as the product of slowly maturing socio-economic processes of the type outlined. For them it was a sudden creation, of divine origin, with a history extending back almost as far as the beginning of the world (Lampl 1968, 7–8). Certainly, the foundation or refoundation of a number of individual towns later in the period under review can be attributed to the creative will of powerful, if not always divine rulers. The best-known examples are the large Assyrian towns of Kalhu (Nimrud), re-built by Ashurnasirpal II (883–859 BC), and Nineveh, refounded by Sennacherib (704–681 BC), as well as Akhetaten (modern Tell el-Amarna) in Egypt, which was established by Akhenaten c.1365 BC and abandon-ed about 15 years later (Lampl 1968). Several other deliberate founda-tions are testified. The planting of 'towns' was both a mark of royal authority and a means of establishing political control. Three examples may be mentioned. Solomon (c.961–922 BC) established 'cities for his chariots, and cities for his horsemen' in his domains (IKings, 9: 17–19). Ashur-dan II (934–912 BC) and his successors attempted to re-establish Assyrian authority by planting 'towns', first in their homeland between the Zagros Mountains and the Tigris and Lesser Zab Rivers, but subse-quently across the whole of Upper Mesopotamia (Opppenheim 1970, 118–19; Postgate 1974). Even Mesha, King of Mārib, claimed to have founded 100 'towns' (Pritchard 1955, 320–1). Clearly, the role of in-dividuals should not be forgotten, even for the dim days of urban origins in the fourth millennium.

Rather schematic, but not entirely fanciful, representations of towns are found in Egyptian and Assyrian reliefs depicting military campaigns during the first millennium BC (Lampl 1968). For the layout of 'towns' we are dependent partly upon fragmentary town plans from Lower Mesopotamia, especially the scale plan of Nippur, which probably dates from the late third millennium (Harvey 1980, 125; Unger 1935), but mainly upon the results of incomplete excavation. Sampling has been in-evitable on individual sites because of their large vertical and horizontal extent, but an early emphasis on major structures has meant that less is known about both the smaller 'towns' and the humbler quarters of larger ones. Available evidence, however, shows that for practically the whole period under review the 'towns' of the region shared several major characteristics (Fig. 5.4).

Massive walls defined the core areas of most 'towns' after the third millennium, though some of them may have remained unwalled, as in the case of the short-lived Akhetaten. The celebrated wall of Uruk, attributed to Gilgamesh, extended for 9.5 km and defined an area of 502 ha. Walls seem to have been intended to separate the town from the countryside and to display the power and authority of the ruler, especi-ally in the height and elaboration of the gates. They also had clear defen-sive purposes which are particularly obvious in the crowded hill-top 'towns' of Palestine and Armenia. Land use outside the walls seems to have combined houses and farms with areas of intensive cultivation, partly

Figure 5.4: Model of Early Mesopotamian Town

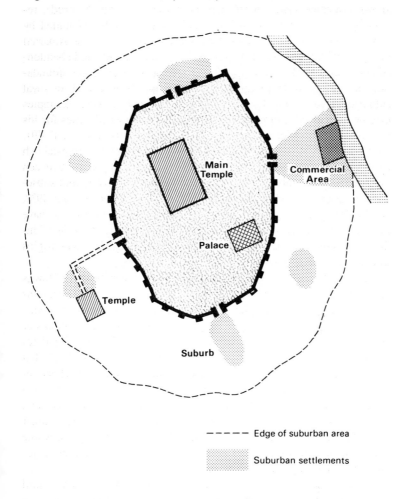

Main Temple

Commercial Area

Palace

Temple

Suburb

– – – – – Edge of suburban area

Suburban settlements

to supply the town with food but also to provide pleasure gardens for the wealthy. Complete ground plans have rarely been recovered from the areas within the walls, but the available information suggests that they were seldom laid out to any overall design. This seems to be true not only of the completely new towns planted by Assyrian kings but also of towns in Egypt, including Akhetaten, even though the mortuary complexes of tombs, temples and service settlements do themselves exhibit clear design principles (Lampl 1968, 20–1, 29–32). It was clearly something worth recording that Sennacherib straightened several streets in Nineveh (Oppenheim 1970, 140).

In general, the built-up areas of towns have the appearance of organic growth, often around ceremonial complexes. Some areas contained densely packed cell-like structures, combining houses, workshops and stores. Others appear to have been open ground. Intensive cultivation has been suggested for some of the spaces, at least in Mesopotamia (Adams and Nissen 1972, 30), while refuse dumps developed elsewhere and against the outside of the defining walls. Abandoned and ruined areas were equally characteristic, providing quarries for mudbrick-makers and enriched earth for spreading as fertiliser. Few market squares have been discovered, though the Old Babylonian version of the *Gilgamesh Epic* refers to Uruk as 'broad-marted'.[4] Throughout the region the town gates and their vicinity often acted as market areas, while along the rivers of Mesopotamia the quays became trading centres, as well as landing places (Oppenheim 1970, 115–16). Colonies of foreign traders have been detected at several 'towns' (Leemans 1960, 57–77, 108). Excavation located the houses of one such group from Assyria outside the walls of Kanish in Anatolia. It was part of a network of factories and agencies. Both archaeological and textual evidence reveals significant groups of craftsmen, concentrated in the town according to their specialism (Larsen 1976, 50–3; Lloyd 1956, 112–26).

The whole urban complex was dominated by temples and palaces. Although built in distinctive period and sub-regional styles (including the ziggurats of Mesopotamian temples and the pillared halls of Egypt), various common elements may be recognised. Compared with other structures, the temples and palaces were massive and elaborately planned. Both combined ceremonial or ritual spaces with living quarters and storehouses, generally comprising cell-like rooms arranged around courtyards. In Egypt and Lower Mesopotamia temple and palace were spatially separate. Individual Egyptian kings frequently created new palaces and usually built their own mortuary complexes, both of which were provided with service settlements. Thus, the built-up area comprised several separate, sometimes scattered, building groups (Kees 1977, 157, 288–307). Across the rest of the region, however, palace and temple were often combined into a single complex, defined by a wall. Separatism was emphasised, especially where such a group occupied an elevated position in the core of the old town and the bulk of the settlement had spread over low-lying land. This was the striking position, for example, at both Asshur, ancient capital of Assyria, and Hattusas (modern Boğazköy), capital of the Hittite Empire.

In addition to their religious and ceremonial functions, temples and palaces were the administrative centres of the complex organisations necessary for the running of supportive estates, the control of trade and the administration not merely of the towns themselves but also of the dependent polities. In fact, it is possible to argue that the material needs of those people with spiritual and temporal power virtually sustained the rest of the urban population and even, in a real sense, created the towns themselves. Many 'towns', in consequence, experienced only one or at most two phases of intense flowering — as indicated in architecture, artefacts and areal extent of their built-up area — before lapsing into drab obscurity and relative neglect when their elite was destroyed or moved elsewhere (Oppenheim 1970, 117).

## State Formation and Warfare

The connection between 'towns', on the one hand, and political power, on the other, raises the question of the development of political states in the region. As with the origins of 'towns', most work has been carried out on Mesopotamia. Research to date suggests that state formation there was virtually contemporaneous with the emergence of the first 'towns'. It is identified with the appearance of clear hierarchies in settlement size and of discrete spatial patterns dominated either by a single, relatively large settlement or, in some cases, by two large settlements but one of them larger than the other. These microcosms appear to have been fairly evenly spaced, roughly equal in extent and, to begin with, relatively autonomous in their functions. They may even have corresponded with 'natural' units determined particularly by local peculiarities of water supply. Their emergence was paralleled by the use of symbols, chiefly on seal-stones and sealings. Clearly meeting administrative and commercial needs, they argue for the control of the movement of goods from points of production to bulking centres and thence to other centres for use and possible redistribution. The degree of control and planning implied suggests the existence of political units (Wright 1977a). We may call them 'city-states'. The original form of government is unclear. There are hints of, on the one hand, public assemblies and, on the other hand, of the notion that kingship was at least as old as 'towns'.[5] Friction and rivalry existed between the different units, as is clear in the texts, once writing had emerged out of the pictograms on seals and sealings into a flexible medium for recording events, as well as assisting administration.

The Sumerian King List suggests that, after the Flood, supreme power in Lower Mesopotamia passed from one town to another, and in some cases back again. But it took place within a recognised territory, 'the land' (of Sumer). However, Lugalzagesi, a ruler belonging to the Third Dynasty of Uruk (about 2460 BC), claimed to have secured the submission of all the people 'from the Lower Sea, along the Tigris [and] Euphrates, to the Upper Sea', that is, from the Gulf to the Mediterranean. An era of imperialism had begun, and the state gradually ceased to be focused upon just one 'town'. Instead, it became territorial in scope, amalgamating the territories of formerly independent 'city-states' into a single unit. Sargon of Akkad (2371–2316 BC) and his grandson, Naram-Sin (2291–2255 BC), were the first rulers in the new mould of whom we have some detailed information. Although Sargon's conquests included the whole of Lower Mesopotamia and claimed to extend to the Sealand and even Tilmun, they and those of his grandson embraced much of the 'Upper Region' as far as the 'natural' limits set by Mt Lebanon, the Taurus Mountains and the Zagros; Naram-Sin may have penetrated the fastness of Anatolia (Finegan 1979, 40–3) (Fig. 5.5). For the most part, the conquests seem to have been made within the orbit of Mesopotamian civilisation, that is, a sub-region marked by such indicators as the use of cuneiform script on clay tablets to write the Mesopotamian languages Sumerian and Akkadian, the transfer of religious, cultural and social terminology and the adaptation not only of literary conventions but also of aesthetic standards and style (Oppenheim 1970, 67–73). Mesopotamian civilisation was extended beyond the 'natural' limits of Mesopotamia by conquest and trade. Khuzestan (ancient Elam) was thoroughly 'Mesopotamianised'. The Hittite culture of central Anatolia was deeply influenced, and even the literature of the Urartians around Lake Van slavishly copied Akkadian prototypes. Despite the influence, however, essentially non-Mesopotamian cultures evolved in the uplands. In touch with Mesopotamia, they were none the less separated from it by mountains or deserts and tended to develop in their own ways. Anatolia and Persia, in addition, were exposed to influences coming from the steppes of Asia.

Indo-European groups entered the Middle East from central Asia during the second millennium, or possibly the third, most probably as an extension of their perennial wanderings in search of pasture.[6] They brought with them the use of the horse, terrifyingly effective in raiding and warfare, and introduced a dynamic element into state formation in the region. The Hittites created a large empire in Anatolia during the second millennium. Devising means to operate beyond the mountain rim in Upper Mesopotamia and Syria, for a short while they clashed with Hurrian principalities there, some of whose rulers were also of Indo-European origin. Another Indo-European group, the Medes, established themselves in the

Figure 5.5: From City-state to Regional Empire: Examples of Different Scale Polities in the Region

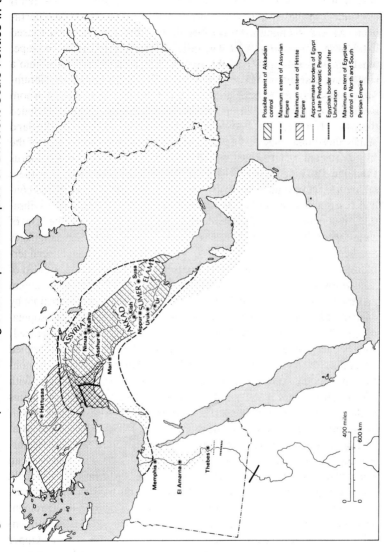

Legend:
- Possible extent of Akkadian control
- Maximum extent of Assyrian Empire
- Maximum extent of Hittite Empire
- Approximate borders of Egypt in Late Predynastic Period
- Egyptian border soon after Unification
- Maximum extent of Egyptian control in North and South
- Persian Empire

northern valleys of the Zagros Mountains and 'on the borders of the salt desert', that is, astride some of the few routes from Mesopotamia onto the Persian plateau. A revived and expansionist Assyria clashed with them during the ninth century BC. Cimmerians and Scythians, horse-riding nomads of similar origins to the other Indo-Europeans, burst into the region during the eighth century and finally settled to the north-east of the Medes, around Lake Urmia. After shaking themselves free from Scythian rule, the Medes went on to capture 'that great city', Nineveh (Jonah 1:2, 3:2), in 612 BC and replaced the Assyrians as a regional power. Their king, Cyaxares II (652–613 BC), ruled much of the northern part of the Middle East, including Anatolia as far as the Kızıl Irmak River (the ancient Halup), which became the border with the Kingdom of Lydia. While these events were taking place, 'Babylon the destroyer' (Psalm 136:8) created a short-lived empire in Lower Mesopotamia, Syria and Palestine. Its rulers were Chaldeans drawn from a group of Semite tribes which had moved into Lower Mesopotamia from the west as nomads but had become settled along the Euphrates since the ninth century BC.

Subsequently, another Indo-European people came to prominence. These were the Persians, who had finally settled in the foothills of the southern Zagros (Ghirshman 1965; Olmstead 1948). Under a series of of vigorous rulers (the Achaemenians), the Persians spread through the Zagros Mountains and eventually established an empire which in the early sixth century came to embrace not only the whole of the Middle East as defined but also the borderlands between it and India, on the one hand, and, on the other hand, central Asia at least as far as the Syr-Dar'ya. The Achaemenians are remarkable in other ways. Building on Assyrian experience, they devised a decentralised and tolerant system of administration which held their vast empire together until the late fourth century BC, when much of it was taken over piecemeal by Alexander the Great in his 'many and great struggles for supremacy' (*Diodorus Siculus* 17.6.13) with Darius III (335–330 BC). They adopted and propagated the teachings of Zarathushtra (Zoroaster, c.630–553 BC), which, in emphasising the transcendence of the supreme god, Ahura Mazda, and the cosmic struggle between good and evil, prepared the way for Islam and, in their popular manifestations as Mithraism, deeply influenced the Christian liturgical and mystical traditions. Achaemenian toleration of other religions, however, allowed the development and spread of Judaism. Through its emphasis on the one, ever-living God and his demand for absolute righteousness, Judaism eventually produced both Christianity and Islam. Attempts have been made to relate monotheism to various aspects of the physical environment, but the implied impoverishment of the human spirit by a crude determinism is clearly out of place. On the other hand, both the ritual forms of Judaism and its calendar of festivals, just as much as Zoroastrianism, owed much to the development of earlier religions which had emphasised fertility and were geared to the seasonal cycle. Nor should we forget that polytheistic systems of belief remained important in the region until at least the fourth century AD.

Although the Exodus provided a vital binding historical experience for the Jews, the Egypt from which they claimed to have emerged was relatively isolated from the rest of the region by the mountains and deserts of Sinai, the sparsely inhabited spaces of the Negev and by the trough of the Red Sea. In addition, Egypt was relatively closed off from the west by the virtually empty Western Desert, though communications with its oases were considered politically important to the Nile-Valley state (Kees 1977, 127–30). Political development in Egypt, not surprisingly, took an individual line.[7]

Many scholars have accepted the ancient view that state development culminated around the middle of the fourth millennium in the unification of the two kingdoms of Upper and Lower Egypt, each with some basis in the ecology of the Nile Valley. The final stage is associated with the semi-legendary Menes, who is also credited with the symbolic act of founding a new capital, Memphis, near the junction of the two parts of the country (Trigger 1983, 44–51). Writing, that important tool of administration, appeared at the same time as, or soon after, this significant political development. The elaborated system was hieroglyphic, but derivation from cuneiform has been suggested.

The unification of the Delta and the whole of the Nile Valley as far as Jebel el-Silsila, where the cliffs close in on the river, must have been preceded by similar unifications to produce the two basic kingdoms. We may speculate that the identity of some of the earlier political units may have been retained in the administrative divisions (*nomes*) of united Egypt, with their distinctive ensigns, emblems and deities. Although little evidence has been produced to link early state formation with urbanism in Egypt, this must remain a distinct possibility. Hints are provided by the emergence of craft specialisation and long-distance trade, as well as evidence for both synoecism and the emergence of several large population centres. These developments were paralleled during the fourth millennium by the spread along the Nile of a uniform culture which is strikingly indicated by wheel-made, beautifully decorated, pottery. Subsequent cultural developments south of Jebel el-Silsila and the First Cataract take a different turn and it has been suggested that the differentiation is related to state formation in Egypt, including the establishment of a definite southern frontier (Baines and Málek 1980, 30–1).

The unity of the northern sections of the Valley and the Delta was always fragile and ultimately dependent upon the personality of the Pharoah. During the so-called Intermediate Periods (2134–2040 BC; 1640–1532 BC; 1070–712 BC) it broke down. The country decomposed into varying numbers of rival states built by control over those primary units, the flood basins. Low-level floods may have caused the collapse which ushered in the First Intermediate Period (Bell 1971), while startling rises in the price of emmer wheat during the period 1770–1100 BC may argue for persistently low flood levels (Butzer 1976, 55–6), which could have precipitated the crises leading into the Third Intermediate Period. Ecological relationships are less easy to postulate in the run-up to the Second Intermediate Period, but similar economic and social hardship could have been created by a period of erratic behaviour in the Nile, which included a number of low floods, as well as some extremely high ones (Bell 1975; Butzer 1976, 52).

When united, Egypt's resources provided the basis for a strong and potentially aggressive territorial state. From the First Dynasty (c.2920–2770 BC) onwards, conquests were made from time to time beyond the First Cataract. Territory was held and even colonised. Conquests in the Middle East, however, followed the final defeat by Ahmose (1550–1527 BC) of the hated Hyksos, foreign rulers who had established themselves in Egypt during the sixteenth century. They were preceded by cultural penetration, as revealed in iconographic themes and symbols. In 17 campaigns, Thutmose III (1479–1425 BC) carried Egyptian arms through Syria, to reach their greatest extent when they rested along the great westward bend of the Euphrates south of Carchemish and possibly penetrated Cyprus. The success was comparatively short-lived. Egyptian rule was periodically restored in Palestine and southern Syria, only to fade away again. Finally, the warring shatter-belt states between Egypt and expansionist Assyria, including united Israel (Israel and Judah), collapsed. Assyria was able to invade and install puppet kings. The final collapse of Assyria, in turn, brought Persian rule to Egypt around 525 BC. With it the whole region was integrated into a single empire.

As well as creating dynamic historico-geographical entities, state formation had several other geographical consequences in the region, in both the long and the short term. Positive, but often short-lived achievements were the organisation of agriculture, the creation of integrated irrigation schemes, the colonisation of empty land, the plantation of 'towns' and the facilitation of trade. Methods were evolved which, together with concepts of law, became part of the received wisdom of the region and were replicated, occasionally advanced by later rulers. The ability to command and organise large amounts of manpower produced the great monuments of the early civilisations which overawed and puzzled subsequent generations familiar with their ruins through dwelling in and around them, and even making their living out of them through such activities as quarrying, treasure-hunting and tourism. The characteristic Mesopotamian monument was the ziggurat, or step-tower, set amidst a complex of associated religious and secular buildings, which began to evolve in the fourth millennium. Some 31 of these brick-built structures are known within the borders of present-day Iraq. Vast temples were built in Egypt, as in an area of about 1.2 km$^2$ at Karnak. But the monuments thought of as characteristic of early Egyptian civilisation were the stone pyramids, built mainly between the middle and the end of the third millennium. Groups of pyramids and associated temples, tombs and cemetaries are found at various points on the edge of the Valley, but the largest and most famous concentrations form a 30-km-long chain between Gīza and Daheir, across the river from Memphis. The building and equipping of these monuments has recently been seen as a highly effective means of sustaining the Egyptian economy (Trigger, Kemp, O'Connor and Lloyd 1983, 87).

Much more negative was the warfare involved in the creation and expansion of states. It had several effects on the contemporary landscape. Most terrifying must have been the deliberate destruction of towns and villages with their inhabitants. Such barbarous actions were intended to weaken rival powers and to prevent rebellion. Related to that were attempts to make farming difficult or impossible and to hinder economic recovery. Thus, Saul (c.1020–1000 BC) was instructed by his god to destroy the Amalekites: 'Spare no one; put them all to death, men and women, children and babes in arms, herds and flocks, camels and asses' (I Samuel 15:3). Around 850 BC the Israelites 'entered the land of Moab, destroying as they went. They razed the cities to the ground; they littered every good piece of land with stones, each man casting one stone on to it; they stopped up every spring of water; they cut down all their fine trees' (II Kings 3:25) — if the text is exaggerated, the intention is clear! Thutmose III boasted how in Palestine he 'took away the very sources of life, [for] I cut down the grain and felled their groves and all their pleasant trees' (*The Barkal Stele*, Pritchard 1955, 240a). The Assyrian kings glorified in destruction: '592 towns . . . of the 16 districts of Damascus I destroyed [making them look] like hills of [ruined cities over which] the flood [had swept]', reported the annals of Tiglath-pileser III (744–727 BC) in characteristic fashion, before going on to note the killing of 1,000 inhabitants, 30 camels and 20 head of cattle belonging to Semsi, Queen of Arabia (Pritchard 1955, 282a, 284). In consequence of such activity, many sites were condemned to 'dust and silence' (Sandars 1972, 28), though it is clear that resettlement and rebuilding often took place, sometimes almost immediately after the destroying forces had departed and perhaps by the three or four people overlooked by those who searched the houses to kill (II Esdras 16: 29–31). Sometimes the destruction was not as complete as the official records claimed.

Closely related to the physical destruction of war was the deportation of populations. When the people of captured settlements were not slaughtered out of hand and their heads heaped up, they were treated as booty and enslaved before being put to onerous and sometimes dangerous work. On other occasions, people were taken from their homelands and resettled in another part of the victorious empire. Most is known of Assyrian practice. Oded (1979, 19–21) has estimated on the basis of surviving texts that, over the three centuries of empire from 1090 BC, at least 4.5 million people were deported. Most single operations known involved fewer than 10,000 people, but in at least 13 others more than 30,000 people were reportedly moved. Lower Mesopotamia was affected most, in terms of both the total deported and the number of times deportation occurred (Oded 1979, 26). Wherever it happened, the aim of deportation was not only to reduce the possibility of rebellion but also, as practised by the Assyrians, especially to restore prosperity to core areas of the state and to colonise strategic territory where manpower was presumably in short supply (Oded 1979, 41–74; Postgate 1974). Deportees often worked on massive construction projects, the costs of which were met by the tribute impose on defeated populations and partly by the booty and loot taken during campaigns. Indeed, the opportunities which it offered for acquiring wealth must have encouraged warfare, which was to some extent self-financing. Another powerful motive was the desire to control both resources and trade routes.

## Commerce

Possibly the earliest recorded attempt to secure vital raw materials by expansionist war was that of Sargon of Akkad, mentioned earlier (p. 106). He is reported to have pushed his frontiers to the Cedar Forest (presumably Mts Lebanon and Amanus) and the Silver Mountain (the Taurus) (Finegan 1979, 39–49). Timber was evidently in short supply in Lower Mesopotamia during the third millennium. Gilgamesh is represented as being enchant by the luxuriance and height of the cedars he found on his journey, perhaps implying that he had never seen large trees or forest before (Leemans 1960, 11; Pritchard 1955, 82b). Gudea, ruler of Lagash about 2100 BC and famous in his day for his building projects, brought massive cedar trunks (25 cubits long) and boxwood from the Amanus, as well as other timber from the mountain foreland between there and the Euphrates (Leemans 1960, 11–12). Silver was not only attractive in its own right, or for what might be made from it. It was also important as a medium of exchange throughout the region (Gurney 1961, 84–7; Leemans 1960, 130–1).

Although states in Lower Mesopotamia lacked many raw materials, not all of them were secured by territorial aggrandisement. The shortages included copper and tin, precious and semi-precious stones, stone for building and sculpture, ivory, oils and essences, as well as wood. Copper was imported from Magan (possibly Oman) through Tilmun (probably Bahrain) in the latter part of the third millennium. In the second millennium supplies may have reached Lower Mesopotamia through Assyrian trading colonies in Anatolia. Tin may have come from interior Persia, but through Assyria and not Susa as was once supposed. Jewel stones came from a variety of sources which are now generally impossible to identify. Exotic wood, often used in making furniture, was imported from Magan and Meluḥḥa (probably northern India). Ivory may have been brought from east Africa, but western India seems more likely, especially when it was imported as manufactured articles (During Caspers 1977; Larsen 1976; Leemans 1960 and 1968; Oppenheim 1954). It should not be forgotten, however, that Thutmose III is reported to have hunted elephants in Syria in 1464 BC and that Ashurnasirpal II (883–859 BC) is credited with killing 30 (Zeuner 1963, 276–8). Perfumed oils and essences seem to have been imported into Lower Mesopotamia from the north-west and this was also the source of wine. Some of these items, notably oils and essences, were re-exported. Lower Mesopotamia's own trade goods, however, were wool, hides, sesame oil and barley (possibly in demand in the more arid stretches of the Gulf coast) and high-value manufactured items, notably garments but also seal-cylinders. Being small, the latter could be sent to a larger variety of markets, including those at a considerable distance.

United Egypt commanded a greater variety of raw materials than Lower Mesopotamia. Various minerals were found in territory accessible from the Nile. The main scarcity was timber. Although techniques of 'fitting together' allowed use to be made of the diminutive local resources (Bass 1972, 19; Kees 1977, 107), imports were necessary. Large timber was required in huge quantities for the interiors of royal tombs and palaces, as well as for sea-going ships and even tall flag-poles. Although Nibbi (1979) has argued that the demand could have been met from southern Palestine, or even from parts of the Delta, most historians have little doubt that from the middle of the third millennium, if not earlier, the larger timbers came from much further north, principally from Mt Lebanon. Byblos seems to have been the main port used by the Egyptians (Drower 1971; Mikesell 1969).

Other scarce materials were obtained in various ways. Expeditions were sent sporadically to the 'turquoise terraces' of Sinai and to the 'incense terraces' of Punt (probably in the Horn of Africa) (Kees 1977, 117–18). The latter involved the dangers of sailing the length of the Red Sea and working back again. Control over territory beyond the First Cataract in Nubia provided additional supplies of wood and gave access to a variety of exotic commodities ranging from ostrich eggs and leopard skins to gums, precious stones and female slaves. Nubian gold was of fundamental importance to the success of Egyptian expansionist dreams in Palestine and Syria during the fifteenth and twelfth centuries BC. Total imports in the 1420s BC averaged about 10,000 oz per annum. Probably no other state in the entire region could then command such resources, especially when it is recalled that Egypt itself produced gold (Hayes 1973). Burial customs, however, took a certain proportion of the scarce metals and precious stones out of circulation, thereby stimulating a demand for more. In contrast to expeditions and colonialism, normal trading relations supplied Egypt with Syrian wines, oils and essences, with copper and with goods manufactured in the Levant towns themselves or even brought from as far away as Crete. In return, Egypt exported gold and cereals, together with its famous specialities, linen cloth and papyrus.

Despite the detail and volume of evidence about commodities, the scale and extent of long-distance trade is difficult to estimate for the fifth to the first millennium. Artefacts recovered on archaeological sites are only a sample of the items traded, and are non-perishable commodities at that. Provenance and destination can often be established but not overall quantities. Single items are just as likely to have been gifts as part of dispersed bulk consignments. None the less, distance-decay effects seem apparent in the numbers of Egyptian and Sumerian artefacts found away from their production areas. The texts giving economic information may be even more unrepresentative than objects. Egyptian sources from all periods often talk in terms of 'tribute' and indicate the existence of state monopolies. Assyrian commercial letters found at Kanish and dated c.1900 BC show something of a policy of commercial penetration of Anatolia

Figure 5.6: Trade Contacts of Lower Mesopotamia in the Late Third Millennium BC

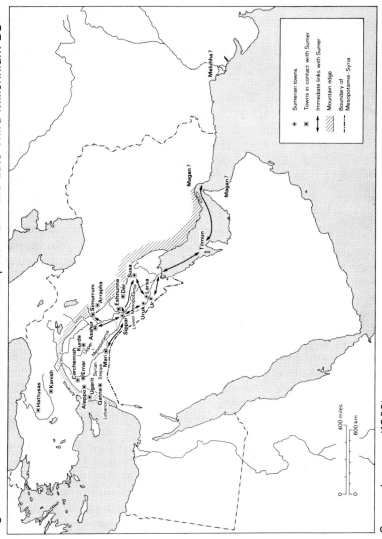

Source: Leemans (1960).

by Assyrian kings of the early second millennium, and reveal the export of large consignments of woollen textiles, tin and copper from Assyria. Some 200 donkey-loads of tin (c.13,500 kg) were sent from Asshur to Kanish in a period of about 50 years (Larsen 1976, 102–5). Roughly contemporary Old Babylonian texts suggest both the shipment of smaller individual consignments and a lesser scale of organisation. They also point to the existence of a set of circumscribed, but overlapping and constantly shifting, trading areas within Mesopotamia which linked with other sets beyond the sub-region through towns at strategic locations (Fig. 5.6). Thus, Eshunna on the Diyālā fan connected Sippar on the Euphrates with a web of routes leading to trading areas lying north-east of the Tigris. Mari and Carchemish were important centres for trade along the arterial route of the Euphrates and converging on its axis (Leemans 1960, 85–96, 102–3, 116–17, 137). Tilmun provided links with the lower Gulf and areas beyond, at least in the second millennium BC (Leemans 1960, 5, 10, 23, 36–53).

During the first millennium 'towns' developed in north-western Arabia at various control points along the routes which supplied the eastern Mediterranean sub-region, as well as Egypt, with myrrh and incense produced in Dhufa (Doe 1972, 50). The Phoenician ports on the Mediterranean façade of Lebanon and Syria similarly flourished on transit trade during the second and first millennium (Culican 1966; Frankenstein 1979). In their case, the trade was in goods destined from interior Syria and Mesopotamia to the 'Upper Sea', some of them for Egypt (Frankenstein 1979). In the opposite direction, trade goods included items from Cyprus and Crete. Phoenician prosperity was assisted by the export of the sub-region's own products. The timber resources of the nearby mountains were exploited, as we have seen (pp. 113–14), and provided a useful basis to the commercial sector of the economy. The Phoenicians were manufacturers and, in addition to carved ivory and purple-dyed cloth, produced such semi-transformed items as wine, oils and essences. Trading posts were established abroad first during the fifteenth and then during the tenth century, perhaps initially in Egypt but later in Cyprus. Under Assyrian pressure in the following century, definite colonies were founded outside the region. The most famous was Carthage (established in 814 BC). Phoenician dependence upon trade may partly explain how the people came to adopt one of the alphabetic systems of writing devised in Syria–Palestine during the second millennium BC. Sheer convenience and flexibility may be reasons why the Phoenician system became so widely diffused in the Middle East under the Assyrian and Persian Empires and spread throughout the Mediterranean world. Its derivatives, of course, remained in use throughout the subsequent history of the Middle East.

Timber was taken from Lebanon and the mountains further north not just because of the intrinsic attractions of cedar wood — in scent, colour, grain and length. Access to the sea, in one direction, and the fast current of the Euphrates, in the other, were of great importance. Although the timber had to be hauled with considerable difficulty between 150 and 350 km to the Great River, water made onward shipment relatively easy. The timber was made up into great rafts, as in more recent times in Anatolia, and these were simply floated down-river on the current. The capacity of the Euphrates for floating large timber inevitably meant that resources deeper and deeper into Armenia were tapped to meet the insatiable demands downstream. One consequence of extensive deforestation within areas accessible from the river must have been to increase the amount of silt washed into the Euphrates. In time, flushing and transportation would have raised progressively the amount of alluviation in stretches of slack water, notably in Lower Mesopotamia (Rowton 1967). On the Mediterranean side of the mountains, deforestation must have increased delta development at the mouths of numerous seasonal streams and could have brought serious local flooding. The timber itself appears either to have been used directly on the coast to build ships or to have been assembled for towing to its destination (Drower 1971). The latter must have been a slow business when the destination was Egypt. Unless the ships went out towards Cyprus before turning south, they would be working against the current and in danger of ending up on a lee shore.

Ships operated on the rivers of Mesopotamia and they must have been prime carriers of barley to the Gulf, since barley is comparatively heavy and bulky and would have to be shipped in substantial quantities for the effort to have been worthwhile. While wind and current assisted downstream movement, haulage was necessary in the opposite direction. Consequently, while the voyage from Nippur to Lagash lasted 4 or 5 days, the return took 16 or 17 days against the current (Bass 1972, 14, 18; Beek 1962, 16). Water transport was vital in Egypt. Here the wind assisted upstream passages on the Nile, though meanders had to be negotiated, while the current carried craft northwards (Erman 1971, 479–88). The cultural and political unity of the Nile Valley depended upon these facilities and, in reciprocal fashion, must have enhanced them. Such massive building operations as the Pharoahs indulged in would have been impossible without the river to transport the stone, some of it in massive blocks.

Maritime expeditions to Punt, and the evidence for contacts along the Gulf (During Caspers 1977; Hayes 1973; Oates *et al.* 1977), point to an awareness of the seasonality of winds and confidence in the seamen's abilities to cope with severe hazards to navigation. Doe (1972, 53–4) has suggested, however, that the land routes across south Arabia flourished precisely because of the dangers of the Red Sea for regular traffic. Direct sailings between south Arabia and western India would have involved less risk and may have developed as early as the first millennium.

Land routes followed river valleys. They were anchored by passes through the mountains and directed by the availability of water at frequent intervals. Abundant supplies helped to turn the Tigris and Euphrates Valleys into arterial routes for north-south traffic. Sparser but frequent supplies from springs, streams and wells dictated that west-east movement followed the edge of the Upper Mesopotamian steppe. Alternative routes crossed to the Mediterranean from the great westward bend of the Euphrates by the shortest negotiable routes. The preferred routes from the Nile to the Red Sea left the Valley where it bends towards the coast between Qena and Thebes, and then followed a set of comparatively well-watered valleys eastwards. Travel across the 'silent land' of the Western Desert to the oases was more difficult and perhaps somewhat irregular (Kees 1977, 121, 127–30). As indicated earlier, donkeys were the main caravan animals of the period throughout the region (Larsen 1976, 102–5; Leemans 1960, 134; Zeuner 1963, 374–7). Their need for frequent watering and individual drivers slowed the progress of the 50 or more animals forming a caravan (Leemans 1968, 134) and probably reduced the number of practicable routes. Dromedaries offered more speed and flexibility. However, the 'ship of the desert' (Qur'ān 23:22) was not readily adopted in the less arid and northern parts of the region. Some of the people there knew of dromedaries from their use on the incense routes through the deserts from south Arabia (Bulliet 1975, 57–78). Widespread use of dromedaries for caravans came later than the period under discussion (p. 87). The overall consequences can be deduced. Long-distance trade by land in the period 5000–500 BC must have been very limited in terms of frequency of movement and the quantities of goods carried at any one time.

Considerably less is known about local trade. While goods were no doubt exchanged in fixed and periodic markets, many of the items transferred from the countryside to the 'towns', and vice versa, were not strictly trade goods. They were the produce of landed estates destined for the storehouses of the owners and those of the temples and palaces. Some of the produce was consumed by the great households directly. In both Egypt and Lower Mesopotamia, varying amounts were redistributed as 'rations' to different grades of official and servant. The population everywhere shared in the communal meals which followed the gargantuan sacrifices offered at major festivals; for some they provided a rare opportunity to eat meat. But sacrificial animals and basic foodstuffs were not the only items moved on a local scale. Fuel and industrial raw materials must have loomed relatively large (Wright 1969, 89–116). Locally manufactured items, such as yarn for dyeing and pottery for everyday domestic use, may have been carried over restricted distances from villages to a nearby 'town'. In Egypt the Nile would have been used for local trading above the single village level. In Lower Mesopotamia the numerous water courses were exploited by simple reed boats (Leemans 1960, 115 n.3; 117 n.2). As with long-distance trade, donkeys provided the principal means of transport away from the waterways (Erman 1971, 430–2, 490). Although chariots were widely employed in warfare from perhaps the eighteenth to the eighth century BC, the use of carts and waggons, generally drawn by oxen (Bulliet 1975, 16–21, 231–5; Erman 1971, 491), would have been restricted by terrain, as well as by the availability of alternatives to these slow-moving, relatively expensive means of transport. Horses were too costly to rear for them to be used on any scale as pack animals, despite their size and speed. Inevitably, the hinterlands of most 'towns' must have been fairly small. On the other hand, the size of the imperial capitals of Assyria (conceivably of Memphis and Thebes, too) implies that when they were at their height in the ninth century, provisions must have been drawn from much wider areas and that considerable organisation was required. Once conceded, this point raises two questions. The first is that of how it was achieved within the technological constraints outlined. The second is the character of the differential effects created by varied access within the context of relatively severe environmental and distance constraints. Booty and tribute were obvious solutions to the supply problem tried by various states, but they were difficult to sustain without continuous conquest and unlimited expansion. Distant frontiers raised considerable logistical problems. The intensification suggested by the construction of canals by Sennacherib (704–681 BC) around Nineveh brought more positive effects (Reade 1975). Although precariously balanced and expensive, increased use was made of this solution after 500 BC, at least in some parts of the region, and helped to sustain a region-wide empire.

## Notes

1. Percentages calculated from data in Waldbaum (1978, 38–58).
2. Unless otherwise stated, the discussion of developments in Egypt is based upon Butzer (1976) and Trigger, Kemp, O'Connor and Lloyd (1983). Egyptian chronology has been standardised as far as possible from Baines and Málek (1980).
3. The trend was set by Childe (1950). See also Wheatley (1972).

4. Sandars (1972, 68): 'Uruk, the city of great streets', 'Uruk, that great market'.

5. The Sumerian King List opens with the phrase, 'When the kingship was lowered from heaven . . .'. In the *Epic*, Gilgamesh is represented as consulting 'an assembly of the elders' and 'an assembly of the men' (Finegan 1979, 23, 30).

6. Political outlines from the *Cambridge Ancient History*, 3rd edn, vol. 1, pt 2 (1971), Finegan (1979), *Power and Propaganda* (1979), and Trigger *et al.*, (1983).

7. On the history of Egypt I have used the *Cambridge Ancient History*, 3rd edn, vol. 1, pts 1 and 2 (1970–71), Emery (1961), Kees (1977), and Trigger *et al.* (1983).

# 6 HELLENISTIC WEST AND PERSIAN EAST

Westward expansion across Asia Minor brought the Persian Empire into contact with Greek-speaking people settled along the coasts and in the islands. Incorporated into the Empire, the *poleis* (city-states) were freed in the great war launched against Persia from the Greek mainland (479–448 BC), only to pass under Athenian hegemony. The Peloponnesian War (431–404 BC) allowed Persian control to be re-established in Asia Minor, though rebellion was supported by Sparta (397–394 BC) and to a lesser extent by Athens. A settlement was reached at Sardis in 387 BC ('The King's Peace') but conflict between the Greeks and the Persian Empire continued to flare up from time to time. In an event rich with symbolism, Alexander of Macedon crossed the Hellespont (Dardanelles) in 334 BC. Within three years he had destroyed the Achaemenid dynasty, going on to seize much of its eastern territory (Fig. 6.1).

Alexander's Empire did not last long after his death at Babylon in 323 BC, though the eastward thrust of Hellenic culture which it launched had repercussions lasting for centuries. When Alexander died, his general, Ptolemy (304–283 BC), early made sure of Egypt, the most compact and unified part of the Empire. By 285 BC he was master of the coastal areas of the eastern Mediterranean and of western Asia Minor. Although most of these possessions were lost within about 50 years, the Ptolemaic dynasty ruled in Egypt until the suicide of Cleopatra VII (31 BC), when it finally passed under direct Roman rule. Alexander's cavalry commander, Seleucus (323–281 BC), acquired the largest part of the Macedonian's Empire in the Middle East. By 301 BC the Satrap of Babylonia ruled over Syria, as well as the whole of Mesopotamia and most of Persia. Within another 20 years he was master of central and western Asia Minor. Persian territory in the north and east had been bypassed by Alexander on his rapid advance across the peninsula and a series of independent states soon emerged to plague the neighbouring Seleucid Empire and to assist in its contraction. Despite strenuous efforts to retain territory, the Seleucid state broke up. Most of Persia eventually proved too distant from the core lands of Syria and Lower Mesopotamia to be held. Much of it was overrun by successive waves of nomadic horsemen breaking through from central Asia. During the first century BC these incursions finally destroyed the most easterly outpost of Hellenism, the Kingdom of Bactria, which had broken away from Seleucid control about 225 BC. One

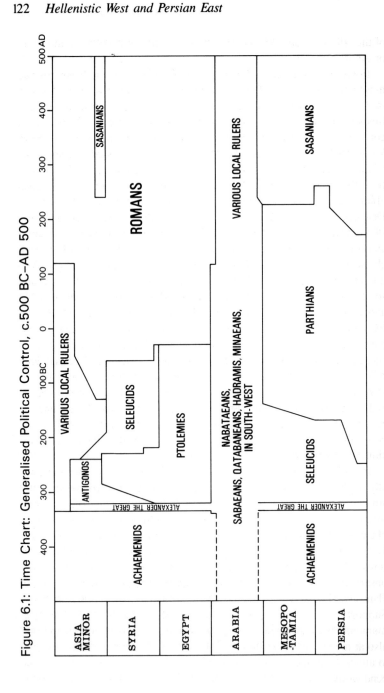

Figure 6.1: Time Chart: Generalised Political Control, c.500 BC–AD 500

of the tribes became known to the Greek world c.250 BC as the Parni or Aparni (the Parthians).

Initially established to the east of the Caspian Sea, the Parthians had moved southwards into Persia and became associated with the Satrapy of Parthia, from which western commentators derived their name. Several times repulsed by the Seleucids but never driven out of the Middle East completely, the Parthians bounced back to take over much of the territory lost by the Seleucids in Iran. By 155 BC they occupied Media. They crossed the Tigris into Mesopotamia in 141 BC and detached the original core from the Seleucid Empire. Meanwhile, Seleucid involvement in Aegean affairs had drawn the rising and ruthless power of Rome into Asia Minor. Defeated there, the Seleucids were forced by the Treaty of Apamea (188 BC) to withdraw south of the Taurus Mountains. By diplomacy, inheritance and conquest, Roman expansion was relentless from then onwards, though checked briefly by the anti-Roman hatred of Mithridates VI, King of Pontus, and the three wars which it engendered (88–64 BC). Central and southern Asia Minor were brought under Roman control (between 133 and 44 BC); and Syria, Cyprus and Palestine were annexed (64BC, 88 BC and AD 6 respectively). Egypt was secured in 31 BC. Eastward expansion, however, brought confrontation with Parthia and later with its successor, Sasanian Persia, the rise of which is conventionally dated from AD 208. Their spheres of influence overlapped in the mountain fastness of Armenia, but met sharply on the open plains of Upper Mesopotamia and Syria. The long and exhausting struggle between the Roman and the Sasanian Empires weakened them both, particularly in the late sixth and early seventh centuries AD, and paved the way for the victory of expansionary Muslim Arabs.

Initially content with a loose network of client states, Rome eventually organised the protection of its eastern provinces by a frontier in depth, stretching from the Black Sea to the Gulf of 'Aqaba (Fig. 6.2). Several times advanced, brought back again and modified, the *limes* consisted of a chain of legionary fortresses and fortified towns linked by paved roads. Forts and observation posts were arranged along the roads and in positions either side of it, where the terrain allowed ease of both defence and observation, as well as control over vital water supplies (Frézoub 1980; Hanson and Keppie 1980; Kennedy 1982; Mitchell 1983; Poidebard 1934). Support troops were kept in the rear, notably at the focal point of communications in Asia Minor, Ancyra (Ankara), and from at least AD 115 also at the regional metropolis of Antioch-on-Orontes, itself vulnerable to attack from the line of the Euphrates where the river takes its great bend westwards. The ruins of desolate fortresses and obsolete town walls,

Figure 6.2: North-eastern Frontier of the Roman Empire in the First Century AD, showing the Line of Forts in Syria

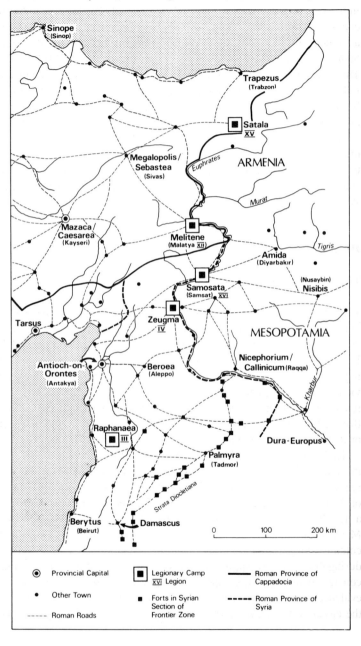

as well as the clear traces of Roman roads, are striking features of the landscape even today. In broad terms, the *limes* system lasted from the initial moves in its creation under Vespasian (AD 70–79) to the Muslim conquests of the early seventh century, when much of it was abandoned as of no use in a state which embraced both sides of the line. For more than 500 years, however, there was a major division in the Middle East. The historical geography of the Hellenistic West and the Persian East followed somewhat different trajectories during this period. There were links, of course. Trade crossed the frontier; some cultural and ideological elements were shared, particularly in the frontier areas of Upper Mesopotamia and northern Syria (Brown 1967). In retrospect, however, the differences emerge as sufficiently strong and sustained to justify the treatment of the two sides of the Roman *limes* as essentially separate culture regions. It is convenient — no more — to begin with the Hellenistic West.

## HELLENISTIC WEST

### The Spread of Hellenism (Fig. 6.3)

With the benefit of a long time-perspective, two features of the history of the lands west of the Euphrates stand out as particularly striking. They are, first, the remarkable degree to which they were influenced by Hellenism and, second, Hellenism's almost total disappearance from the region, leaving ruined theatres and peripheral temples as the most obvious landscape memorials of its passing. The flow and ebb are so remarkable, and the physical evidence so obvious in particular localities, that Hellenisation should be discussed as a preliminary to other aspects of the evolving geography on which it had a large influence.

Hellenism is 'the civilisation, language, art and literature, Greek in its general character, but pervading people not exclusively Greek, current in Asia Minor, Egypt and Syria, and other countries, after the time of Alexander the Great' (Harvey 1955, 198). However, its spread from a hearth in the Aegean basin began much earlier (Toynbee 1959, 16). Mycenaeans, presumably Greek-speaking, established trading posts or colonies along the shores of Asia Minor and Syria, as well as in Cyprus, during the early fourteenth century BC (Boardman 1980). While trading contacts with the Greek lands of the West may have continued, the evidence suggests a definite break in the occupation of many sites around the eastern end of the Mediterranean during the twelfth and eleventh

Figure 6.3: Hellenisation of the Middle East

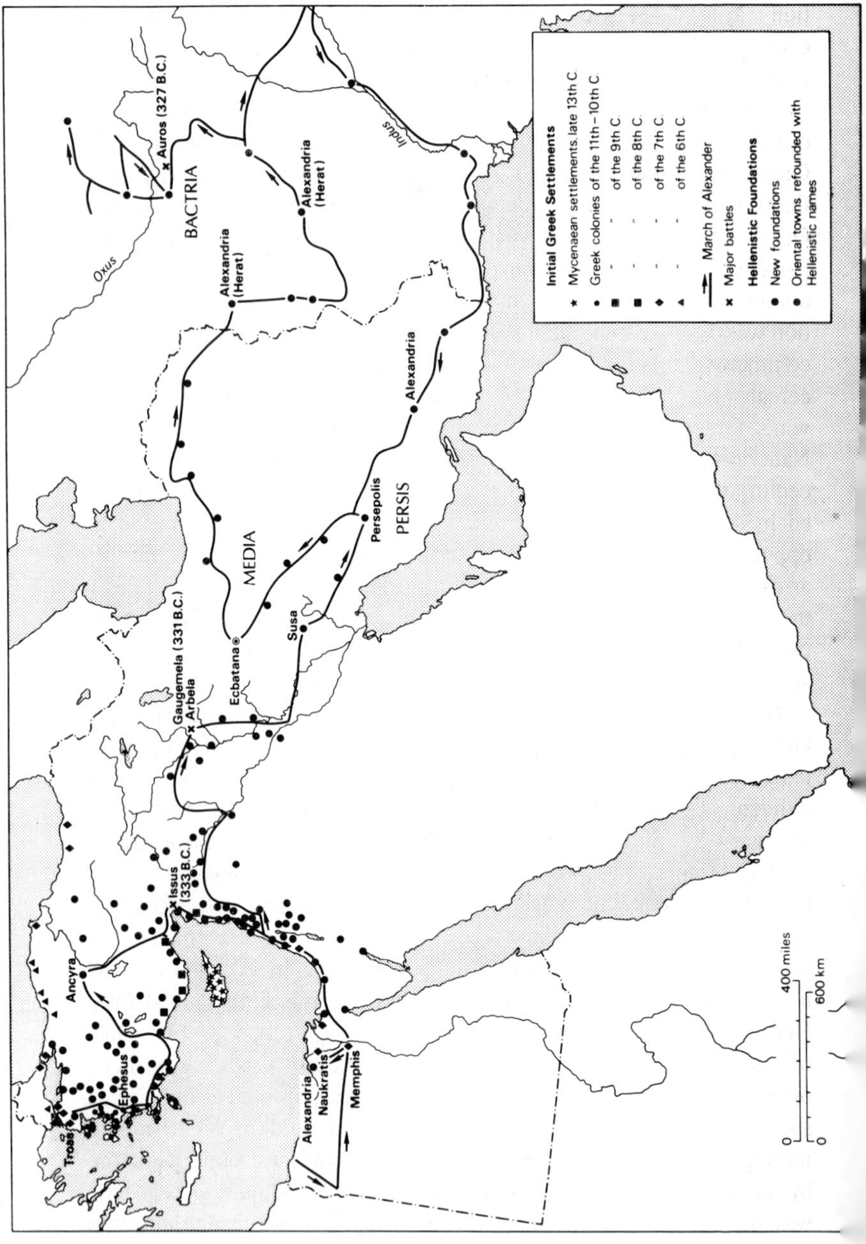

centuries BC. By contrast, a number of sites in western and south-western Asia Minor and the fringing islands suggest continuity. Some colonisation may have added to the number of Greek settlements during the tenth century BC. In subsequent centuries, they were all regarded as integral to the core of the Greek *oekoumene*.

The eighth and seventh centuries BC saw a further coastwise spread of Greek settlements northwards and eastwards (Fig. 6.3). Nowhere did Greek settlement penetrate very far inland, though Greek cultural influences spread along the great river valleys of western Asia Minor. New Greek settlements may have been founded in Cyprus, but the Greek language of the island always remained distinctive. Trading posts were established along the Syrian coast and down into Palestine; inland penetration was purely commercial (Waber 1969). From the late eighth century commercial relations were developed with Egypt, probably to obtain cereals and possibly linen and papyrus, as well as to sell olive oil and wine. Greek traders of diverse origin appear to have congregated at Naucratis in the Delta from the late sixth century onwards. In the fifth century they were joined by Greek mercenaries, and others were stationed in different parts of the country. Persian expansion increased the opportunities for mercenary service by allowing recruitment into the armies of the Great King. War between Persia and the Greek states in the fifth century BC did not prevent the growth of contact between the two sides, and these interchanges prepared the way for a major expansion of Hellenism in the fourth and third centuries BC.

In the early phases Hellenism spread because it was the culture of Greek settlers. In the last great phase of expansion, under Alexander and particularly his successors, Hellenism was adopted on a large scale by indigenous populations. As the culture of the conquerors, it first spread spontaneously amongst the upper classes who were, to some extent, already favourably disposed. Elites had the wealth and leisure to adopt the entire Greek educational programme and they went on to strengthen its literary roots by establishing their own schools, some of them ultimately very distinguished. But Hellenism was also officially promoted at all levels by the state. Various rulers, whether themselves Greek or more-or-less Hellenised, saw Hellenism as a means of giving ideological cohesion to otherwise rather disparate peoples. Popular Hellenism was encouraged, on the one hand, by the contagion which resulted from settling Greek-speaking veterans as military colonies and, on the other hand, by promoting Greek cults, physical training, games and even plays, as well as by providing the facilities which these demanded. Also influential was the continued patronage of Greek styles in art and architecture originally fostered by the Achaemenids.

The degree to which Hellenisation was generally successful depend-ed upon the extent of bureaucratic control from the centre, the vitality of indigenous civilisations, the proportion of urban to rural population and the willingness of wealthy individuals to provide the facilities re-quired by an Hellenic style of life (Jones 1966a). Egypt possessed the most centralised administration in the Hellenistic world, but though the bureaucratic use of Greek penetrated to village level and Greek methods of organisation were widely adopted, the culture did not go very deep. The European element in the urban population outside Alexandria was probably too small to be influential against a powerful and millennia-old civilisation. Even the elites were never completely won over. Coptic re-mained the everyday language of the country. Hellenism spread more widely and effectively where Greek style urban communities (poleis, pp. 130–4) were numerous, and the Greek-speaking populations larger. The greatest success was achieved in western Asia Minor, where Greek com-munities were virtually indigenous and the contagion effects particularly strong. More striking, though, was the spread of Hellenism on the plains of northern Syria and Upper Mesopotamia. These areas were transform-ed into a second Macedonia through a comparatively dense system of planted *polis* communities. These were named after places in the Greek 'homeland' or carried the honorific titles of the ruling dynasty. The whole organism was centred in the metropolis of Antioch-on-Orontes, itself situated in 'a Syrian replica of the Vale of Tempe' (Toynbee 1959, 24). By contrast, Hellenism's influence on the rural populations of central and eastern Asia Minor came comparatively late and its major vehicle was not the state as such, but Christianity. Even in Syria and Upper Mesopotamia, however, Syriac continued to be used and it became the vehicle of serious literature under the influence of plebeian, popularis-ing Christianity during the first and second centuries.

While local cultures reasserted themselves, Hellenism itself was transformed from within to become Byzantinism — 'the combination of Roman political institutions, Hellenic cultural ideals, and the religious aspirations of Christianity' (Peters 1970, 682). Two forces were crucial. One was the progressive decay of the old style of urban life from the third century onwards as the wealthy, leisured elites on which Hellenism depended failed to shoulder their traditional civic burdens and a centralis-ed form of administration was imposed across the Roman domains (pp. 132–3). The effects interacted with those of the second force, Christianity.

Christianity developed within Palestinian Judaism with its centre in Jerusalem, but it soon spread to neighbouring, strongly Hellenised areas

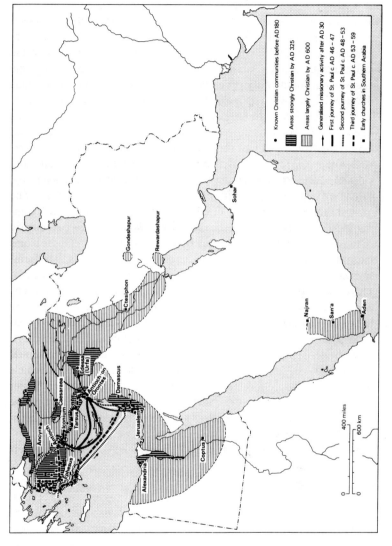

Figure 6.4: The Spread of Christianity, c.AD 30–500

(Fig. 6.4). Although Christianity was assimilated to Hellenism, its adherents felt increasingly uneasy about the contents of pagan literature, the study of which was fundamental to the spirit of Hellenism. Some ancient authors continued to be read, but Christian tracts were often preferred. The well-springs of Hellenism dried up completely in some areas and became less clear in others. More obvious changes followed as Christianity became politically strong. The theatres were closed as promoting lewdness and immorality. Physical training, so important in Hellenic education, was viewed as improper and the games, for which it was a preparation, either were formally ended or died from lack of support. Consequently, there was no need for elites to maintain those quintessentially Hellenistic buildings, the gymnasia, even if they were willing to do so. Gymnasia fell into disrepair or were converted to other purposes. Spiritual athletes became the inspiration of the emergent Christian culture of the third and fourth centuries AD, and the ascetics won their victories in the solitude of desert caves or country ruins rather than in the stadia. In some places 'the desert was made a city' (*Vita Antonii* col. 14, quoted Chitty 1977, 5), while monasteries became both the guardians and the transmitters of the new ideals. Greek, however, continued as the official language of the Eastern Roman, or Byzantine, Empire down to the Muslim conquests. Coptic and Syriac survived the Arab conquest for a few centuries but, as related languages, were virtually replaced by Arabic by the twelfth and fourteenth centuries respectively. The Turkisation of Asia Minor after AD 1070 (pp. 183–5) progressively reduced the areas where Greek was still used as a popular language almost to those from which it had originated. An unsuccessful bid (1919–22) to restore the Greek Empire after the First World War resulted in the removal of the last living vestiges of Hellenism from these shores, though not from the neighbouring islands. Now only the small indigenous Orthodox communities of Lebanon and Syria, still celebrating a Arab Liturgy, as well as a host of ruins, testify to Hellenism's former extent in the western half of the Middle East.

## Towns and City-states

Hellenism is virtually identical with the polis (city-state). On the one hand, the polis provided the environment for the expression and nurture of Hellenism. On the other, it was the main vehicle for the spread of this distinctive culture across the region. A polis was essentially an autonomous community, living in a particular territory and focused for administrative, religious and economic purposes upon a single, comparatively large settlement which, in discharge of these functions, took on recognisable urban characteristics.

Poleis had a variety of origins. They appeared first in the off-shore islands of the Greek-speaking areas (Snodgrass 1977) and were later replicated in the Middle East by the colonisation movements already outlined (pp. 125–7). As Hellenising influences gradually spread to the inland part of western Asia Minor, for example in Caria and the upper Cayster (Küçük Menderes) Valley, federations of villages focused upon local sanctuaries agreed to amalgamate their populations and to create poleis (Jones 1966a, 28). The early stages of similar synoecisms were found elsewhere in the region at various dates but, as with autonomous communities east of the Jordan rift during the Roman period, the final concentration of population never took place. Synoecism was often forced upon particular communities by the Seleucids and the Romans. The objectives were to provide for effective administration and to encourage Hellenisation. Their achievement usually involved the concentration of population to create an urban settlement and perhaps the instigators had in mind the notion of Plato (*Laws* V. c.10) and Aristotle (Barker 1948, 339–43) that an optimum size was some 5,000 citizens. Somewhat similar was the bestowing of polis-status and institutions upon garrisons, colonies of veterans and indigenous towns (Cohen 1978). Although in east-central Asia Minor and on the fringes of Arabia some territories had still not been brought within the structure of poleis, by about the fourth century AD the Roman half of the Middle East was organised in 'a mosaic of city territories' comprising more than 900 separate units, some of them resulting from the amalgamation of smaller poleis (Jones 1966b, 237–8). The whole cellular structure was held together, as earlier under the Persians and the Seleucids, by a shared culture, as well as by officials appointed from the centre, a centrally promoted cult and, in the last resort, by force.

Although the original poleis were autonomous, they lost control of important aspects of their life — foreign affairs, the raising of troops and taxes — with the victories of Alexander. Subsequent foundations never enjoyed much independence, but, like their predecessors, they were allowed privileges of self-government. Most poleis managed to retain these for several hundred years, but by the third century AD these rights had vanished. The loss was largely a consequence of the refusal, or the inability, of wealthy individuals composing the poleis councils or senates and occupying the major offices to carry the burdens always expected of them by the state but increased with the expansion of the imperial regime.

In earlier centuries governmental and cultural needs blended with civic

Figure 6.5: Model of the Hellenistic Town

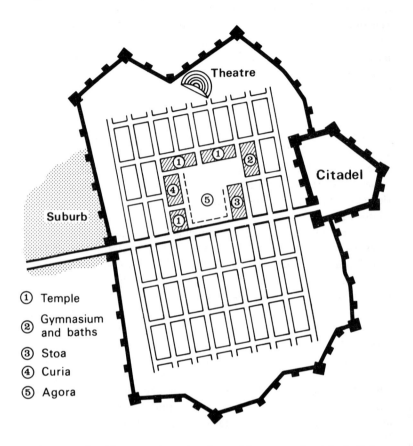

① Temple

② Gymnasium and baths

③ Stoa

④ Curia

⑤ Agora

and personal pride to produce the planned layouts and public buildings characteristic of the focal settlement of the polis, the immense ruins of which puzzled subsequent generations. Old-established towns had their ground plans transformed in one major phase of building activity, while new ones were laid out as complete entities. Most towns received subsequent additions and experienced modifications.

The focus of the whole urban complex, indeed of the polis itself, was the *agora*, or forum (Fig. 6.5), though a city as large as Antioch-on-Orontes acquired several *agorai* over the centuries. The agora was not simply a paved open space similar to a European market-place of recent times. It was a 'gathering place' for the polis community, for the assembly of citizens where such a body was part of the constitution. On the sides of the agora, or very close to it, were grouped such buildings as the council or senate house and the law courts, often with the principal temples. Colonnades were frequently added around the edge of the open space — a striking example still survives at Jerash — and these gave access to the lock-up shops in which much of the trade of the polis was concentrated.

Leading from the agora were the principal streets. These were comparatively wide, paved thoroughfares which, like the agorai, were often provided with colonnades at some time in their history. At Palmyra there were at least 375 columns along each side of a street more than a kilometre long (Rostovtsev 1932, 126). The colonnades were a definite amenity. They provided shelter from sun and rain; from them access was gained either to small lock-up shops on the outer side or to booths and stalls between the pillars. At Antioch-on-Orontes they were lit at night. In the Hellenistic and Roman periods the main streets were usually planned to intersect at right angles and to cut minor streets in the same way, creating *insulae*, or blocks. At Antioch-on-Orontes they measured 112 x 58 m (Downey 1961, 70). Although of considerable antiquity, such thoroughgoing geometrical planning was thought by Greek scholars to have been applied first at Miletus and to have been popularised by one of its citizens, Hippodamus (Barker 1948, 83). The ideal, however, had to be accommodated to the qualities of particular sites. Miletus itself was laid out on a peninsula. Neighbouring Priene was situated on a mountainside so that the new town was built in the middle of the fourth century BC on a series of giant steps (Schede 1964, 11). At Pergamum the whole town was closely fitted into a large concave hollow in the impressive citadel hill and a visual climax was provided by the vast theatre. Antioch-on-Orontes was organised along a strip of relatively level ground between the left bank of the river and the steep slopes of Mt Silpius (Downey 1961, 15–16).

Within the framework of the streets were houses, more shops and workshops, gymnasia and baths, *stoas* (covered markets), inns and taverns, and temples; and in many places from the third century BC onwards there were also synagogues. Churches appeared later. When the temples were officially closed at the end of the fourth century, churches became the principal centres of public worship. Christianity, in fact, had spread as an urban religion along the main highways of the Roman Empire. Paralleling secular administration it, too, became organised into town-centred dioceses.

The whole urban complex was often surrounded by a towered wall, remodelled from time to time according to current military thinking (McNicoll 1972). A strong-point was often provided by a citadel. In some

Table 6.1: Maximum Extent of Selected Walled Towns

| Town | Walled area (ha) |
|------|------------------|
| Antioch-in-Orontes | 2,100 |
| Constantinople | 1,200 |
| Alexandria | 920 |
| Smyrna | 640 |
| Bostra | 400 |
| Ephesus | 345 |
| Apamea | 250 |
| Ancyra | 225 |
| Palmyra | 220 |
| Laodicea-on-sea | c.220 |
| Antioch-in-Pisidia | 130 |
| Sardis | 126 |
| Miletus | 117 |
| Heliopolis (Baalbek) | 91 |
| Aphrodisias | 90 |
| Dura Europus | 42 |

Sources: Areas taken mostly from Russell (1958), but Sardis from Foss (1976) and Aphrodisias from Marchese (1976).

of the earliest Greek foundations this was a fortified hill or mountain and may have been the site of the original colony. Where natural strength was lacking, the citadel was simply a fortress. The walled areas obviously varied greatly in size (Table 6.1). Population densities must also have differed. Average figures suggested by Russell (1958) range between about 125–160 per ha for some more or less open-planned towns to about 200 per ha or over for dense agglomerations such as Alexandria. If these figures are approximately correct, they suggest populations higher than those attained subsequently, at least until modern times.

Along the main roads approaching the town gates were the tombs of the wealthy, stadia for track events, hippodromes for more violent sports and chariot-racing, and sometimes triumphal arches. More or less extensive suburbs were common. Vast aqueducts marched above the houses and gardens, delivering water from distant sources to baths and public fountains, though must of the domestic water supply was probably provided by wells and local springs.

## Local and Long-distance Trade

The town was the administrative and cultural focus of its polis. It was, at the same time, the principal market and industrial centre. Baking and simple metal-working, perhaps the production of coarse cloth, were found in every town to meet the needs of the local population. Certain towns, however, were recognised by ancient authors, such as Pliny the Elder (AD 23/24–79) and Strabo (c.64 BC–AD 19), as having developed manufacturing specialities, mainly in high-value goods which could stand the costs of export. For example, the ancient writers associated glass-making especially with Sidon. Silk-finishing was centred in Tyre and Berytus (Beirut) (West 1924). The manufacture of linen was the speciality of Byblus, as well as of Tarsus (Broughton 1938; West 1924). Miletus was famous as a woollen town in Hellenistic times, but during the Roman period the inland neighbours of Laodicea, Hierapolis and Colossae grew into rival centres (Broughton 1938). State factories were established in various towns somewhat later, mainly for the supply of military equipment.

Local trade consisted chiefly of the food, fuel and raw materials transported to the towns from subordinate villages and hamlets, either as payments in kind for land leased from the wealthy families whose scions formed the poleis councils or as 'surpluses' to be sold. Specialist villages provided coarse pottery and cloth for everyday use, at least to some towns. In Hellenistic times, each polis tried to be broadly self-supporting (Rostovtsev 1941, 1241). Even under the Roman Empire, when the opportunities for interaction were greater, the polis was forced to live for the most part from its own territory. Cereals could not be moved economically by land further than about 50 Roman miles (75 km) (Jones 1966b, 311–12). Only those towns which lay within that radius of each other or that critical distance from navigable water could hope to relieve the severe famines brought by periodic bad weather (Brunt 1971, 703–6), though most poleis did their best by maintaining an emergency fund for the purchase of cereals and electing an official to buy supplies (Jones 1966b, 249). Continued polis foundation may have made adequate subsistence even more precarious by reducing the agricultural area controlled by each (Marchese 1976, 198).

The comparatively low cost of water transport, first along the Nile and its distributaries and then by sea, gave Egypt considerable advantages in the grain trade over and above those derived from the famed productivity of its soils and its highly centralised administration. Before the Roman occupation, much Egyptian grain found its way through the bulking and exporting centre of Alexandria into the deficit region of the Aegean basin. Here it was in competition with grain from Thrace, the northern shores of the Black Sea and, after the late second century BC, from Numidia (roughly modern Algeria) (Rostovtsev 1941, 1249–52). Under

Roman rule cereal production in Egypt was organised to supply Rome. We are told that the grain convoys from Alexandria shipped 20 million *modii* (180 million l) annually during the Principate of Augustus (30 BC–AD 14), or an amount sufficient to feed the population of Rome for four months (Aurelius Victor quoted by Johnson 1936; Josephus II.386). North Africa subsequently became a major supplier, but in the fourth century AD Egypt was said to send 8,000,000 *artabae* (240 million l) annually to Constantine's new city (Charlesworth 1926; Jones 1966b, 238).

Many towns traded at a higher level than the purely local, though the duplication of many modest urban and sub-regional specialities, such as cloth and wine, points to the importance of transport costs in limiting trading spheres, except for the most valuable items. Some towns were seaports, while others were stopping places on caravan routes. A few were significant nodes in the web of long-distance communications.

Coastal towns in the Aegean basin and around the eastern end of the Inland Sea were favourably placed for bulking and distributing such locally produced, semi-perishable commodities as dried or salt fish, wine and olive oil. Although comparatively expensive, they appear to have been in high demand and could be shipped long distances by sea. Egypt was a major market, though it produced these commodities itself. Stamps on the handles indicate the place of origin of *amphorae* and from this information it is clear that wine, oil and possibly preserved fish, pitch, cheese and honey — all of Aegean origin — had a wide distribution in Syria and Palestine during Hellenistic times (Waber 1969, 39–89). Imports from Rhodes seem to have been particularly important. The island exploited its position as the hinge between the Aegean and the eastern Mediterranean to develop as a major bulking and distributional centre in the sea trade of the region during the later third century BC (Waber 1969, 14–16). It retained its pre-eminence in maritime affairs until, in 167 BC, a vengeful Rome undermined its position by removing the mainland possessions in Asia Minor and declaring the barren island of Delos a free port (Peters 1970, 318). In general, seaborne trade was facilitated under the Romans. New ports were constructed, and old ones were improved. Access inland was improved by road-building. Piracy, which had flourished in Hellenistic times, was suppressed (Ormerod 1924).

The Ptolemies, followed by the Romans, made various attempts to capture the valuable trade in incense and luxuries which either originated in or passed through southern Arabia (Fig. 6.6). These involved the creation of some new ports in Egyptian-controlled territory on the Red Sea, the clearance and improvement of the fifth-century canal from the Nile towards the salt lakes on the isthmus of Suez and at least one major

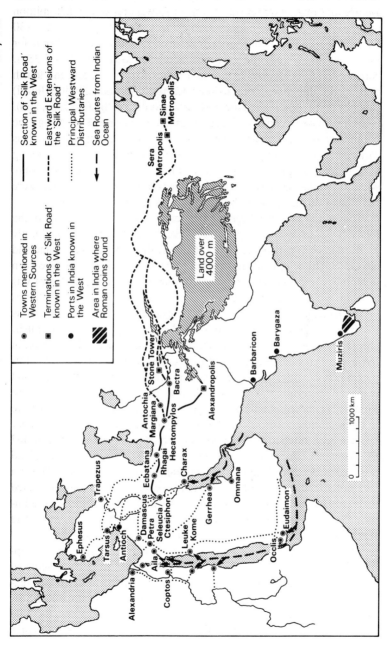

The Principal Trade Routes Connecting the Middle East to the Far East in the First Century AD

military expedition into Arabia, as well as the annexation of the Nabataean Kingdom centred on Petra (Cary and Warmington 1963, 192–3; Charlesworth 1926, 20–1; Murray 1967; *The Periplus of the Erythraean Sea*, 77–9). The real breakthrough came when the rhythm of the monsoons was discovered, allegedly by a Greek called Hippalus, in the first century AD (*The Periplus of the Erythraean Sea*, 57; cf. Raschke 1978). This opened up the sea routes to India (Cary and Warmington 1963, 95–7). Finds of coins indicate that direct trade flourished in the first and second centuries, though contacts were maintained down to the sixth century. Alexandria became the great clearing-house for the eastern trade in the Mediterranean. As ships could carry more consignments than caravans and at less cost, prices fell, and spices became better known and more widely used in the western world (Miller 1969, 200–1, 218). None the less, fears were expressed that the loss of an estimated 55 million *sestertii* each year to India was draining the wealth of the Roman Empire (Pliny VI. 101; Raschke 1978).[1]

Alexandria was not the only terminus for the eastern trade. There was also a succession of towns between 'the desert and the sown' on the western fringes of Arabia (Petra, Bostra, Jerash, Damascus) handling caravans which either crossed the desert from the middle Euphrates, and less frequently from the Gulf in the neighbourhood of Bahrain, or else came north from the Red Sea ports and even south Arabia. In Strabo's opinion, 'the greatest mart [$'εμπόριον$] of all the cities of Asia west of the Taurus' in the first century AD was the expanding city of Ephesus (Strabo XIV. 1.24). Almost certainly rivalled by the neighbouring city-ports of Miletus and Smyrna, as well as inland Apamea (Strabo XII. 8.15), Ephesus was a major outlet for the land routes which crossed Asia Minor from the Euphrates frontier. Altogether Roman land trade with China, India and southern Arabia formed at least 45 per cent of the estimated total value of 100 million sestertii (Miller 1969, 223–9; Pliny XII. 84; Raschke 1978). More than on the sea routes, it was a trade in 'treasures and rarities which contributed to the ostentation of courts but did not alter the character of societies' (Lattimore 1951, xiv). Silk was the most valuable single commodity brought to the West in any quantity. Its use increased considerably from the time of Augustus until, in the fifth century AD, officials of the Eastern Roman Empire set aside their woollen togas in favour of long coats of stiff brocade (Charlesworth 1926, 109–10; Ferguson 1978; Raschke 1978; Runciman 1961, 188). The opening of the 'Silk Road', however, forms part of the historical geography of the Persian East (pp. 147–8).

**Economic Expansion and Decline**

Although the Roman domains contained important extractive and manufacturing industries, many of the goods traded within the Empire and exported from it either came straight from the land or were agricultural products subjected to minimal processing and transformation, such as wine and textiles. Local agriculture was basic to the economy in other ways, too. It produced the food on which essentially labour-intensive productive systems operated; it was the source of most taxes; and it generated some substantial private fortunes. Output fluctuated according to inter-annual variations in weather, but some episodic expansion of the cultivated area can be inferred down to perhaps the middle of the second century AD (Heichelheim 1956 and 1970; Rostovtsev 1926 and 1941). Methods and implements changed little from earlier times and, while the organisation and planning recommended by the agricultural manuals may have improved yields marginally, increased output must have depended chiefly on cultivating more land. Labour was certainly available from a growing population which may have peaked at between 13.2 and 19.5 million people, perhaps equivalent to 21–36 per cent of the total for the whole Empire (Durand 1977).

Syria was considered equal to the Nile Valley for the production of cereals but the famed forests of the western mountains had been so depleted by the second century AD that Hadrian (AD 117–38) attempted to restrict access to the surviving timber (Mikesell 1969). Apples, pears, plums and figs were grown in the sub-region. Vines were found near the coast and wine was exported (p. 136). The remains of dams, barrages, cisterns and canals reveal a concern with irrigation and its probable extension in areas marginal for successful dry-farming. As in Palestine, however, the systems could only be maintained with a high input of labour (Heichelheim 1938). Cicero (106–43 BC) remarked that 'Asia is so rich and fertile that it easily surpasses all lands in the fruitfulness of its soil, the variety of its products, the extent of its pastures, and the number of items for export' (Cicero, 'Pro Lege Manilia' VI. 14). Cereals and grapes were produced practically everywhere, but there were some regional specialities in other crops. For example, the fruits of the lower Meander (Büyük Menderes) Valley were celebrated, as were the olives and figs of the south-west (Caria), while large flocks of sheep were remarked on the western edge of the central plateau, in Phrygia. Much of the interior was bare of forest by the first century AD (Brice 1955; Roberts 1982b). Agricultural expansion in river catchments during Hellenistic and Roman times, which must have involved the clearance of forest, has been

Table 6.2: Progression of the Cayster (Küçük Menderes) Delta

| Time period | Apparent extension (km) |
|---|---|
| 750–300 BC | 1.0 |
| 300–100 BC | 5.0 |
| 100 BC–AD 200 | 2.0 |
| AD 200–700 | 1.5 |

Source: Eisma (1962).

blamed for the acceleration of delta progression by the Cayster (Küçük Menderes), the neighbouring Meander and probably also the Sarus (Seyhan) and Pyramus (Ceyhan) in lowland Cilicia (Brice 1978a; Eisma 1978; Erinç 1978).

Agriculture in Egypt was particularly profitable, with yields of 70:1 claimed for cereals, but it was also exploitive and labour-intensive (*Descriptio Totius Orbis* quoted by Jones 1970; Johnson 1936). The early Ptolemies orientated it completely towards the export of wheat, for which there was a ready market in the Aegean basin and Italy. To meet demands, a new wheat (*Triticum durum* Desf.) was introduced and reached 65–99 per cent of the winter-sown crops (Crawford 1979; Kees 1977, 74). Olive groves and vineyards expanded, initially to meet the needs of the Greeks and Macedonians settled in the country on modest farms and large estates by the Ptolemies. Cultivation probably reached the physical limits of basin irrigation in the Nile Valley. The Delta was the scene of considerable activity, partly because of the northward shift of power consequent upon the foundation of Alexandria and its development as the capital and leading port. Reclamation was launched in swampy areas and greatly assisted by a slight fall in sea-level. Settlement numbers increased (Butzer 1976, 95). In both the Delta and the Valley the amount of perennially irrigated land, often in areas usually beyond the reach of normal floods, increased with the introduction of the sāqiya, or animal-powered waterwheel (Rostovtsev 1941, 363). Cultivation expanded in the Faiyūm depression, especially during the third century BC. During the first century BC the population of the whole country may have been just under 5 million (Butzer 1976, 85, Fig. 13).

When Egypt passed under their direct control, the Romans found that the later Ptolemies had neglected the irrigation system. Augustus set troops to cleaning silt from existing canals and the digging of new ones, a practice followed by some later emperors, too (Johnson 1936). The government, however, was initially unwilling to invest its own capital in further agricultural development. This was left to private individuals who were granted large estates. Under Vespasian the policy was changed and many of the large estates were broken up. This helped to stimulate the development of secondary towns throughout the country (Peters 1970, 396–8). None the less, the regime remained one of colonial exploitation for the benefit of Rome (Butzer 1976, 92). In addition to the wheat destined to feed the population of the City, an annual revenue of 100 million *drachmae* was extracted in the middle of the fourth century AD. By comparison, Judaea yielded 8 million drachmae (Johnson 1936). Population topped 5 million at about this time (Butzer 1976, 85, Fig. 13), a total sustained in part by the continued importance of gathering in the swamps (*Diodorus Siculus* I 80. 5–6; cf. Herodotus II. 92). It plunged thereafter. Population was down to about 3.25 million by the end of the first century AD and continued to fall (Butzer 1976, 85, Fig. 13).

The third century AD was one of profound crisis for the Roman Empire. There were many facets to it, but the one which most directly affects the historical geography of the Middle East was the appearance of *agri deserti* (abandoned fields, unused lands). This is a controversial subject, and Whittaker (1976) has stressed both the unsatisfactory quality of the evidence and the differential nature of local experience. None the less, a 10–15 per cent reduction in the agricultural area has been claimed (Jones 1964, 812). Many of the agri deserti may have been found in areas which were physically marginal for successful dry-farming, and it is perhaps debatable whether they were out of use for more than a season or two of reduced rainfall. Others were located in areas like the Faiyūm, where neglect of the irrigation system and inadequate water supplies would force the abandonment of land (Johnson and West 1949, 7–18). The phenomenon, though, was widespread, and this has induced modern scholars to search for 'universal' explanations (Jones 1966b, 307–8). Most can be dismissed. There is little evidence to support the hypothesis of progressive or pulsatory desiccation of climate in the period concerned (Huntington 1911; McDougall 1956; Sperber 1974; Tchalenko 1953, T.1, 66–72). The erosion hypothesis has more strength, for the resulting silts have buried ancient structures and led to delta progression, but the chronology and mechanism are problematic (Butzer 1974; Vita-Finzi 1969a and b; Wagstaff 1981). Labour shortage may have more to commend it (Boak 1955; Jones 1953a), but this hypothesis must be linked to controversial evidence for massive inflation and punitive levels of taxation, both arguably produced by large increases in the size of the army and the civil service, as well as by expensive building projects (including fortresses on the eastern frontier) and persistent warfare (Jones 1953b; Whittaker 1980).

At least localised recovery came during the fourth and fifth centuries. This is exemplified again by the Faiyūm, a particularly sensitive area (Johnson and West 1949, 9–10). It may have been part of a general revival of the economy, as indicated perhaps by evidence for road maintenance in western Asia Minor, and urban renewal both there and at Antioch-on-Orontes (Downey 1961, 403–7, 434–5; Foss 1972) and associated with administrative and fiscal reforms initiated by Diocletian (AD 285–305). Agricultural expansion, however, is at present evinced only from the Belus massif in northern Syria, where olive plantations appear to have spread rapidly (Tchalenko 1953, T.1, 381–2, 422–6), and the explanation may lie in local labour shortages rather than in rising demand. In fact, it seems unlikely that the basic structural problems of the imperial economy had been solved. The growth of large estates, notoriously inefficient to run, and increase in state requisitions made matters worse, except possibly in remote, mountain valleys. Some large towns (Ephesus, for example) lost so many people that not only were large tracts of built-up area unoccupied (Foss 1972, 215), but farming in the neighbourhood must have been, at best, depressed. In any case, agri deserti remained a problem for the imperial government until as late as the reign of Justinian (AD 527–65) (Jones 1964, 812), itself a period of expensive confrontation with the other 'superpower', Persia.

## THE PERSIAN EAST

South-east and east of the developed *limes* of the Roman Empire lay the heartlands of successive Persian states (*Cambridge History of Iran* 1982 and 1983). Under the influence of resilient local cultures, developments here diverged from those in the western part of the region. The differences may be displayed best by adopting a plan similar to that used in the first half of the chapter. Attention will be given to the advance and retreat of Hellenism, then to the development of towns, and finally to economic change.

### Hellenism

Hellenism infiltrated the lands east of the Euphrates even before

Alexander's victory at Gaugemela, near Arbela in the Assyrian hills (331 BC). The Achaemenids became enchanted with the Greek art and architecture revealed by their expansion into Asia Minor. Greek masons and sculptors were accordingly employed at Susa and the ceremonial centres of Pasargadae and Persepolis in the dynasty's homeland. Xenophon's 10,000 were simply the best-known of the mercenaries attracted to Persian service and, even at Alexander's victories at Issus (333 BC) and Gaugemela, Greek soldiers were in the centre of Darius' army. Force of arms carried Hellenism as far as the Oxus (Amu-Dar'ya) in central Asia and the Hypharis (Bear) in India under the leadership of Alexander, but its hold was consolidated after his death (323 BC). Alexander's eventual successor in the East was Seleucus. As in their western territories, he and his descendants planted colonies of Greek and Macedonian veterans and pursued a policy of Hellenisation (Fig. 6.3). The primary objective was to hold territory and to unite a diverse empire. Greek, of course, was the administrative language in the East just as much as in the West. It remained so even when the Parthians had taken over all the Seleucid domains up to the Great River, and was used in inscriptions on coins and reliefs, as well as in recording comparatively minor legal decisions (Colledge 1967, 73–4). Parthian kings adopted Seleucid titles and styled themselves 'Philhellene'. Ordoes II (c.57–38 BC) is reported to have been listening to an actor reciting from the *Bacchae* of Euripides when the head of Crassus was brought to him following the resounding victory near Carrhae (Harran) in Upper Mesopotamia in 53 BC (Plutarch, *Crassus* XXXIII. 2). Greek was one of the three official languages used in inscriptions by the early Sasanian rulers, but the quality deteriorated and gradually it went out of use (Colledge 1967, 97).

Hellenism as a complex of institutions and behaviour seems to have taken root particularly in two areas. One was the fertile and strategic area of Media, where a number of Greek settlements were founded and some adorned with purely Greek temples. The other was in Bactria, on the middle reaches of the Oxus (Amu-Dary'ya), where nearly 20,000 of Alexander's soldiers were reportedly settled to protect Persia from nomadic incursions (Bernard 1967; Ghirshman 1954, 221). A tenuous chain of Greek and Macedonian settlements linked the two areas together and led on to Seleucia-on-Tigris (founded in 274 BC), which contained a relatively large Hellenic population. The total number of Greek and Macedonian settlers is unlikely to have been substantial, despite the difficulties of their homeland and the attractions of the East. Many veterans must have returned home. The culture of those remaining was that of a privileged minority. It was largely confined to the towns. The majority of the population in the immense territories of Persia was completely unaffected. The lingua franca of the highly mixed populations of Mesopotamia, a legacy of Assyrian transference policies (p. 112), was not Greek but Aramaic. Akkadian was still studied in priestly circles, and much of what is known of ancient Babylonian culture is due to a kind of renaissance which began under the Seleucids. Even the philhellenism of the Parthian upper classes was little more than a fashionable affectation. They remained nomadic horsemen at heart and retained the dress of loose blouse and trousers (Colledge 1967, 93). They used their own language, Pahlavi, among themselves and occasionally in bilingual inscriptions. It is hardly surprising, then, that the 'Iranianisation' claimed by the Sasanians actually began under the Parthian ruler Vologases I (c.AD 51–80). Inscriptions on his coinage are in Pahlavi written in an Aramaic script. The text of the *Avesta*, the Zoroastrian scriptures, may have been established in his reign. The components of the national epic, the *Shahnāma* ('Book of Kings'), were probably being sung (Meuleau 1965), though they were not given fixed form order until the eleventh century. Towns began to re-use their Iranian names — Susa, instead of Seleucia-on-Eulaios; Merv, in place of Antiochia Margiana (Frye 1966, 225–6).

The Iranian revival was continued and extended by the Sasanians. Not surprisingly, the dynasty sprang from Persis (Fars), which was one of the provinces scarcely touched by Hellenism but with strong nomadic traditions (Briant 1979; Colledge 1967, 51, 94). With the Sasanians came the restoration of a purer form of Zoroastrianism as the state religion (Frye 1966, 247–53). Christianity, spread from north Syrian centres such as Edessa and reinforced by prisoners of war, remained as one of the most resilient manifestations of Hellenism in the state. Its adherents were repressed and persecuted as potential fifth columnists until the authorities realised that the Nestorian bias could be politically useful against the Roman Empire, where Orthodoxy was the official ideology. The last traces of Hellenism were found in the use of the Greek script in Bactria as late as the eighth century AD (Ghirshman 1954, 267), in a residue of Hellenistic motifs in Persian art (including, arguably, in carpets) and in the Iskandur (Alexander) romances.

## Development of Towns

Alexander and the Seleucids are credited with establishing towns in Mesopotamia and Persia, as elsewhere in the Middle East. They were almost certainly fewer than either the total claimed or the number actually known from the Hellenistic West (Cohen 1978). For the most part, the 'new' foundations were *katoikiai* (military colonies) planted on the sites of Achaemenid administrative centres in strategic localities (principally Bactria and Media), and along the major routes, including the rivers flowing to the Gulf. Voluntary associations among the former soldiers provided social amenities and a basis for self-government, but relatively few of the katoikiai developed the institutions of the fully-fledged polis. Perhaps there were too few Greek and Macedonian settlers and relatively little interest by local elites to make this development possible. Among the few places where polis organisation developed were Ragae Europus (near modern Tehrān), Dura Europus (on the middle Euphrates) and, above all, Seleucia-on-Tigris. Seleucia was laid out on a grid-iron plan. At the height of its prosperity it covered about 4 km² and had a population estimated by Pliny the Elder (VI. 122) at 600,000 but probably nearer 20–25,000 (Adams 1965, 175 n.7). It was destroyed by the Romans in AD 165 and lay abandoned when Septimius Severus reached the site some 33 years later. A small settlement in the area was destroyed during the Roman foray of AD 283 and the site was totally deserted when reached by Julian 80 years later (Ammianus Marcellinus XXIV. 5.3; Dio Cassius 71.2; 76.9.3). While it lived, Seleucia-on-Tigris was a stronghold of Hellenism, but the new ways did not take root in many other Greek foundations.

In addition to establishing Hellenic settlements, some of which became towns in terms of function and size, the Seleucids helped to revive some of the ancient towns of Mesopotamia and Assyria. Nimrud appeared deserted to Xenophon and his companions in 401 BC and Nineveh was unidentifiable (Xenophon III. 4.6–7, 10), but both revived in the third century. Babylon was half-ruined and largely deserted when Alexander arrived, partly as a consequence of Achaemenid reprisals against persistent rebels. Attempts were made to restore it as a Greek town, notably by Antiochus IV (151–143 BC), who endowed it with a gymnasium and theatre. Competition from nearby Seleucia, however, tended to weaken its economic vitality. Tablets and sealings from Orchoi (Uruk) reveal the presence of a large Greek community; impressive new monuments were built there. Ur, however, languished after reaching a peak in its fortunes under the Achaemenids. Decline was due partly to a shift in the course of the Euphrates which affected its canal network and partly perhaps to commercial competition from the new settlement of Alexander-Charax nearby (Adams and Nissen 1972, 55–7; Roux 1966, 369–81).

The Parthians left the institutions of the Greek towns alone, while installing their own governors (Pigulevskaya 1963). To some extent they encouraged further developments, allowing the formation of vassal states based on particular towns or groups of towns. These included Osrhoene (focused on Edessa in Upper Mesopotamia), Adiabene (around the revived towns of Assyria) and Araba (focused on the desert town of Hatra). Adiabene (Assyria) seems to have been particularly fortunate. Long-ruined towns were revived, often in new form. They were given straight streets lined with columns, a citadel (usually built on the mound of an ancient ziggurat) and a market-place. Fortifications were added in stone instead of the mudbrick used in earlier times (Roux 1966, 382–4). The Parthians, however, added some new towns of their own. The most important was Ctesiphon. It was established as a military camp to overawe Seleucia-on-Tigris, but gradually took over from it as the major urban centre in Lower Mesopotamia. Like other Parthian foundations (e.g. Hatra), it was roughly circular in shape. The form may have been derived from military camps, but the circuit would have been cheaper to build and easier to defend than a rectangle. Within the walls the most important buildings were palaces and temples, but *suqs* (rows of shops) may also have been provided amidst the typical courtyard houses (Colledge 1967, 116–18).

The first Sasanian ruler, Ardashir (c.220–240 BC), was credited with founding at least six towns and his successor, Shapur I (240–c.272 BC), with several more (Pigulevskaya 1963, 122). It is difficult to substantiate all of these claims because of the confusion created by multiple names. The same town may have a Greek, Parthian and Sasanian name, but be a still older foundation. The major new foundations appear to have been circular in shape, like the Parthian towns, but there were exceptions. Bishapur-in-Persis (Fars) was laid out on a grid-iron plan presumably copied from Hellenistic towns, and Hellenistic influence is also apparent in the mosaics found on the site (Frye 1966, 244–5; Ghirshman 1954, 320). The shape and plan of Gundeshapur and Weh-Antioch-i-Khosrau are not known in any detail. The former was one of a number of towns reportedly settled by Roman troops under Valerian's command who were captured in AD 260 and it may have been laid out in a fashion 'reminiscent of a Roman military camp'. The latter was built much later near Ctesiphon, but on a site as yet undiscovered. It was reported to have been a painstaking reconstruction of Antioch-on-Orontes which Khusru I (Chosroes) had ordered to be razed in AD 540, for the benefit of the original city's captured population (Adams 1965, 70; Downey 1961, 533–45). A large number of settlements in the vicinity of Uruk in Lower Mesopotamia attained 'urban size' during the Sasanian period. Most of them lay along the main trunk canal, and at least three bear traces of some degree of planning in their layout, though possessing an overall sprawling form (Adams and Nissen 1972, 62). They were probably more characteristic of urban development than the individual examples mentioned above. At the same time, the towns were part of a comprehensive attempt at regional development associated with an expanding economy.


## Economic Expansion

Writers on the Hellenistic West have interpreted the erection and proliferation of public buildings as evidence for economic prosperity. The same has been true for the Persian East. Despite a history of warfare, with its associated destruction, the Seleucid, Parthian and Sasanian eras down to the third century AD have been viewed with some justification as years of economic growth. The basis lay in trade and agriculture.

Several trade routes crossed the Persian East to the Hellenistic West, but the most romantic to later generations was the so-called 'Silk Road' (Fig. 6.6). Although trade between the Middle East and northern China via the oases of central Asia was always a possibility, the potential for through-trade was not realised until around the beginning of the Christian era. The release mechanism was political unification. The way was prepared by the migration southwards into eastern Persia of the Ta Yueh Chi nomads in the second century BC and the creation, under the leaders of the Kushan tribe, of a state which stretched from central Asia as far as northern India. Diplomatic relations were opened with the Kushans by the former Han Emperor, Wu Ti (140–71 BC), with the aim of outflanking the Huns pressing on his own frontiers. The ambassador returned with reports of An-hsi (Parthia) and T'iao-Chih (Babylonia), as well as of lands further west. A military offensive then brought many of the central Asian oases under Chinese control and soon after 100 BC the first silk began to arrive in the West. Although the Chinese were not able to retain their conquests continuously, an army did penetrate to the Caspian Sea in AD 93 and an embassy reached Antioch-on-Orontes four years later. The Silk Road was open. It was largely Roman demand for silk which sustained trade along it, rather than any desire on the part of the Chinese to find markets for expanded production or even any great desire on their part for western goods, though these may have been acceptable to the populations of the more westerly oases (Ferguson 1978; Lattimore 1951, 492–3; Miller 1969, 121–37, 205, 234–5, cf. Gray 1970; Peters 1970, 521; Raschke 1978).

Although western commentators were reasonably well informed about the stages of the Silk Road, they knew next to nothing about the major rival, the route along the Gulf (Cary and Warmington 1963, 80–6, 93, 96; Gray 1970). Despite the hazards of large waves, rough seas, eddies and whirlpools, *The Periplus of the Erythraean Sea* (8–12) makes clear that trading ships used the route. In the first century AD cargoes were brought from India, through which there were connections to the Silk Road and, perhaps by stages, even from the Far East. Seleucia/Ctesiphon was the major entrepôt for the seaborne as well as for the overland trade, but it is probable that goods were also landed at Gerrha in eastern Arabia to be carried overland to Petra.

Control of the routes through Mesopotamia, and of the Silk Road across northern Iran, was obviously lucrative to the Parthians and Sasanians. They monopolised the supply of a very high-value product in demand from the West. Customs duties were levied; merchants, carriers and inn-keepers collected their charges. It is no wonder that silk was very expensive by the time it arrived in the Roman Empire or that the Chinese formed the impression that the Parthians were anxious to prevent direct contact (Miller 1969, 121–37). None the less, both the Parthians and the Sasanians were aware that the continued flow of goods depended not just on maintaining the roads but also on providing such basic facilities as water points and caravanserais. The income raised from silk and other transit goods contributed to the capital invested in the extensive irrigation systems of Mesopotamia, traces of which are still apparent in the landscape.

Study of the lower Diyālā Valley and of the territory around ancient Uruk has revealed the emergence in both areas of similar patterns of large-scale irrigation during the Seleucid and early Parthian periods (311 BC–AD 125) (Adams 1965, 58–68; Adams and Nissen 1972, 55–9). Instead of earlier systems, which simply used canals to enhance and extend the natural waterways, long main canals were cut to feed complex branching networks through brick-built take-off points. Comprehensive planning is indicated. While the Uruk district may have experienced some economic decline as a result of diverting water to supply the twin city of Seleucia/Ctesiphon, the lower Diyālā Valley saw a large, if uneven growth in settlements until the built-up area was some 15 per cent larger than under the Achaemenids (Fig. 6.7). Wet rice cultivation was established in the area (Adams 1965, 69–83). During the same period, Susa was re-established (p. 144) and there is literary and documentary evidence for extensive irrigation systems in its vicinity and throughout Khuzestan, also related at least in part to wet rice production. These developments in Khuzestan were accompanied by an increase in the number and size of settlements until the built-up area was more than three times its extent at the beginning of the period. During later Parthian times (AD 125–225)

Figure 6.7: Generalised Irrigation and Settlement Pattern
Development in the Lower Diyālā Valley, c.300 BC–AD 637

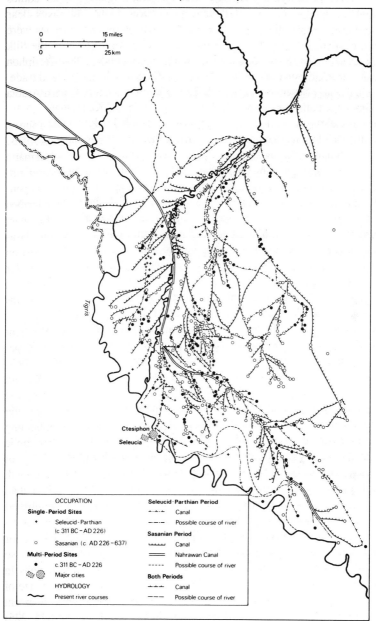

Source: Adams (1965), Figs. 4 and 5.

the built-up area of the district increased by another 7.6 per cent, though the irrigation structure remained much the same (Wenke 1975–6). In the Uruk area, however, large-scale irrigation appears to have advanced even further and cultivation began to expand to something like the maximum extent which was physically possible, a position probably attained in early Sasanian times. Most settlements, though, were concentrated along the main trunk canal (Adams and Nissen 1972, 62).

The early Sasanian period (AD 225–400) brought still further expansion in the cultivated land of Lower Mesopotamia and Khuzestan. Again, it was clearly associated with large-scale, integrated irrigation networks. In the Deh Luran (Dehlorān) plain of Khuzestan, water was obtained partly through qanāts fed by seepage from the two principal rivers and partly by tapping springs with contour canals (Neely 1974). Near Susa a unified canal system was imposed upon a variegated topography to supply relatively small, scattered pockets of cultivable land which the circumstances of the time made worth cultivating. Sugar-cane, as well as wet rice, was produced (Adams 1962). Stone and brick weirs built on the Karkheh, Dez and Kārūn Rivers are still apparently described as 'Roman' (Adams 1962), and this may give a clue as to how the irrigation works were constructed and for what purpose.

A major objective of the Sasanian monarchs may have been the economic development of territories under their direct control, with a view to building up their power so that they could dominate the powerful territorial nobility of the more distant provinces of Persia (Adams 1974). It was done by resettling captives brought from Roman-controlled Syria and Upper Mesopotamia, as well as transferring people from the Persian plateau (Adams 1974; Morony 1976). This would account for several apparently contradictory developments. One was the contraction of the built-up area in Khuzestan at the same time as the emergence of at least two substantial towns. Another was the growth of large towns in the lower Diyālā Valley against a diachronic background in which the number of occupied sites of all types appeared to double compared with the Parthian period (Fig. 6.7). At the same time, failure to complete the plans for resettling population may explain why the new towns in Khuzestan contained large unoccupied areas within their walls, as well as the seemingly incomplete state of some of the irrigation works in the area. The engineering works themselves may have been created by captured Roman soldiers, especially those from the ill-fated army of Valerian, but evidently under the direction of Sasanian irrigation experts. The Sasanian state, in fact, appears to have been highly centralised. Such authority was necessary to the contemplation, let alone the successful execution, of the large-scale irrigation systems outlined above. Nothing similar seems to be evinced from the Hellenistic West. The achievement was impressive. For example, almost the entire area of the lower Diyālā Valley with any potential for agriculture (some 8,100 km$^2$) was brought into cultivation, using water transferred to the area from outside the catchment. However, the whole structure was very vulnerable.

The large-scale integrated systems were completely dependent upon the state, not only for planning and organisation but also for their maintenance. Any weakening of interest, resolve and direction would obviously put them at risk. In addition, the networks of water courses were highly artificial. They would be threatened, if not necessarily over-whelmed, by the tendency of hydraulic systems to establish their own con-figurations through such events as larger-than-average floods. Output was certainly increased, but the cost was considerable strain on ecological relationships which, in Lower Mesopotamia particularly, had earlier in-itiated an increase in salination followed by falling yields (pp. 96–7). Altogether, then, the productive system was delicately poised. After about AD 400 its equilibrium was disturbed. The instruments were the growth of conflict between the region's 'superpowers' and natural disaster.

# Note

1. This sum is put into perspective when compared with the fortunes of contemporaries. That of Seneca (c.4 BC–AD 65) was estimated at 300 million sestertii (Miller 1969, 229), while Cicero (106–43 BC) left the governorship of Cilicia with more than 2 million sestertii in the hands of Ephesian bankers (Peters 1970, 339).

# 7 ISLAM AND THE ARAB EXPANSION, AD 500–1500

**The Ancient Empires, AD 500–630** (Brown 1978; *Cambridge History of Iran* 1982; Frye 1966; Peters 1970)

The recovery of prosperity in the Roman provinces was sustained by a relatively efficient administration down to roughly the reign of Justinian I (527–65). Before then the role of the military had been diminished and the Empire resumed a more civilian character. Its nodal points, the towns, had lost their Hippodamian purity (p. 133). Large estates had again increased in number and extent, while during the fifth century the once disreputable and impoverished Church had suddenly become a sizeable landowner. Monasteries were numerous in some districts. All this was to change over three or four generations of almost uninterrupted war with Persia (528–628).

East of the Roman *limes*, the Sasanian Empire experienced a series of crises during the fifth century. Wars were fought over the succession. Oppression sparked a series of revolts by those inspired through the communistic teachings of Mazdak. Outlying provinces made a bid for independence. Nomads pressed on the eastern frontiers and were able to exact tribute from the weakened Empire. To crown it all, famine appears to have struck on several occasions.

Reference was made at the end of the previous chapter to what seems to have been an attempt by the later Sasanian monarchs to develop a power base in Lower Mesopotamia (pp. 148–51). Implementation of their plans required treasure and manpower, as well as skills in administration and engineering. Captives seem to have supplied at least the necessary labour (Adams 1965 and 1974). It has even been suggested that Khusru I (531–79) renewed the wars against Rome specifically to promote his development plans (Brown 1978, 167). Sasanian successes in terms of the expansion of cultivation and settlement numbers have already been outlined, together with their dependence upon elaborate but ultimately vulnerable irrigation systems (pp. 148–51). An impressive as well as a vital element in the Lower Mesopotamian system was the Nahrawān Canal, designed to supplement the flow of the Diyālā. It took water from the Tigris in the vicinity of Sāmarrā, the later palace town, and carried it for about 150 km. The canal was then conveyed across the Diyālā and ran parallel to the river for another 30 km before cutting south-eastwards for more than 100 km to rejoin the Tigris (Adams 1965, 76–9).

The price for success was high. Development in Lower Mesopotamia was promoted to the neglect of Persia, and as Persia was neglected, so dependence upon the resources of Lower Mesopotamia increased. The Arab conquest revealed near-anarchy in the Persian plateau (Brown 1978, 170). The wars against Rome did not pay for themselves. Taxes were raised. The income from the Sawād (an area roughly equivalent to ancient Babylonia and including the lower Diyālā Valley) rose by some 12 per cent between the reign of Kavadh (488–531) and 608, and by a further 42 per cent (2 per cent per annum) in the next 20 years. While the increase can be explained in part by the expansion of the cultivated area, the amount of disssatisfaction and internal strife which became apparent under Khusru II (590–628) suggests that the rates of taxation were raised to unacceptable levels. Almost certainly, burdensome taxation was part of the explanation for a 42 per cent reduction in the settled area of the lower Diyālā Valley which occurred before, or very soon after, the end of the Sasanian period (Adams 1965, 71, 81). Other factors included the destruction wrought by the Roman army when it penetrated to Ctesiphon in 627–28, as well as such Persian defensive measures as the breaching of dykes to create floods and the blocking of waterways to impede the enemy advance. The destruction was difficult and expensive to repair, especially since the foreign captives settled by Khusru II appear to have fled as the Roman army withdrew. Salination resulted from impeded drainage. Silting-up developed because of administrative failure under the succession of weak rulers who followed Khusru II and because of the local fighting created by disputed successions (Adams 1965, 81–2). A further complication, particularly in the area of Uruk, was unprecedented flooding by the Tigris and the Euphrates in 629. Although vigorous attempts may have been made to repair the dykes, political instability and then the onslaughts of the Muslim Arabs prevented full restoration. The consequences were the further abandonment of land and the expansion of the swamps along the lower Euphrates, a development assisted by a westward shift of the main channel of the lower Tigris which was probably occasioned by the floods (Adams and Nissen 1972, 59–62).

The great war was ultimately disastrous for the Eastern Roman Empire. By the time of Justinian I it was no longer a state organised for war and, though the old militarism revived during the emergency, reliance was put on diplomacy and the fortification of the eastern frontier. Both were expensive. However, they were generally successful down to the reign of Heraclius (610–41), despite the penetrating and destructive raids into Syria by the Persian heavy cavalry (*cataphracti*) accompanied by Arab allies. The fall of Antioch-on-Orontes in 540, and its subsequent destruction, was a great blow to Roman prestige, but perhaps no more. With hindsight we realise that, with its fall, an era had passed, and that the city was entering on a protracted and painful decline (Downey 1961, 533–52). The villages of the neighbouring Belus massif appear to have been little affected by these attacks (Tchalenko 1953, T.1 426–9). Away from the frontier provinces, however, the archaeological evidence suggests that little public building took place after the reign of Justinian. Relative impoverishment is suggested, and it was probably caused by the expenses of the protracted war and the Emperor's own extravagance. Population seems to have declined in the larger towns of western Asia Minor, no doubt partly as a result of the spread of plague from Egypt in the 540s, when perhaps 20–25 per cent of the total population of the region may have died (Procopius 22.1–24.12; Russell 1968). Paradoxically, social and economic life appears to have flourished (Foss 1972).

Resumption of the war in 572 brought a change in its character. Whereas it had previously been something of a hit-and-run affair, the Persians now began a war of conquest through the slow but systematic reduction of the fortress towns of the Roman *limes* in Upper Mesopotamia and Syria. By 606 and 607 they had 'overrun' Syria and taken large numbers of captives (Downey 1961, 571–5). They broke into Armenia and Cappadocia in 610–11 and occupied Antioch-on-Orontes yet again. An attempt by Heraclius to prevent any further advance was defeated in 613 near Antioch. This opened the way to Asia Minor and Egypt. The Persians advanced on Constantinople and made thrusts towards the west coast. The effects were uneven, of course, in such a large territory. Some long-established towns, such as Sardis, were virtually obliterated (Foss 1976, 53–5). Many more were so threatened that their inhabitants concentrated into fortresses surrounded by a scatter of cowering villages. These included Ancyra, Pergamum, Priene, Laodicea and Miletus. A few relatively large centres, notably Ephesus and Smyrna, survived reasonably well. Districts away from the main lines of communication and out of the way of foraging parties were relatively untouched (Foss 1975). The Holy City of Jerusalem fell to the Persians in 614 and the Holy Cross — preserved there since its alleged discovery in 326 by the dowager Empress Helena — was taken to Ctesiphon. Egypt was occupied as far as the First Cataract during the years 616–18. Persian rule lasted until 627 or 628 both here and in Syria. It meant the effective end of the shipment of corn to Constantinople and may have encouraged production in the relatively secure coastal districts of Asia Minor to meet the needs of the beleaguered capital. In Syria, the Persian occupation brought an end to construction in the Belus massif and ruined the olive-growers, already suffering the effects of a tree-killing drought in 599, by cutting them off from their overseas markets (Downey 1961, 571; Tchalenko 1953, 431–6). Desertion may have begun (Tchalenko 1953, 431–6, cf. Sourdel-Thomine 1954).

Heraclius was eventually able to collect an army in the capital estimated at 120,000 men. He had these moved by sea to Cilicia in 622, the momentous year in which Muḥammad migrated from Mecca to the oasis of Yathrib (later Medina). Heraclius was thus able to place his troops at the weakest point in the Persian communications with western Asia Minor and to begin driving a wedge which would split their forces (Butler 1978, 119–27). The move was successful. Troop reinforcements were brought in through the Black Sea port of Trapezus (Trabzon) in 623 and within six years Heraclius had recovered his territories in the east, struck hard at Ctesiphon and recovered the Holy Cross.

## Arabia and the Rise of Islam

Such a titanic struggle between superpowers was sooner or later bound to involve the lands south of the Euphrates, in Arabia. Several interdependent socio-economic systems existed here in the late sixth century. The northern margins between the steppe and the desert were the domains of pastoral nomads and semi-nomads. Different tribes had drifted in and out of these zones for millennia, but the development of an efficient, framed camel saddle (*shadād*) during the last half-millennium before Christ initiated a series of changes which came to fruition in the third and fourth centuries AD (Bulliet 1975, 87–104). The range of the nomads was increased immediately, making them less dependent upon the water and grazing of settled communities on the steppe margins. As a consequence, northern Arabia became increasingly 'bedouinised' (Dostal 1959). At the same time, the nomads had an effective means of attacking or protecting caravans which crossed their territories, while they enjoyed a virtual monopoly in the supply of the now principal draft animal, the dromedary (Bulliet 1975, 87–104). Nomadic activities thus became integrated with those of settled communities at a level which rose above mere subsistence. It was probably as a result of closer contact that stable forms of political organisation began to emerge among the nomads in the third and fourth centuries AD (Peters 1978). These were exploited by the Persians and Romans, first, to provide auxiliaries for their armies, and then to create buffer states along their southern flanks where aridity prevented the deployment of large regular forces. On their side, the Romans recognised the dominant position of the Christian Banū Ghassān and used them in the wars against Sasanian Persia. The Sasanians, for their part, worked through the Banū Lakhm who dominated the tribes along the middle and lower Euphrates. Some of the associated tribes here had become Christian, but conversion did not prevent them from raiding deep into Syria during the wars of the early seventh century. Much of central Arabia was dominated until the 540 by the Kinda tribe, which was loosely allied with the Romans. Their confederation, however, broke up and was not effectively replaced before the rise of Islam (Hitti 1964, 78–86).

Pre-Islamic conditions among the nomads were characterised by later Muslims as 'The Days of the Arabs' (*Ayyām Al-Arab*) (Gabrieli 1968; Hitti 1964, 88–104), when the uneventful but harsh life of the nomad was punctuated by the exciting rituals of raid and counter-raid (p. 68). Incidents from these were long remembered in complex and lengthy odes (*qasīdahs*) celebrating the manly virtues of courage, loyalty, generosity and honour (Gibb 1963, 13–31). 'The turning planets and . . . the stars that rise and set' (Qur'ān 81: 15–16) regulated the seasons of migration, but some of them were worshipped as deities; Judaism, as well as Christianity, had made some advance.

Both the Banū Ghassān and the Banū Lakhm controlled villages and towns. Oasis-communities flourished throughout the rest of Arabia, providing nomads with basic foodstuffs, assisting local exchange and facilitating long-distance trade by caravan. Cultivation and a settled way of life were particularly well developed in the misty mountains of the south (Doe 1972). Here the arduous building of terrace systems and the tapping of both mountain springs and unpredictable flash floods for irrigation underpinned a succession of states which had waxed, waned and then disappeared ever since the early part of the first millennium BC. Additional wealth was derived from the export of salt from a number of salt domes along the northern edge of the mountains and, more spectacularly, from the control of both the sources of incense and myrrh in Dhufa and the overland routes which took those valuable commodities towards the demand areas of the north (Bowen and Albright 1958, 35–41). Trade in local products was supplemented by transit trade in goods from east Africa, the Gulf and India. An elaborate and sophisticated culture developed.

This happy corner of Arabia was drawn into the great war between Persia and Rome in two ways. Its transit routes bypassed the major zone of conflict between the two great powers in Syria and Mesopotamia. At the same time, a power with a foothold in Yemen could command the

sea route through the Bāb al-Mandab, while effective control of south-
west Arabia would clearly threaten the flank of its rival. Thus, the occupation of Yemen by Abyssinian forces in 525–75 at the instance of local
Christians (Doe 1972, 27–9) can be seen as a Roman advance by proxy.
The Persians intervened directly in 575 and conquered the area. Neither
great power, however, was able to use its advantage. On the one hand,
the intervening terrain presented enormous difficulties while, on the other,
the logistical problems of mounting complex, sea-based operations at a
long distance were immense (Doe 1972, 15, 27–30, 80).

The caravan routes to the north ran either through a series of rain-fed
mountain basins or across the zone transitional to the Rub' al-Khali. One
of the stopping-places, about half-way between Mārib in Yemen and Gaza
in Palestine, was Mecca, known to Ptolemy (6.7,32) in the second century AD as Macoraba. This was the birth-place of Islam, the principal
socio-religious system in the Middle East today. Mecca lay in a narrow
rocky valley, 'unfit for cultivation' (Qur'ān 14:40), and just to the north
of the limits for dry-farming. The name means 'sanctuary'. In pre-Islamic
times it was the focus of a pilgrimage made at a specific time in the year
and associated with both a truce in the perennial fighting and a fair of
perhaps more than local significance. Its object was the *Ka'ba*, built according to tradition by Adam and restored by Abraham and Ishmael (his
son by Hagar). Here they set the Black Stone presented by the Archangel
Gabriel, which is still in position on the south-east corner of the structure, and organised a cycle of ceremonies (Burton 1893, vol. 2, 300–3;
Hitti 1964, 100–4). From these activities crystallised a commercially based
urban community, a development traditionally associated with the
Kuraish tribe, who became custodians of the shrine, and possibly dated
as late as the 440s (Eickelman 1967). The mainstay of the community
was the transit trade, for which it provided caravans. This would have
brought its members into contact with neighbouring nomads who must
have supplied many of the necessary dromedaries. At the same time,
Mecca was in touch with various settled communities, especially, it would
seem, in the oases of al-Yamamah in central Arabia, from which it drew
its food supplies (Donner 1977).

It was in a cave under the grey and black crags near Mecca that a
respected, 40-year-old merchant with a contemplative nature named
Muḥammad received the first of a series of divine revelations: 'Recite
in the name of your Lord' (Qur'ān 96:1; Cook 1983; Watt 1961). Muḥammad was convinced that there was only one God, not the pantheon
acknowledged by his contemporaries, and that God was the All-powerful,
the Creator, before whom each individual person would stand alone at

a Last Judgement. He felt compelled to declare these truths to his contemporaries, to give them warning, as the ancient prophets had done,[1]
so that they too could submit to God, fulfil his commandments and reach
Paradise.[2] Western commentators in particular have pointed out the
derivative nature of the Prophet's message (Southern 1962, 92–4), and
suggested that its origins lie in contacts with perhaps ill-instructed converts to Judaism and Christianity whom Muḥammad met during his
journeys with the Meccan caravans (Qur'ān 16). There are certainly
numerous parallels between the Qur'ān and the Bible, especially with
the Pentateuch. Muslims believe that the revelations received by Muḥammad over the next 22 years complete the earlier ones given to the Jews
and Christians but distorted by them. The Jews attempted to reserve the
love and mercy of God to themselves alone, while Christians had erred
by worshipping Jesus ('Isa), son of Mary and one of the great prophets,
as if he were the Son of God.[3] In the end, the authenticity of the Prophet's message is demonstrated by its continuing ability to satisfy for many
people their spiritual hunger for righteousness and their need for a rule
of life (Hodgson 1960). Its spread has been of the utmost importance to
the life and geography of the region since the early seventh century and,
accordingly, is given more extensive treatment here than that allowed to
other religions.

The Prophet's first converts were in his own family and among his
immediate kin. Despite Muḥammad's efforts and considerable personal
charisma (his face shone),[4] further conversions came slowly. Scorn and
persecution developed, forcing some of the Muslims to find refuge in
Abyssinia (AD 615). A sympathetic hearing in 620 from some pilgrims
from Yathrib (Medina), an oasis 400 km away to the north and off the
main caravan route, prepared the way for the secretive, but carefully
organised withdrawal of most of the Muslims from Mecca some two years
later, followed a few months afterwards by the Prophet himself. His migration (*hidjra/hegira*) was accepted subsequently as the decisive step which
inaugurated the Muslim era (Hitti 1964, 116).

Free from the constraints of Mecca and faced with the need to provide for and organise the Muslim community, further relevations from
the Qur'ān allowed Muḥammad to consolidate and extend his teaching.
Prayer was regulated, its direction changed from Jerusalem to Mecca and
the simple 'call' instituted in place of trumpets, gongs and bells to sum-
mon the faithful. The sabbath was moved to Friday. Ramaḍān, the month
in which Muḥammad received his first revelation, became the period of
fasting, while pilgrimage to the Ka'ba was maintained. Rules for alms-
giving and social behaviour were promulgated. The poverty of the

Meccan Emigrants (*Muhājirūn*) turned them to somewhat unsuccessful raiding. Their attempt to intercept a Mecca-bound caravan in January 624 provoked a battle with a relieving force from Mecca encountered at the wells of Badr, about 136 km south-west of Medina. It was the first Muslim victory (Glubb 1963, 61–7). The new faith was plainly vindicated (Qur'ān 3:19; 8:42–3).

Although defeated and wounded at Mt Uhud during the return battle in the following year, Muhammad was able to build up his political base in Media, in a few of the captured oases to the north, and among some of the nomadic tribes. His methods were those of expulsion (Jews — probably converts — from Medina), assassination (of those who disputed his authority), small-scale raiding and — arguably — the control of vital food supplies to the nomads (Donner 1977 and 1979; Glubb 1963, 67–8). The interruption of food supplies to Mecca provoked an unsuccessful attack on Medina ('The Battle of the Trench', AD 627). In the following year Muhammad was prevented from making the pilgrimage but he negotiated the Truce of Hudaibiya which, in effect, recognised his authority and the existence of the Muslim community, while allowing him to make the pilgrimage unhindered in 629. Influential converts were made, including the hereditary guardian of the Ka'ba. Muhammad's triumph came in 630, when he marched on Mecca with, it is claimed, 10,000 men. After performing the prescribed rituals he ordered the idols defiling the shrine to be smashed and those in private hands to be destroyed. The whole population was summoned and swore loyalty to God and his Prophet. Further military success followed. Delegations came from different parts of the peninsula to acknowledge the authority of the Prophet, perhaps to secure vital food supplies. In 632, Muhammad again led the pilgrimage. It was the last time, for in June he died at Medina.

## The Muslim Expansion, c. 632–60 (Fig. 7.1)

After a noisy incident created by despair and uncertainty immediately after the Prophet's death, his old companion Abdulla (or Atiq) abu Bakr was recognised as Successor (*Khalīfa*/Caliph) (632–34). Hijāz had been the scene of the Prophet's life and it remained loyal to Islam when he died. People in the more distant areas, whose submission was always incomplete, refused to pay the tax levied by the Prophet and, in the view of later Muslim commentators, turned apostate. Some local and tribal prophets came forward, though it is not clear to what extent they were imitating Muhammad or following an established tradition (Eickelman

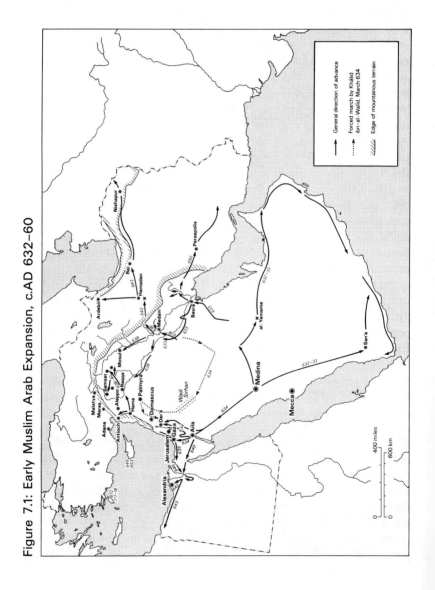

Figure 7.1: Early Muslim Arab Expansion, c.AD 632–60

1967). The Caliph would brook no secession. War was unleashed against the apostates. The Muslims' success against their most serious opponents, the followers of Muslama ibn Ḥabīb, based in al-Yamamah, not only secured all of central Arabia for Islam but led, within two or three years, to the submission of the whole of the south as well. In addition, success in central Arabia brought the Muslims into contact with the largely pagan Banū Bakr confederation. Its members had been raiding the Euphrates fringe of Mesopotamia for at least a generation following the rebellion of the Banū Lakhm against their Sasanian masters (Donner 1980). At the same time, the defeat of the apostates allowed the Caliph to turn his attention back to the Prophet's own favourite project, an attack on the outskirts of the Roman Empire in Palestine and Syria. The time was ripe. Withdrawal of their subsidy by the Emperor Heraclius had undermined the power of the Banū Ghassān and made them resentful and disloyal. Almost without wishing it, the Muslims found themselves fighting on two fronts in northern Arabia. None the less, within 20 years or so of the Prophet's death, 'the long-haired Saracens' (George of Pisidia *De Expeditione Persica Acroases* 11.209, quoted Butler 1978, 151) had seized practically the whole of the region except for Asia Minor and the Caspian lowlands (Fig. 7.1). Armenia was isolated and raids were made deep into Asia Minor. Constantinople, the Roman capital, was besieged for the first time by the Arabs in 668–69. The military success is staggering, especially for comparatively small armies. How was it attained?

While the Muslim will see the hand of God at work, the basis of success was undoubtedly geographical (Glubb 1963; Hill 1975). Djazīrat al-'Arab ('The Island of the Arabs') is a wedge of largely arid territory penetrating the better watered lands of Syria–Palestine and Mesopotamia. Expansion within Arabia was almost bound to carry the Muslims into such 'soft', attractive areas, along whose margins were established people similar in background and life-style to the majority of their armies. By contrast, steppe and desert rendered inner Arabia virtually impenetrable to armies unaccustomed to its rigours. The Arabs could move at will in this environment, living off the country in the early days, with their families and herds accompanying them. These were the twin bases of 'desert power', where 'range was more than force, space greater than the power of armies' (Lawrence 1962, 203; Lewis 1964b, 55). It was effectively exploited by the Muslim leaders, probably intuitively. Their forces could strike without warning at any point they chose and Khālid ibn al-Walīd's lightning attack on Palmyra and Damascus in March 634 is simply the most dramatic instance of a common tactic. Equally important, Muslim forces could rapidly withdraw to safety and regroup, as in the case of Muthanna ibn Ḥāritha's retreat from the line of the Euphrates after defeat at the Battle of the Bridge in October 634. The Muslims could build up their forces out of sight and range of their enemies, as for example during the winter of 637–38 before Sa'd ibn abu Waqqās' successful advance into Lower Mesopotamia (Glubb 1963, 131–6, 161–5, 189–93).

No physical obstacles impeded the advance of largely cavalry armies to the Euphrates, though the river itself and the ditches and banks of irrigated farmland presented problems to horsemen and camel-riders. Similarly, the way was open for mobile forces to advance up the slopes into Upper Mesopotamia and Syria until they reached the hills and mountains which run parallel with the coast in the west and set limits to the steppe in the north. Hills and mountains checked the Arab advance there. Like walled towns, they were difficult to subdue by the types of forces available. Movement was channelled into narrow passes. There was no room for cavalry to manoeuvre. Advance was difficult, especially when the armies were dependent upon soft-footed dromedaries for transport and were accompanied by dependent families with livestock. These difficulties help to explain several peculiarities in the pattern and pace of Muslim expansion. One is the importance assumed in the campaigns for Palestine and Syria of the Der'ā Gap, which carries the road from the south-east towards Damascus between the almost impenetrable lava fields of the Hauran and the deep, steep-sided gorge of the River Yarmuk (Glubb, 1963, 140–2). A second problem which the existence of mountainous terrain may help to explain is the relative slowness of the conquest of eastern Persia, which was not complete until the eighth century. Although the Elburz Mountains seem to have formed a major obstacle to northwards advance, it was not so much the terrain of interior Persia which was responsible for the delay as the difficulty of getting forces there from Lower Mesopotamia through the passes in the Zagros Mountains. A somewhat similar explanation may apply to the lack of success in conquering the Anatolian plateau from the south-east, despite the subjugation of lowland Cilicia and the maintenance of fortress towns west of the upper Euphrates and north of the Taurus-Anti-Taurus Mountains. An additional factor in both sub-regions must have been the snow and extreme cold which made winters harsh for dromedaries and people not used to them. The fact that Persia was overrun, while most of Asia Minor was not, however, emphasises the human element in the Muslims' military success.

Most of the Sasanian Empire in Persia had effectively disintegrated before the Muslims crossed the mountains and it required only a determined push to topple it altogether. Asia Minor was much more united. It was the core of the Eastern Roman Empire, which the emperors were determined to hold. No Arab population was already settled there whose affiliation with the enemy might undermine this resolve. It was quite otherwise in Syria-Palestine and Mesopotamia. People related to the Muslim Arabs had been drifting into these areas for generations and their descendants seem to have been sympathetic to the newcomers. Moreover, the people had been alienated from their imperial masters by, on the one hand, Sasanian attitudes to Christians and Manicheans settled in Mesopotamia and, on the other hand, Eastern Orthodox persecution of the monophysite and monothelite Christians concentrated in Egypt, as well as in Syria–Palestine. They had all suffered in the recent great war between Persia and Rome. Additional human elements assisted Muslim success. Muḥammad's teaching had succeeded in creating a new sort of 'tribe', the Muslims. Among its members, Islam had reduced the claims of ordinary tribal loyalties and transmuted traditional animosities. Promise of a vivid and sensual Paradise provided a substantial incentive to personal valour and sustained effort, but, as an Arab poet observed, it was backed by 'love of bread and dates' (quoted by Brown 1978, 192) — in other words, expectations of booty which raiding and warfare in the settled lands brought to the nomads and semi-nomads who formed the bulk of the Muslim armies. Their natural exuberance and native fighting spirit were also important. Finally, we should recognise that the Muslim leaders, for the most part from Mecca, were exceptional men. They made skilful use of 'desert power', and had learnt from the Prophet the patient diplomacy which secured them such key towns as Damascus without resort to expensive sieges, though they were actually in no position to mount such sophisticated operations.

**Political and Military Developments: an Outline to c. 1520** (Fig. 7.2)

Military success created an empire. It was ruled at first (632–56) by Caliphs chosen from the Prophet's own tribe and generally acknowledged by all Muslims in the conquered territories. The proclamation (656) of 'Ali, the Prophet's first cousin and son-in-law, brought division. It not only introduced armed conflict in the short run, but also created a permanent cleavage in the House of Islam between *Shī'a* and *Sunni*. This is marked by disagreement over the legitimate line of authority in Islam and by certain practices. The Shī'a (derived from *Shī'at 'Ali*, 'The Party of 'Ali') believe that the true line of 'leaders' (*imāms*) runs through 'Ali and his two sons, Hasan and Husain, all of whom are recognised as martyrs (*maḵtail*). They developed secondary pilgrimages, notably to Karbalā, where Husain is buried, and instituted passion plays (*taʿziyāt*). These re-enact the engagement in which Husain was killed in 680 and are performed during a period of ritual mourning which is often accompanied, as in modern Iran, by flagellation. Immediately, though, the vengeance killing of 'Ali (661) opened the way to dynastic rule by the descendants of Umaiya.

The Umaiyads (661–750) consolidated the Muslim Empire and continued its further expansion, notably by incorporating north-eastern Iran, where many Arabs subsequently settled. Their rule, however, was marked by a reliance upon Syrian personnel, often Christians, and increasing, if culturally brilliant secularisation. Opposition smouldered. It was fuelled from various directions. Pious Muslims wished to see a return to the austere religious purity of earlier days. Non-Arabs were eager to end social and economic discrimination. 'Nationalism' began to revive in Persia. Finally, the diverse opposition groups gradually crystallised around the descendants of 'Abbās, the Prophet's uncle, who was promoted as a secret early Muslim, though he had not openly declared his 'submission' until comparatively late. Revolt broke out in June 727. Within three years the 'Abbāsids were in control of the Muslim lands of the Middle East, with the exception of parts of Arabia. The new rulers renewed the war against the Eastern Roman Empire, but after 838 no serious attempt was made to conquer territory north of the Taurus. 'Abbāsid rule saw the flowering of the civilised arts, as well as of philosophy and science, and created canons and traditions which persisted in Muslim lands under subsequent rulers.

Another major change came with the revolt of the Zandj (868–83). These were predominantly, but not exclusively, workers of African origin apparently employed in the reclamation of salt-infested land in and around the swamps of southern Mesopotamia. Some scholars have seen their revolt, with its religious undertones, as a classic slave rising, while Shaban has argued recently that it was part of a struggle to control the terminii of the lucrative African trade (Ashtor 1976a, 115–21; Shaban 1976, vol. 2, 100–12). Be that as it may, the central government's preoccupation with a revolt so close to the new capital, Baghdad, prevented it from dealing with the centrifugal forces which, now that Islam was securely established, increased the autonomy of the more distant provinces. From the tenth to the sixteenth century the political history of the Middle East is a kaldeidoscope of locally based principɐ'ities, expanding and contracting over more or less short periods of time, but complicated by intrusions

Figure 7.2: Time Chart: Generalised Political Control, AD 622–1600

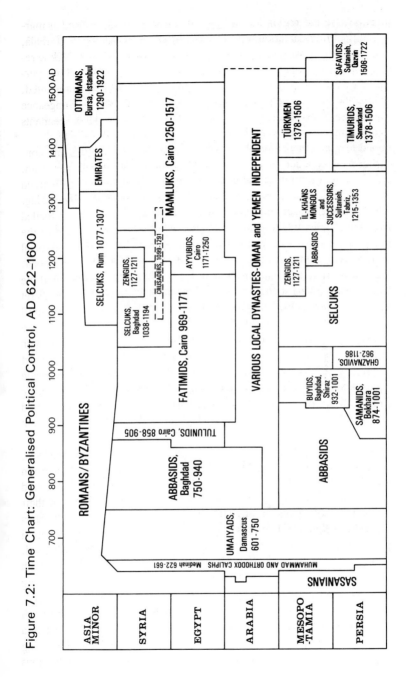

from outside the region.

Political fragmentation was advanced by two developments. The first of these was the weakening of the Caliph's authority, consequent upon the growing influence of the military establishment and the increasing role played by Turkish mercenaries, first recruited by Mu'taṣim (833–42) (Hitti 1964, 466). The second was the arrival of the Turks from central Asia in the 1030s (Cahen 1968, 1–137; Shaw 1976, 1–9). They came not as detachments of mercenaries but as 'hordes' of migrant nomads determined to graze their animals at will and to conquer territory. Like the earlier Arab conquerors, they had the advantages of familiarity with arms and fighting, of mobility and being able to live off the country. In addition, the Turkish leaders were able to field comparatively large armies which made devastating use of mounted archers. They created a short-lived but efficiently administered empire out of a confederation of semi-independent emirates and tribal territories. The defeat of the Byzantine army at Manzikert near Lake Van in 1071 opened up Asia Minor and began the 'Turkisation' and 'Islamisation' of the whole peninsula (pp. 184–86) Harassment of Christian pilgrims, following the capture of Jerusalem in the same year, helped to provoke that extraordinary event, the First Crusade (1096–99), and thus the establishment of relatively ephemeral Frankish principalities in Syria and Palestine. These were: the Kingdom of Jerusalem (1099–1244), the Principality of Antioch (1098–1168), the County of Edessa (1098–1144) and the County of Tripoli (1102–1289) (Benvenisti 1970; Cahen 1940; Prawer 1972; Richard 1978; Runciman 1965; Setton 1962–77). In addition, the way was prepared for a dynasty of Kurdish descent to take over Egypt (1171) and to be succeeded in turn (1253) by the slave troops 'bought' (hence *mamlūk*) to maintain the army (Glubb 1973). It was the Mamlūks who eventually stopped the expansion of the 'inhuman and beastly' Mongols (Boyle 1970) in Syria at the battle of 'Ayn Djālūt (1260). Periodic incursions of Mongols into the region (1218–22, 1240s, 1253–58) incorporated Persia under an Īl-Khān (tribal lord) into a vast empire stretching across central Asia to the East China Sea (Boyle 1970; Saunders 1971, 55–70; 128–39). The Mongol Empire disintegrated, to be revived briefly under white-haired Tīmūr Lang (Tamerlane) (1336–1405) (Hookham 1962). His death brought further fragmentation which persisted until the sixteenth century, when cohesion was restored by the rival Ottoman and Safavid Empires.

This succession of political events cannot be viewed as totally detached from other aspects of the region's historical geography. Warfare and changes of political regime meshed very closely with the whole of the socio-economic life of the people over the 36 or so generations sketched above, and the entire interacting complex produced significant changes in the region's landscapes. These are outlined in the following sections. They deal, first, with the long-term consequences of the Arab conquests and then with towns and economic activity in the Muslim parts of the region. Asia Minor will be discussed in terms of the reactions to Arab pressure. Its 'Turkisation' forms the prologue to the next chapter.

## Consequences of the Arab Conquests

The conquests of the Muslim Arabs were so rapid that initially few changes were made to socio-economic life in the region. Arabs dominated the power structure but everyday life went on much as before for most of the inhabitants of the conquered territories. Their status as 'protected people' (*dhimmīs*) was new, but the definition of their communities in terms of religious allegiance had already emerged under the Roman and Sasanian Empires. Some Christians fled from Syria. Many of the Persians whose ancestors had been introduced to Mesopotamia by the Sasanians were among those people who abandoned areas east of the Tigris and concentrated in the southern parts of the sub-region in and around the new towns of Kūfah and Basra (Ashtor 1976a, 13; Morony 1976). Of greater long-term significance were the immigration of Arabs and the 'bedouinisation' of the steppes of Syria and Upper Mesopotamia. The movement of people out of Arabia into neighbouring lands had been going on for generations, forced by the search for pasture in years of relative drought and facilitated by military weakness in the invaded polities (Butzer 1957). The incursions which accompanied and followed the military advances were predominantly of the same type. Military success, however, took some Arab tribes outside what might have been regarded as their normal range. Some went to lowland Cilicia, in the west, and more to the Persian plateau, where Khurasan in the north-east (Cahen 1975) proved particularly attractive. While many of the nomads turned to a settled life of cultivation, some continued in their traditional ways. The dromedary became common throughout the Muslim parts of the region. Wheeled transport vanished (Bulliet 1975, 6–19, 217–23). Except in the more impenetrable hills and mountains, the seasonal movements of flocks and herds were often organised with total disregard for the rights and needs of cultivators. Some land must have fallen out of cultivation, while olives and other fruit trees were probably consumed in innumerable camp fires (Ashtor 1976a, 16–18; Kramers and Wiet T1, 203–22). However, it is impossible to estimate either the extent or the permanence of the devastation.

Dislocation, devastation and loss of population were the immediate consequences of Arab raids into Asia Minor and on the Aegean Islands during the period from the middle of the seventh down to the end of the ninth century. They were launched almost annually from lowland Cilicia and the valley of the Euphrates, in both of which substantial, fortified garrison towns were maintained, against territory which had suffered from the ravages of Persian armies during the great war. Major attacks were launched on Constantinople (676–77, 717–18). The larger towns of Asia Minor were not only difficult for the raiders to capture, but they were also to some extent ignored by them in the search for easy plunder and prisoners to ransom or sell, and because of the ritual and propaganda elements in carrying on war against the infidel. Consequently, many of the large towns survived in truncated form, especially along the main lines of communication with the East from Constantinople. Scattered villages sometimes emerged among the extensive ruins. Smaller towns found their vitality sapped by the disruption of farming in their hinterlands, the inadequacies of their own defences and the cutting of aqueducts. They were either deserted altogether or else contracted to hill-top fortresses, perhaps with a scatter of villages nearby (Ahrweiler 1962; Brooks 1898; Foss 1972 and 1976; Miles 1964). The reduced populations of Asia Minor were organised into, and administered through, military districts (*themes*) responsible for their own defence (Ahrweiler 1962). People were transferred from other parts of the Byzantine Empire to make up for losses (Charanis 1961). Small freeholdings were instrumental in maintaining agricultural and military manpower and appear to have increased in number, while great estates declined. The emergence of iconoclasm in the eighth century may have been an attempt, at the ideological level, to secure support from people perhaps to some extent sympathetic to the austere claims of Islam. The village community became collectively responsible for paying taxes and attempts were made to revive the cultivation of abandoned land (Ostrogorsky 1962; Setton 1953). Military and political necessity, however, collaborated with insecurity and debt to encourage the re-formation by the tenth century of another generation of large estates. Both a form of feudalism and a class of frontier warrior (the *akritai*) emerged in what the Arabs continued to call 'The Land of the Romans' (Bilād ar-Rūm) (Setton 1953; Vasiliev 1933).

Beyond the Roman frontier, Arabic (*'Arabiyya*) was introduced by the conquerors but made comparatively little advance at first among the indigenous population. It was simply the language of the military and religious elite and, though Arabic began to develop a written literature from the middle of the eighth century, local languages continued in use, even for most administrative purposes. Syriac translations from the Greek were the main vehicle for transferring Hellenistic science and philosophy to the Arabs: the languages are related. The production of the first fully Arabic coins in 695 marks the beginning of a change-over from local languages to Arabic for administrative purposes. Significant advance at the popular level, though, came only with conversions to Islam. The Qur'ān is an Arabic Qur'ān (19:97; 39:27–8; 42:7; 43:3), to be read and recited in the original tongue. Arabic provided the formulations of prayer. Even then, Arabic replaced only languages in the same related group — Syriac and Coptic. Coptic was virtually 'dead' as a spoken tongue by the twelfth century, while Syriac was not finally superceded until some 200 years later, and then a few relatively remote communities in Syria and Upper Mesopotamia managed to retain it in everyday, as well as liturgical use down to the present century. Similarly, non-Arabic Semitic languages survived until recent times beyond the deserts and mountains in parts of southern Arabia; Arabic was originally indigenous to the north-central parts of the peninsula. Persian, as an Indo-European language, was the most resistant among the tongues of the conquered peoples. It remained the vehicle of ordinary speech, even when Arabic became for a while the main literary language, and enjoyed a renaissance which started as early as the ninth century.

The spread of Islam was undoubtedly the most important long-term consequence of the Arab conquests. Not that mass conversion took place — as it often did in contemporary northern Europe — at the point of the sword (Keen 1969, 31–2). Conversion to Islam was more of a social than a political process. A plot of the number of conversions against time produces a logistic curve similar to that described by the adoption of other innovations (Fig. 7.3) (Bulliet 1979). As measured by the giving of Muslim names, Persia was the first of the conquered territories to have more than 80 per cent of its population submit to Islam. This was achieved before the end of the ninth century. The early date helps to explain the extent of Persian influence on the evolution of Islamic civilisation often remarked on by commentators. Egypt and Iraq were about a century behind and, somewhat curiously in view of its location, Syria a little later still. Everywhere the first converts ('innovators') may have become Muslims to enhance their social status or, as prisoners of war, to avoid enslavement. The 'late majority', on the other hand, may have converted as the result of social and economic pressures, amounting at times to persecution (Lapidus 1972). For the 'early adopters' and the 'early majority', however, Islam had other attractions. First, by insisting on the

Figure 7.3: Cumulative S-curves for the Conversion of
Sub-regional Populations to Islam

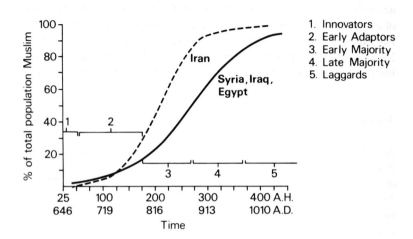

Source: Bulliet (1979).

transcendent unity of God, it resolved the intellectual problems of the
Godhead which had proved so difficult to many Christians and
Zoroastrians and produced major divisions in society on both sides of
the *limes*. Second, the revelation given to Muḥammad showed in unequivocal fashion how man should obey the God under whose final judgement he stands (Hodgson 1960). Both must have come as relief to many
troubled spirits.

Islam's claim to guide behaviour in every aspect of life was gradually
realised through a pattern of interlocking institutions developed principally
in the towns. The *Sharī'a* ('Holy Law') was basic and all-embracing.
Private prayer was obligatory five times in the day, while community
prayer at midday on Friday was an expression of the community's sense
of shared obligation before the mercy and compassion of God. The 'place
of prostration' (*masdjid*, mosque) was at first either a simple structure,
such as the Prophet's own house at Medina, or a fenced enclosure, as
at Basra (Cresswell 1958, 3–9). Conquest brought the Muslims control
of churches, fire-temples and palaces. Some were converted directly to
Muslim use; others were shared initially with local non-Muslim wor-
shippers. More than a generation passed before elaborate, purpose-built
mosques like the Dome of the Rock in Jerusalem (691–92) and the

'incomparable' Umaiyad Mosque in Damascus (714–15) were erected (Wiet 1937, 174). From then onwards, mosques began to dominate the townscapes of the region. Architectural styles varied from area to area according to local building traditions, but the basic form remained that of an open space with arcades for prayer. Minarets, from which the muezzin could call the faithful to prayer, were perhaps first introduced in Lower Mesopotamia in 665 or 673 (Rogers 1976, 87) but soon became increasingly elaborate and distinctive symbols of Muslim dominance.


## Towns

As well as introducing the mosque as an element into the townscape, Islam has been credited with increasing the degree of urbanisation in the Middle East and with the emergence of a distinctive form of town (Benet 1963; Eickelman 1974; Fischel 1956; Grunebaum 1955a, 141–58; Hourani and Stern 1969; Ismail 1972; Landay 1971; Lapidus 1969). This identification of Islam with urbanism may seem paradoxical in view of the role played by nomadic Arabs in the initial spread of Islam, but a strong case has been made for it as the religion of townsmen *par excellence*. The Qur'ān itself appears to disparage the nomadic life as conducive to impiety, hypocrisy and apostasy (Qur'ān 9:98). Public prayer at midday on Friday, which was meant to unite the Muslim community (Qur'ān 60:9–11), could only be regularised and guaranteed in a permanent settlement. While a minimum community of 40 adult males was needed to authenticate the service (Benet 1963), the thirteenth-century geographer, Yākūt, recognised the possession of a mosque where Friday midday prayers were said as a major criterion for distinguishing a town (Grunebaum 1955b). In addition, the religious scholars and jurists who interpreted and sustained the faith, collectively known as the *'ulamā* (the learned; religious leaders and scholars) could only function at nodes in local communications and information systems, that is, in the towns, where perhaps the necessary leisure could be guaranteed. Thus the argument for a close relationship between urbanisation and Islam is at least plausible.

　　Since the Muslim conquerors clearly took over existing towns, sometimes creating new quarters alongside them, urbanisation should be examined in terms of possible increases in both the number of new towns and the proportion of the urban population. The latter is difficult to establish directly. Bulliet (1979, 53–7) has argued, however, that early converts to Islam tended to migrate to the towns, especially the large towns, because they found a style of life suited to their new status in these politically important centres. Archaeological evidence from Lower Mesopotamia and Khuzestan, while reflecting local circumstances, indicates a decline in the number of settlements after about 800 but a parallel tendency for the built-up area of surviving towns to expand. Measured in areal terms, then, urbanisation increased in these areas. None the less, most of the major Sasanian towns in the Uruk area disappeared, probably undermined by the establishment of new Muslim towns at Basra (635), Kūfah (639) and Wāsit (702–05) (Adams and Nissen 1972, 62–5). Further north, Baghdad (762) and Sāmarrā (835) in their heyday occupied areas considerably greater than the aggregate of all the other urban settlements in the vicinity of the lower Diyālā. With areas of about 7,000 ha, they covered 13 times the area of Ctesiphon and 5 times that of tenth-century Constantinople (Ashtor 1976a, 89). The picture elsewhere in the Muslim lands is not as clear, but a lesser degree of urbanisation can be postulated.

Relatively few new towns were established in the Middle East after the Arab conquests (Grabar 1969; Lapidus 1973, cf. Hamdan 1962). The exceptions are various military camp and palace towns. The Arab commanders established substantial base camps at nodal points in the web of local communications and at places where they could be in touch with routes back to Hijāz. Some of these, like Dabik north of Aleppo, were ephemeral establishments which vanished once the military frontier rolled on. Others were transformed into substantial permanent settlements. The best known are Basra and Kūfah, on the edge of Lower Mesopotamia, and Fuṣṭāt (640), the kernel of later Cairo, almost at the apex of the Delta. Two factors were involved in the establishment of permanency. One was the importance of the new centres in the reorganised administrative and economic patterns of the conquered lands. The other was the attraction of such purely Muslim enclaves for indigenous converts, together with the multiplier effects which relatively large numbers were able to generate.

If anything, probably fewer palace towns took root. A palace town was created when a ruler decided to build a new residence for himself, with associated accommodation for administrators and soldiers. Such a concentration of wealth and socio-political importance immediately attracted a substantial service population. The most famous examples are the Round City of Baghdad (Lassner 1963a and b, 1966 and 1969; Le Strange 1900), Sāmarrā (Herzfeld 1948; Rogers 1969) and Cairo (Abu-Lughod 1971). After its foundation, each of these palace towns showed amoeba-like development. They focused and transformed themselves; they moved location as well as changed configuration, and all within a relatively circumscribed area. The basic stimulus was undoubtedly the building of new royal residences, which acted successively as points of crystallisation within the urban mixture. But flood damage, changes in the location of the nearby river which exposed or reduced building land, and the destruction of war were also important morphogenetic agents. Extensive areas of ruins were left behind as refocusing took place, and these restricted subsequent redevelopment. The three cities shrank when the locus of political power shifted elsewhere and when the economic base, widened by empire, finally contracted to the local area or, at best, the immediate sub-region. Baghdad and Sāmarrā survived on residual politico-economic and religious functions respectively. The geopolitical importance of inland connections ensured that Cairo remained the capital of Egypt.

Islamic religious and cultural values have been advanced as generating a specific form of town (Grunebaum 1955a, 141–58). In model terms, its focus is the Friday mosque (Fig. 7.4). Nearby is the principal commercial area of the town, with its shops, workshops and warehouses, *khans* (combining residences with commercial premises) and *qaysāriyyas* (lock-up markets). Localisation was never complete, however. Bakers and grocers tend to be found in residential areas and noxious trades (like tanning) are forced to the periphery of the built-up area. Public baths may be centrally located because of the importance of ritual ablution. The citadel is the centre of political and administrative power, its walls serving to accentuate the division between the mass of the population and the often alien governors and soldiers. Streets through the town are described as narrow and meandering; culs-de-sac are common. Residential quarters are distinguished on the basis of community affiliation, particularly religious allegiance.

Such a complex has been contrasted with the planned lines and openness of the type Hellenistic town, which is seen as expressive of quite a different cultural spirit (Grunebaum 1955b). In fact, there are several weaknesses in this view. Most fundamental is its neglect of morphogenetic development. The centuries immediately before the Muslim era saw the transformation of the model Hellenistic town. Buildings encroached upon the streets and debris was not removed. Sharp turns and dead-ends must have emerged very soon afterwards. Colonnades collapsed or were incorporated into the built-up fabric. Whole areas became derelict. The demise of local government was partly to blame, but so too was the disappearance of wheeled traffic. Earthquake and flood brought devastation to some towns which could only partly be made good because of lack of funds and patrons. Society changed, as well. One consequence was,

Figure 7.4: Model of the Islamic Town

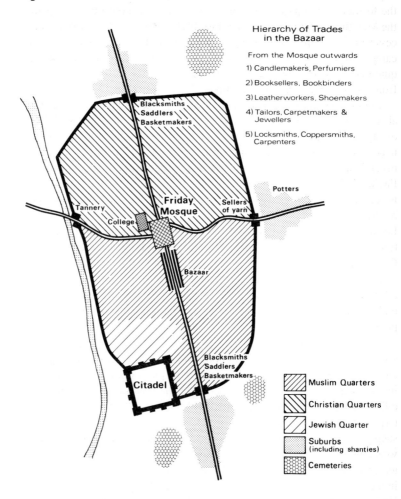

Hierarchy of Trades
in the Bazaar

From the Mosque outwards

1) Candlemakers, Perfumiers

2) Booksellers, Bookbinders

3) Leatherworkers, Shoemakers

4) Tailors, Carpetmakers &
   Jewellers

5) Locksmiths, Coppersmiths,
   Carpenters

Potters

Blacksmiths
Saddlers
Basketmakers

Tannery

Friday
Mosque

Sellers
of yarn

College

Bazaar

Blacksmiths
Saddlers
Basketmakers

Citadel

Muslim Quarters

Christian Quarters

Jewish Quarter

Suburbs
(including shanties)

Cemeteries

as we have seen, the loss of such characteristically Hellenistic institutions as gymnasia and theatres (p. 130).

Social and morphogenetic development continued under Islam. The original Muslim conquerors brought only the 'germ of Islamic society' (Bulliet 1979, 1–2) and its full flowering did not occur until the ninth or tenth century. It is unlikely, therefore, that a distinctively Muslim stamp could have been put on the whole townscape much before then. As Muslim society matured, so further distinctly Islamic elements were added. The larger cities soon possessed more than a single Friday mosque, despite the formal limitation to one. This was partly a response to the size of the Muslim community, but it was also partly to display the glory of successive rulers. While community or tribal mosques existed in the military camps, during the twelfth century small mosques proliferated in the quarters of most towns. *Madrasas* (religious schools/colleges) appeared from the eleventh century onwards.

Finally, the usual formulation of the Islamic town model implies the absence of conscious planning and the importance of a relative informality in the ground plan. Reality is rather different. The Prophet himself may have stipulated a minimum width for streets of 7 cubits (about 3–4 m) and such a module was used at Basra, Kūfah and the Round City of Baghdad (Planhol 1959, 15–16). Another indication of planned design is rectilinearity in the layout of streets and buildings in towns founded by the Muslims. This is still apparent at 'Anjar (714–15) in present-day Lebanon and it was a feature of the layout of the original al-Qāhirah section of Cairo (Abu-Lughod 1971, 18–19; Rogers 1969). The lost Round City of Baghdad was laid out on different principles, but to a deliberate and long-established design (Lassner 1969). Indeed, deliberate design was a feature of most, possibly all, palace towns. The palace itself was often build on a *meydan* (square or esplanade) reached by a ceremonial approach and, in some cases, lay near pleasure gardens.

Towns experienced various vicissitudes during the Middle Ages. Most seem to have flourished down to perhaps the tenth century. During the tenth and eleventh centuries physical decay was reportedly widespread in Upper and Lower Mesopotamia; much of Baghdad was in ruins by c.1260. Throughout Mesopotamia the problem was due to a complex interplay of warfare, local tyranny, neglect of communications and water supply (for irrigation, as well as for drinking), epidemics and a progressive weakening of the agricultural base. Towns in Egypt and Syria fared better, at least down to the arrival of the Selcuk Turks and the Crusaders in the eleventh century. The invaders brought upheaval and destruction. Under the 'Aiyubides, however, the larger towns of Syria and Upper Mesopotamia increased in population and new suburbs grew at Damascus and Mosul. Mongol incursions in the thirteenth century brought further depopulation; the people of captured towns were often massacred in cold blood. In many specific cases, however, wanton destruction completed a long process of decay resulting from such human factors as factional strife and the diversion of caravan routes but, notably in Lower Mesopotamia, also from uncontrollable shifts in river channels. Nor were the Mongols simply destroyers. The Īl-Khān capital of Tabrīz expanded, while various administrative centres were promoted in the rest of Persia. Relative peace followed the Mamlūk victory over the Mongols in 1260. This allowed growth to resume in at least some of the Syrian towns. The Mamlūk capital, Cairo, resumed its expansion and on the eve of the Black Death (1347) its population possibly exceeded 600,000 people (Dols 1977, 198). Stagnation and internal decay then afflicted Cairo, like Alexandria and most of the larger cities of Syria. Aleppo, with its flourishing trade to the West, was something of an exception and continued to expand into the fifteenth century (Ashtor 1976a, 303–4; Kramers and Wiet 1964, T1; Lapidus 1967).

**Trade and Agriculture**

As Lombard (1975, 10) observed, the Muslim Middle East during the Middle Ages was a series of urban islands linked by trade. Trade looms so large in contemporary sources that it is possible to think in terms of a 'merchant economy' (Cahen 1970). The amount of activity in the system seems to have increased under the Umaiyad Caliphs and a golden age of inter-regional commerce emerged with the 'Abbāsids. It was largely stimulated by the unification of much of the region into a single state, as well as by the circulation of relatively large quantities of precious metal. The bullion was initially released as booty but subsequently became available with access to important sources of supply, notably the gold of sub-Saharan Africa and the silver of eastern Persia (Ashtor 1976a, 80–6). Improvements in credit facilities, a commercial mail service and the construction of caravanserais and bridges on major routes doubtless helped. Political fragmentation after the middle of the tenth century brought realignments of trade (Ashtor 1976a, 144–5; Goiten 1967, vol. 1, 32, 211–14; Lombard 1975, 102–14, 218–37) and assisted an overall decline in production.

Much of the inter-regional trade consisted of the industrial specialisations of individual towns and districts. Despite the large variety of goods traded,[5] the most important, in terms of value and probably quantity, were textiles. These were produced in almost bewildering variety, especially for the upper end of the market (Lombard 1975, 181–6). Linen cloths were the specialism of towns in the eastern Delta, while the production of cottons spread from Lower to Upper Mesopotamia and then to Syria and lowland Cilicia. Silks continued to be produced in Syria and Khuzestan using local cocoons, but became important also in the mountains of Armenia, in the Caspian lowlands and north-eastern Iran

(Khurasan) (Frantz-Murphy 1981; Lombard 1975, 181–6).

The insatiable demands of the towns for wood were increased by the expansion of glass-making and the development of sugar-refining. While a variety of purely local resources were tapped for fuel, they were probably inadequate to meet all industrial and domestic demands. Coupled with the need for construction timber, these sustained an important timber trade across the eastern Mediterranean (where production of large timber had virtually ceased from Lebanon) and down the Tigris and Euphrates from the Armenian mountains (Lombard 1969 and 1975, 174–8). Considerable areas must have been ravaged, if not always denuded. Further devastation was caused by the spread of Maronite Christian and Druze villages into the higher reaches of Mt Lebanon to find security. Not only did the new mountain communities require fuel themselves and ship it to neighbouring towns, but they also needed to carve out fields and pastures. Other refugee communities did much the same elsewhere in the region.

The wealth of urban elites encouraged trade beyond the region, though the pattern and intensity of activity changed considerably down to 1500. Doubt has been cast on Pirenne's celebrated thesis that the Arab Muslim expansion destroyed trade with south-western Europe (Pirenne 1939; Riising 1952). Certainly, it now appears that Ifrīkiya (Tunisia and eastern Algeria) and Sicily played an important role as middlemen during the tenth century (Goiten 1973, 31). Italian merchants were established in Egyptian and eastern Mediterranean ports by the end of the next century and negotiating the first trade concessions (Ashtor 1976a, 195–7). The volume of trade with Europe increased. At first interested in the exchange of raw materials and slaves for luxury goods, the Frankish merchants gradually changed the character of their Levant trade. By 1500 they were marketing western textiles so successfully that local manufacturing was in decline (Ashtor 1976a, 306–11, cf. Frantz-Murphy 1981). They were, in addition, shipping raw cotton and flax westwards, together with large quantities of dyestuffs and spices from further east. Land routes in that direction were first reopened by Muslim penetration of central Asia and contact was further expanded with the foundation of the vast Mongol Empire in the thirteenth century. But the rise of the Middle East's own silk production, together with the perennial political and technical difficulties of the central Asian routes, made the seaways much more important. The Indian Ocean was virtually an Arab lake from the eighth to the fifteenth century (Goiten 1973, 175–270; Hourani 1963). For a brief period there was a Muslim factory at Khānfu (Canton) (Sauvaget 1948). In the eleventh-thirteenth centuries more than half of the commodities entering the

Mediterranean trade were imported from India (Goiten 1973, 175). During the 'Abbāsid period the chief means of entry was along the Gulf, but changes took place in subsequent periods. The Fatimid conquest of Egypt led to the promotion of the Red Sea route during the late tenth and early eleventh centuries and involved some revival of the ancient caravan routes from Yemen to Arabian ports with access to Suez. After a lapse in activity, revival of the Red Sea route took place during the late fourteenth and fifteenth centuries under the Mamlūks. Meanwhile, the Īl-Khāns succeeded in diverting the Gulf trade on to a land route which ran from opposite the island port of Hormuz to Tabrīz and then westwards, initially through Trabzon and Lajuzzo (a small port in lowland Cilicia), but from the late fourteenth century through Aleppo and its port, Alexandretta (İskenderun) (Ashtor 1976a, 80–6, 147–8, 195–7, 263–7, 297–301; Lombard 1975, 204–37).

The brilliant opulence of urban elite life and culture in the Arab golden age (Hitti 1964, 240–78, 297–316, 332–428) was founded on agriculture. Changes in the routes and commodities of trade affected the vitality of the towns; reciprocal developments in the countryside shaped their long-term survival. The precise trends here are difficult to discern and date, mainly because of the poor quality of the evidence. The foundation of new administrative and military centres in the seventh and eighth centuries seems to have been associated with schemes of land reclamation and development. These were rather scattered. Elsewhere the sparce evidence suggests at least a concern on the part of administrators to maintain the existing infrastructure of agriculture (Lapidus 1981). Watson ((1974 and 1983) has argued that the first two or three centuries of Muslim rule saw an 'agricultural revolution'. It was characterised, first, by the spread of new, or at least formerly localised crops such as sorghum, rice, hard wheat (*Triticum durum* Desf.), sugar-cane, citrus, spinach and aubergine (eggplant). Many of these were summer crops which required irrigation. Their cultivation thus necessitated increased use of relatively sophisticated means of obtaining water, principally qanāts and wheels (pp. 60–2, Fig. 4.7b,d). The new crops and techniques increased production by allowing more intensive use of land, as well as by facilitating the spread of cultivation to soils and areas which had previously been considered unusable. Villages increased in numbers and size. Although the new crops introduced by the 'agricultural revolution' survived, by the late tenth century the expansionary trend had come to an end. Decay set in.

Despite grave difficulties of interpretation, reported tax yields (Fig. 7.5) and statements by chroniclers and geographers, as well as limited and problematic archaeological surveys, all seem to indicate that a widespread secular decline in the extent of the cultivated area had been initiated and that this was accompanied by the abandonment of villages. The evidence for decay is most striking for Mesopotamia, particularly from the lower Diyālā Valley. Reduction in the cultivated area of the lower Diyālā is estimated to have reached 25 per cent (200 km$^2$) by the middle of the ninth century. Before the end of the tenth century, 62 per cent of all the recorded settlement sites outside Baghdad had been abandoned. The downward spiral continued, though there were clearly areas — as along the middle reaches of the great Nahrawān Canal — where cultivation continued for rather longer and others, notably around the palace towns, where some short-lived expansion took place (Lapidus 1981). The techniques of irrigation changed in the lower Diyālā area. Greater reliance was put on a weir and branch canals than had been the case in the Sasanian period (Adams 1965, 97–111). Further south, the situation is less clear. A combination of neglect and 'natural' disaster ensured that the areas of swamp expanded in the vicinity of ancient Uruk and much of the drainage reverted to an unregulated condition. Occupation sites contracted (Adams and Nissen 1972, 63–5). Parallel developments took place in Khuzestan (Adams 1962; Neely 1974; Wenke 1975–76, 138–39). However, the foundation of Basra and Wāsit was accompanied by land reclamation and the cutting of new canals (Lapidus 1981). Unfortunately, it is not clear how far these developments compensated for decreases in cultivated land elsewhere or how long they were sustained, though date plantations continued to flourish along the Shatt al-ʿArab.

Less scholarly work has been done on Upper Mesopotamia, but it seems clear from the geographer Ibn Ḥawḳal's account of his home district (last redaction 988) that its appearance had deteriorated: fruit trees had been cut down, irrigation works were neglected, villages were deserted and land was abandoned (Kramers and Wiet 1964, T1, 203–22). In Egypt, the cultivated area may have declined by as much as 30 per cent during the eighth century and first half of the ninth. The number of villages continued to fall, despite some oscillations in the trend (Fig. 7.6) (Toussoun 1924, quoted Ashtor 1976a, 60; Bianquis 1980). A similar picture begins to emerge for Syria and Palestine, though al-Mukaddasī (c.946–c.1000) depicts generally flourishing conditions there in the late tenth century (Le Strange 1896) and the first Crusaders were pleasantly impressed with what they found (Benvenisti 1970, 213, 389; Cahen 1950–51). Despite fluctuations in tax yield (Fig. 7.5), the downward spiral may not have become too serious until the destructive campaigns of Ṣalāḥ al-Din (Saladin, 1138–93) in the 1180s.

While the Arab conquest of Persia does not appear to have been

Figure 7.5: Reported Tax Yields for Six Areas, c.AD 630–890 (standardised in dinars)

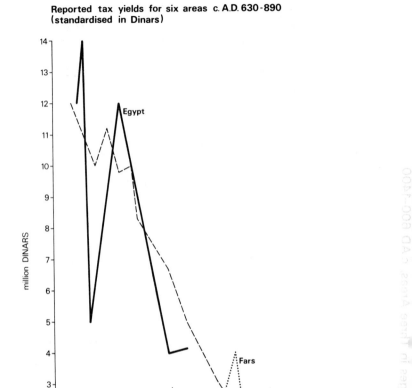

Reported tax yields for six areas c. A.D. 630-890 (standardised in Dinars)

accompanied by any widespread recession in agriculture, land abandonment does seem to have taken place during the ninth and tenth centuries as centralised authority broke down. Some recovery may have occurred once the Selcuk Turks established control. The whole country suffered

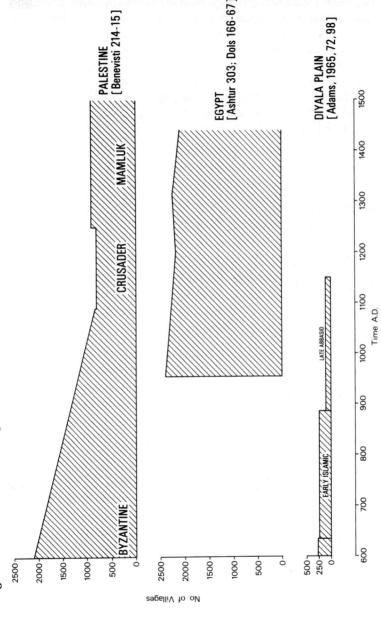

Figure 7.6: Number of Villages in Three Areas, c.AD 600–1400

severely during the Mongol conquests of the thirteenth century. Systematic extermination seems to have been practised by these fierce horsemen, though it is impossible to evaluate either the figures or the atrocities reported by the chroniclers. The north-eastern province of Khurasan suffered most, but neighbouring Mazandaran took a heavy toll and the plain of Gorgān was abandoned. In most districts the vital links between town and village were severed. As urban populations declined and town economies faltered, so contraction affected the land cultivated from the surrounding villages. Arbitrary and rapacious taxation reinforced the downward spiral in population and output. Revenues appear to have declined (Fig. 7.5). Deserted villages were not restored. The degree of nomadism must have increased, partly because it was the life-style of the conquerors. Ghazar (1295–1304) regularised tax collection and encouraged the resettling of abandoned land, but these measures did not completely restore prosperity (Latham 1959, 27, 33; Le Strange 1919; Petrushevsky 1968). Restoration was particularly difficult in much of Persia because of agriculture's dependence on qanāt-irrigation, which requires regular maintenance of the underground aqueducts, as well as considerable capital investment for renewal.

Against the dominant trend of decay and decline after about 1000 AD must be set evidence for recovery and reclamation, in addition to that for the continuance of new areal specialisations in commercial crops initiated during the Islamic 'agricultural revolution'. The evidence for checks in the secular pattern of decline, however, seems to indicate only localised, episodic activity by a few powerful individuals (Ashtor 1976a, 42–9, 127–8, 259–60, 316–18). This had little effect on the long-term, regional decline. Indeed, the downward movement probably continued until the end of the fifteenth century. Specialisation is well documented. Particularly important was the emergence of cotton-, rice-, silk- and sugar-producing districts (Lombard 1975, 162–8; Watson 1983). Cotton-growing became well established in Upper Mesopotamia (particularly for a while in the Khābūr Valley) and Syria in the eighth century and continued to be important in these sub-regions. Rice cultivation spread from the Caspian lowlands and Khuzestan into Lower Mesopotamia by the second half of the ninth century, then into Upper Mesopotamia and lowland Cilicia, and finally to Egypt (Canard 1959). Silk-production became more widespread (p. 176). Plantations of sugar-cane were initially confined to Khuzestan, but by the tenth century they were well established in Yemen, Lower Mesopotamia, parts of Syria and Palestine (where the Crusaders continued to maintain them for some time) and the eastern Delta (Ashtor 1976a, 157; Benvenisti 1970, 253–60; Lombard 1975, 167). They were introduced to Cyprus somewhat later (Christodoulou 1959, 134–5; Thiriet 1959, 274, 329–30). Such commercial crops were almost invariably found on large estates and in districts where the number of villages seems to have declined. These associations suggest possible reasons for the expansion of cash cropping. They amount to a desire to maximise the return from a given unit of land by producing for the urban market and concentrating both capital and labour. The trend is thus not as paradoxical as it may seem when set against the background of a contracting arable area. Explanations for that will be explored in Chapter 9.


## Notes

1. Qur'an 7:59–206; 11:50–123; 19:41–58; 21:7; 26:10–208; 28:59; 29:18, 36–9; 36:13–14; 37:79–148; 38:13–15; 51:24–49.

2. Qur'an 2:25, 82; 3:15, 136, 198; 4:124; 22:14, 23; 37:42–9; 38:51–4; 47:5–6, 12, 15; 52:17–24; 56:10–40.

3. Qur'an, Jews: 2:109–13; 3:48–83; 4:46–8; 5:12–13; 9:29–32; 17:2–8; 43:21–4; 62:5–8. Christians: 2:253; 4:48, 116, 156–8, 171; 5:14–17; 10:18–19; 19:34–5; 23:50; 43:57–63.

4. In pictures illustrating incidents in the Prophet's life he is frequently represented with his face covered.

5. Goiten (1967, 209–10) notes that about 200 items appear in the Geniza Letters, of which 40 were of major importance.

# 8 OTTOMAN WEST: PERSIAN EAST

Chapter 7 left developments in Asia Minor at the stage where Arab raids had virtually ceased (pp. 164, 168). This chapter resumes the story and then, through an outline sketch of the rise of the Ottoman Empire, broadens into a review of changing historico-geographical patterns across the whole of the western section of the Middle East between 1500 and 1800. It concludes with a brief description of developments in Persia over the same period.

## THE OTTOMAN WEST

### 'Turkisation' of Asia Minor (Cahen 1968; Vryonis 1971)

In the eleventh century most of Asia Minor was still part of the Eastern Roman, or Byzantine, Empire. Its population, perhaps increasing in the tenth and eleventh centuries, was predominantly Christian and heir to the traditions of Hellenism. Greek was widely used. The annexation of Christian and Muslim principalities in the mountains of Armenia during the late tenth and early eleventh centuries appeared to demonstrate renewed political vitality. However, the Empire's strength was fatally weakened by an internal power struggle. Centred in Constantinople and long-drawn-out, it involved the bureaucrats, on one side, against the military commanders, often with power bases in the provinces, on the other. The outcome was a lessening of provincial control and the dislocation of territorial defence systems. These, in turn, opened the way for the remarkable 'Turkisation' of the peninsula, that is, the replacement of an apparently mature Hellenic Christian ethos by one which, despite recent efforts at Westernisation, still survives and is distinctly Turkish and Muslim.

The first Turkish nomads were reported in Armenia during the early eleventh century. They formed the advancing edge of the Turkish expansion across the Middle East which had been in progress for 100 years. Like the Arabs before them, the nomads (Türkmen) converted subsistence mobility into military success, but depended upon the 'hard white-thonged bow' shot from horseback (Lewis 1974, 123). By the middle of the century, they were ranging far into central Asia Minor, with the Byzantine authorities virtually powerless to stop them. The resounding Byzantine

defeat at Mantzikert (Manāzgird), near Lake Van, in 1071 effectively opened up the whole interior. Fighting continued throughout the twelfth century around the mountain rim of the central plateau. Success alternated
between the two sides, with the Crusaders on their way east able to help
the Byzantine Empire to some short-lived territorial gains. Defeat in the
pass of Tzybritze, near Myriocephalum, in 1176 marked the end of Christian attempts to reconquer the peninsula for some 700 years, though not
of a Byzantine presence. Parts of western and north-western Asia Minor
remained in Byzantine hands until the fourteenth century and, in the case
of Trebizond (Trabzon) on the Black Sea, until 1461.

Much of the advance was achieved by Türkmen, and they were the
major agents of 'Turkisation'. Pattern and pace obviously varied over such
an enormous area, but a model proposed by Vryonis (1975) probably has
general validity. Türkmen casually entered a particular area with their
families and herds during the spring and summer, seeking pasture and
taking any plunder and slaves which they chanced upon. This was not
raiding of the earlier Arab type (pp. 161, 168), and the number of people
involved was probably relatively large. The Türkmen generally withdrew
with the first snows. A sequence of such events, over several years, would
dislocate agriculture and produce depopulation through enslavement, flight
and starvation. Total desertion is unlikely. Türkmen were probably always
a minority in local populations. Refugees strained local urban economies
and these were weakened further by constant disruption of farming. The
nomads gradually occupied abandoned territory and began to stay
throughout the year. Localised patterns of transhumance were then
established, often using abandoned villages as winter bases (Lewis 1974;
Planhol 1966). The degree of pastoralism increased, not only through
the extension of the Türkmen's activities but also as a result of the insecurity of cultivation. Some surviving woodland was probably cleared
to extend the upland grazing and provide winter fodder. If Vryonis is right,
such a series of events was repeated as a spatially advancing succession,
a rolling frontier, until checked by change within the system itself or by
superior force. It had progressed sufficiently far in south-central Asia
Minor by the time of the Third Crusade (1189–92) for contemporary
writers to describe the territory as *Turchia* (Cahen 1968, 145).

Turkish spread as a spoken and administrative language as the Türkmen
took over. During the twelfth and thirteenth centuries, agriculture began
to revive behind the military zone, in what Marco Polo described as *Tur-
comania* (Latham 1958, 15), and flourishing conditions were reported
by travellers, at least in some areas (Ibn Sā'id, *Kitab al-Djughrāfiyā*,
quoted by Cahen 1968, 157–9; Markham 1859, 72). Revival involved the

sedentarisation of some of the nomads, though Christian villagers were also transferred and resettled with a view to redeveloping abandoned lands. Later, dervish convents (*zâviyes*) proved an effective way of recolonising land and founding villages (İnalcik 1973, 149–50). Towns began to recover, partly as a direct result of the better fortunes of local agriculture but partly, too, because Turkish leaders settled in them. Corrupt forms of ancient names emerged (e.g. Ancyra became Ankariya, then Ankara; Iconium became Konya). Some shifts of site took place (e.g. Denizli replaced Laodicea, 7 km to the north) and changes in local urban hierachies may be postulated. As town life recovered, so the caravans, previously disrupted by Türkmen attacks, were re-established. The towns of Asia Minor soon became linked with the Muslim heartland. Caravanserais were first built in the second half of the twelfth century, but soon spread along the main routes (Cahen 1968, 163–8). Through-journeys, like those of the Polos (c.1260–69, 1271–95) and Ibn Battuta (1304–68/69), became possible again. Exports to the West revived.

The Türkmen had been carriers of Islam ever since they were converted in central Asia during the ninth and tenth centuries. They brought the faith with them to Asia Minor, where it spread in various ways. Subjects tend to convert to the religion of the dominant group. This was difficult to prevent in Asia Minor. On the one hand, the Türkmen's advance disrupted the life of the normally influential monasteries. On the other hand, many of the bishops were absentees, partly through the impoverishment of their sees by the confiscation of church property and partly through an inability to reach their seats because of the disturbed nature of the country. The absence of bishops had several effects. Priests could not be ordained; doctrine became corrupt; social and cultural cohesion relaxed. In contrast, the Islam of the Türkmen was syncretic. Popular Christian piety was easily accommodated, while the dervishes, who began to spread through Asia Minor in the thirteenth and fourteenth centuries, were very popular and brought about many conversions. The advance of Islam was helped, of course, by mixed marriages, tax assessments which discriminated in favour of Muslims, and finally by the pressures of the *gazi*[1] tradition adopted by the Turks and which emphasised war against the infidels. From the end of the thirteenth century, the majority of Asia Minor's population was Muslim.


**Rise of the Ottoman Empire** (İnalcik 1973; Shaw 1976; Wittek 1938)

Turkish military success brought progressive political control, and with

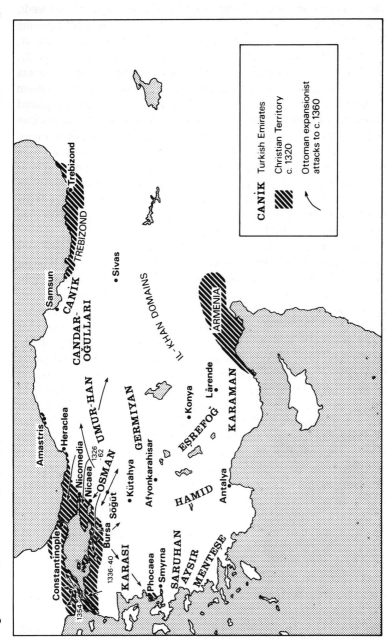

The Emirates in Anatolia in the Thirteenth and Fourteenth Centuries: Early Ottoman Expansion

Figure 8.2: Growth of the Ottoman Empire, AD 1359–1683

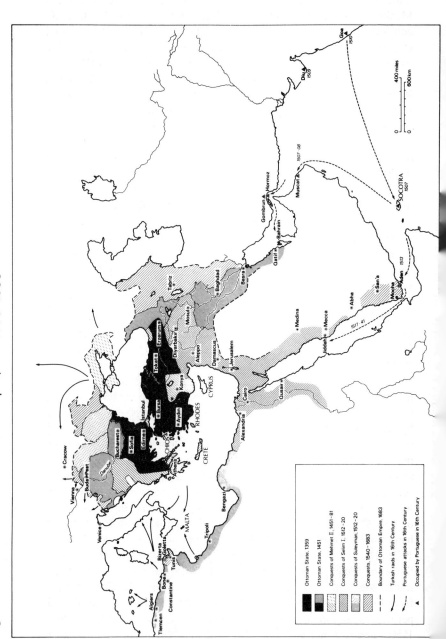

it the introduction of administrative and social institutions learnt in other Muslim lands. Always unruly, the Türkmen were at first under the loose control of the Selcuk sultans based further east, but the breakup of their empire allowed the emergence of a number of separate principalities. Among the most important were those of the Dānişmenids (c.1055–1178), centred around Amasya, Tokat and Sivas, and the Selcuk Sultanate of Rum (1205–43), controlled from Konya. Collapse of Selcuk predominance in Asia Minor under Mongol battering (battle of Kösedağ 1243) atomised political authority, especially in the frontier areas of the West. When it refocused, it was in a number of emirates, one of them the creation of Osman Gazi (lived 1258–1324) (Fig. 8.1). This was the beginning of the *Osmanlı* or Ottoman Empire.

Osman's emirate straggled the passes leading from central Anatolia into the Sakarya Valley and the coastal plains of the north-west. Its initial expansion seems to have conformed to the Vryonis model outlined earlier (p. 184). The nomads grazed the uplands in summer and moved into the lowlands in the winter. As gazis, they fought the Byzantines when they could. Territorial control gradually grew (Fig. 8.1). The capture of Bursa in 1326, however, marks the change from a 'nomadic border principality' into a settled, though dynamic, polity with a fixed capital. Once established on the Straits (Scutari/Üsküdar fell in 1338), major expansion took place in Europe (Fig. 8.2). Although growth continued there for more than 200 years, the high spot was 29 May 1453, when Mehmet II (1451–81) finally plucked 'the Red Apple' of Muslim aspirations by taking Constantinople and making it the Ottoman capital. Substantial gains in Asia Minor were delayed until the late fourteenth century and then Ottoman expansion suffered a major, almost crippling reverse with defeat by Timūr in the plain of Çubak, near Ankara, in July 1402. Unexpected recovery came in just over ten years and the Ottomans went on to make major conquests in the western parts of the Middle East (Fig. 8.2). Here, as in Europe, the basis of their success was a professional army of exceptional loyalty and enthusiasm. It consisted of rather traditional cavalry formations wrapped around a nucleus of crack infantry (the 'New Force' or *Yeni Çeri*/Janissary), equipped from the mid-fifteenth century with hand guns and working closely with cannon (İnalcik 1954 and 1975; Parry 1975). The success of this force brought considerable rewards (Hess 1973). Syria and Egypt gave the Ottomans control over the region's major trade routes, together with substantial taxable populations. The annual tribute from Egypt alone started off at 400,000 ducats in 1525, an amount 33 per cent greater than the annual yield of the Spanish New World before 1550, and it increased steadily (Shaw 1962, 283–312). There were costs. The acquisition of Egypt, and with it responsibility for the Holy Cities of western Arabia, brought conflict with the Portuguese in the Red Sea and the Indian Ocean during the sixteenth century (pp. 197–8). Roughly contemporary expansion in the north-east initiated the first of a series of wars with Safavid Persia (pp. 205–6) which continued sporadically into the eighteenth century and weakened the Ottoman Empire by frequently forcing it to fight expensive wars on two fronts.

## Ottoman Lands in the Sixteenth Century

As well as being a time of wars and territorial expansion, the sixteenth century saw the full flowering of Ottoman civilisation. Its maturity is associated especially with the reign of Süleyman I (1520–66), called in Europe 'the Magnificent' or 'the Grand Turk' but known to his subjects as *Kanuni* ('Law-giver') from his attempts to standardise the administration of his vast, heterogeneous empire. Administrative needs generated a rich documentation, notably the series of statistical registers (*defters*) which, in due time, will be used to produce a reconstruction of the evolving geography of the Ottoman territories during the classical age. Only a coarse and imperfect outline can be attempted here.

The bases of Ottoman wealth and power were the control of land and an effective system for exploiting its 'surplus' to support soldiers and administrators. Most of the land was considered to belong to the state (*miri*) (Gibb and Bowen 1950, 46–52; İnalcik 1955; Lewis 1979). The revenues from about half of it were allocated as strictly revocable, non-hereditary 'livings'[2] of varying size and type, generally known as *tīmārs*. Each allocation was meant either to support one or more mounted soldiers (*sipahis*) or to reward some official. About 577 separate individuals benefited in this way in Palestine–Transjordan, according to surveys of 1596–97(Hütteroth and Abdulfattah 1977, 47–110). The farmers working the land simply had rights of usufruct, but were not apparently overburdened by the simplified, regular and ordered system of payments introduced by the Ottomans (Gibb and Bowen 1950, 236–44, 258–9; Lewis 1979). In general, the tīmār system applied to much of Anatolia, Syria and Palestine. Earlier systems of exploitation were left untouched in eastern Anatolia, while in Egypt and Lower Mesopotamia (the provinces of Baghdad and Basra) tax-farming arrangements were allowed to continue, on the understanding that the Ottoman governor met all local administrative and military expenses and returned an annual tribute to Istanbul (İnalcik 1973; Shaw 1976, 122).

Several important changes in land use took place within this administrative and fiscal framework. The sedentarisation of nomads, both voluntary and coerced, proceeded in parts of Anatolia. Some of the pasture was ploughed up and new villages established, probably on the sites of earlier winter quarters (Jennings 1978; Orhonlu 1969). Even so, the nomadic element fell only from about 18 to 16 per cent in the period 1520/30–1570/80 and there was probably little change in the next century (Fig. 8.3) (Barkan 1970). The proportion of nomads in the population was probably smaller in much of Palestine, though their mobility and warrior qualities gave them disproportionate control and influence (Hütteroth and Abdulfattah 1977; Sharon 1975). Beyond the Jordan, nomads formed nearly 40 per cent of the population of 'Ajlūn province. There, and in the south of Palestine, the frontier of permanent settlement more or less coincided with the 250 mm isohyet, indicating the controlling influence of precipitation (Hütteroth and Abdulfattah 1977, 45–54). Important but unstable tribal polities emerged in Lower Mesopotamia not later than the seventeenth century. They were supported by a combination of pastoral nomadism and small-scale irrigated farming organised around the ramified distributaries leading from a surviving canal or a branch of one of the rivers. Settlements were impermanent. All this was basically the result of past neglect and continued insecurity. It was exacerbated by the frontier status of the area under the Ottomans, though an easterly shift in the lower course of the Tigris, dated to 1500–1650, would have been important, at least in shaping the pattern of irrigation (Adams and Nissen 1972, 67–83; Longrigg 1925, 2).

Cereals were the most important crops in all the cultivated areas. Wheat was twice as valuable as barley in late sixteenth-century Palestine, though progressive aridity ensured the increased dominance of barley in land use towards the south (Hütteroth and Abdulfattah 1977, 84–105). In ten widely scattered areas in Anatolia during the early sixteenth century wheat formed 50 per cent of total production, with a range of 21–66 per cent, while barley formed 30 per cent (range 11–42 per cent) (İslamoğlu and Faroqhi 1979). Both fell slightly over the century, despite the importance of supplying Istanbul and the emergence of a significant, but clandestine grain trade with the western Mediterranean (Aymard 1966). At least part of the explanation lies in the expansion of certain cash crops.

An upsurge in cotton production began in Syria and Palestine towards the end of the fourteenth century (Ashtor 1976b), and seems to have been sustained until after the Ottoman conquests. Cotton production grew from 12 to 15 per cent of the total output in the Tire area of the Küçük Menderes Valley in western Anatolia during the sixteenth century, while around

Figure 8.3: Population Distribution in Anatolia, AD 1520–30

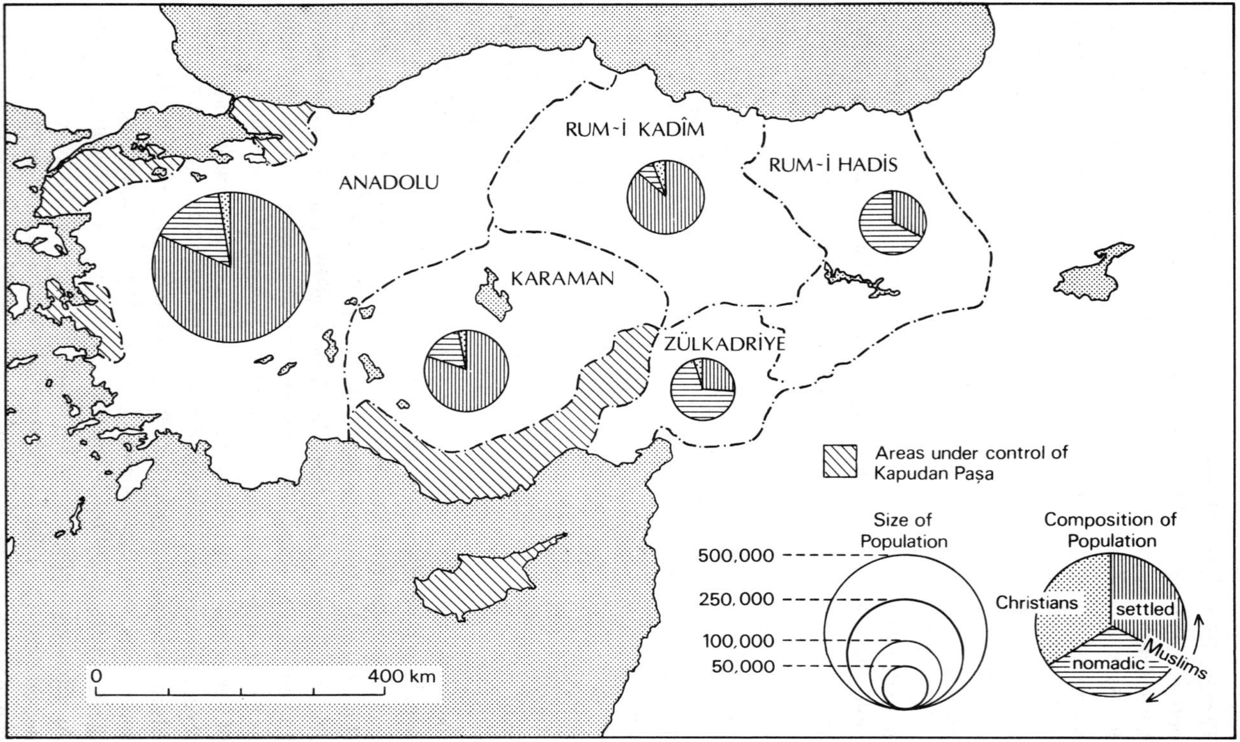

Adana in lowland Cilicia it rose from 24 to 36 per cent (İslamoğlu and Faroqhi 1979). Something similar may have happened in Palestine, for cotton was the main crop in several villages in the plain of Acre and in lower Galilee by the end of the century (Hütteroth and Abdulfattah 1977, 84–5). Maize and tobacco were introduced to the Empire during the sixteenth century and became locally important in the seventeenth (Braudel 1972, 762; Stoianovich and Haupt 1962). Coffee-growing expanded in the mountains of Yemen, acquired by the Ottomans for strategic reasons after 1538, as the habit of coffee-drinking spread during the 1540s and reached Istanbul in 1555 (Braudel 1972, 762; Shaw 1976, 198).

İslamoğlu and Faroqhi (1979) have suggested that the large increase in barley production over the course of the sixteenth century in the ten areas studied points to the cultivation of marginal land. It is difficult to verify, though Cook (1972, 20–1) found positive evidence for small-scale clearance in the mountains near Tokat. The cultivation of marginal land could be explained by the general increase in population apparent in most parts of the Ottoman Empire (Bakhīt 1972; Barkan 1957 and 1970; Cohen and Lewis 1978; Erder and Faroqhi 1979; Göyünç 1969; Sahillioğlu 1965). The mean rate of increase between 1520/30 and 1570/80 was 1.98 per cent per annum in the provinces of Anatolia and 1.1 per cent in Syria; in some places it was as high as 2.4 per cent (Barkan 1970; Erder and Faroqhi 1979). Cook's study of 700 villages in three different areas of Anatolia over the period c.1450–1575 revealed that population growth was more rapid than the expansion in cultivated land. References to deserted villages, to villages newly entered and to uncultivated land disappear from the Anatolian registers studied. The average size of farm decreased and population pressure mounted (Cook 1972).

Urban populations also grew during much of the sixteenth century (Fig. 8.4) (Cohen and Lewis 1978; Erder and Faroqhi 1980; Faroqhi 1980; Jennings 1976). Growth rates were highly differential within the Empire and obviously reflected local circumstances, but where they were as high as 3 per cent per annum,[3] then substantial immigration is implied. Some of it was forced by the Ottoman authorities, notably in the case of the repeopling of Istanbul after the conquest (Barkan 1946–50; İnalcik 1973). Elsewhere, immigration had the character of a drift from the land. In some cases, it was doubtless occasioned by population pressure, as indicated above, but in others the towns themselves exerted a strong 'pull'. As urban economies improved under the stimulus of a large, interconnected empire, so the number of jobs must have increased in industry, commerce and administration. Possibly more important still were the differential rates of tax — low on urban households but considerably

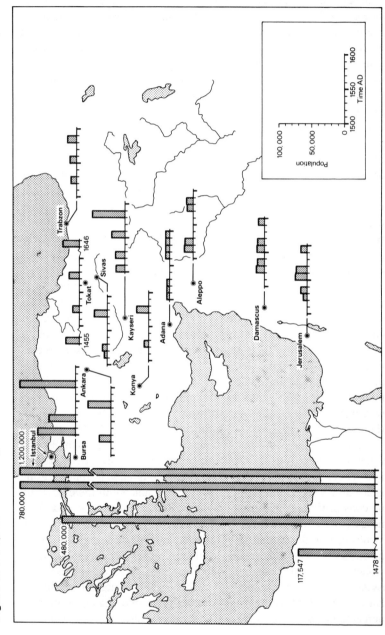

Figure 8.4: Population Change in Selected Ottoman Towns and Cities during the Sixteenth Century AD

higher on those in the villages.[4]

In addition to the revival of old towns, new towns and market centres were created, most notably in Anatolia. The explanation lies less in the free play of market forces in conditions of peace and security created by Ottoman rule behind the frontiers, than in the needs of the imperial system itself. Ottoman administrative and political priorities required an organisation analogous to a command economy (Cvetkova 1970). The backbone of the administration was provided by the *kadis* (judges). They not only dispensed justice according to the *Şeriat* (Sharī'a) and interpreted the provincial *kanunname* (law code agreed at the time of conquest), but also supervised the implementation of the Sultan's administrative and fiscal decrees (*kanuns*) (İnalcik 1973). For efficiency, these officials were located in convenient central places. At the same time, agricultural produce had to be sold for cash, either by those sipahis whose incomes were largely in grain or by the peasants whose obligations were discharged in money (Faroqhi 1979). The state itself needed bulking centres where food and raw materials could be gathered to ensure the tranquillity of the capital and major provincial cities, particularly in times of shortage, and to maintain its regular troops. All this implies an hierarchical urban structure, especially at provincial level. The sources appear to confirm the supposition. Often the linear variant of the Christaller model of urban hierarchies prevailed as a result of the alignment of towns and market centres along river valleys and major routes, while relative ease of access to a major provincial city, such as Damascus, disrupted the full development of expected patterns (Erder and Faroqhi 1980; Faroqhi 1980; Hütteroth and Abdulfattah 1977, 86–92). Istanbul increasingly dominated the urban system, with a population which grew from an estimated 400,000 in 1520–30 to perhaps 1 million by 1600 (Barkan 1970; Mantran 1965, 62–3).

'Turkisation' gradually gave an Islamic character to the towns of Asia Minor, though they often retained important Christian and Jewish populations. Ottoman rule was stamped on the larger cities of the provinces by palace architects, such as the great Mimar Sinan (?1490–1588), through the construction of mosques and especially *imârets* (a combination of mosque, *medresse*, hospital and *han*) in a distinctive style (Goodwin 1977). Population growth led to the development of extensive new quarters beyond the old walls, while the earlier ones often lost any original homogeneity (Cohen and Lewis 1978, 37; Jennings 1976; Raymond 1979). Tax assessments suggest that the food-processing industries flourished and that in many towns textile manufacturing prospered and may even have increased output for a while (Cohen and Lewis 1978, 59–62; Erder and

Faroqhi 1980; Faroqhi 1980). A lively commerce is evinced. The number of market centres grew and sales tax assessments (*bac-ı pazar*) rose, while the actual number of shops taxed increased (Cohen and Lewis 1978, 46–50, 68–9; Faroqhi 1980). Indeed, trade was a necessity for communities specialising in cash cropping. The commercial areas of even large cities, such as Aleppo and Cairo, expanded enormously during the sixteenth century.[5] The existence of periodic markets, markets on the summer pastures (*yayla pazarları*) and occasional fairs shows that even pastoralists and the inhabitants of remote villages shared to some extent in the exchange economy (Faroqhi 1979; Hütteroth 1980).

The provisioning of major cities and the armed services created extensive flows of food and raw materials across the Empire. Istanbul was supplied with cereals chiefly from the plains along the Danube and on the western side of the Aegean, but lowland on the southern shores of the Sea of Marmara was also tapped; occasionally, interior Anatolia was drawn upon (Güçer 1953). Little is known about the timber trade, but considerable quantities must have been required as Istanbul expanded, for many of its houses were built of wood and fires were frequent (Lewis 1963, 111). Similarly, the maintenance of a large high-seas fleet, rebuilt in haste over the winter of 1571–72 following destruction at Lepanto (Braudel 1972, 409–14, 1,120), must have had considerable effects upon the forests accessible from the sea.

As a set of wealthy markets, the Empire was a major force in world trade. Military needs drew in munitions and base metals (including scrap) from western Europe (Parry 1970). Popular cuisine required spices (especially pepper),[6] and these were brought from India and the East Indies by the traditional interconnecting land and sea routes (pp. 177–8). The court and wealthy families were supplied with furs and other luxury goods from all directions, while their domestic slaves (probably several thousand in aggregate) came from the Sudan, east Africa, Russia and Poland (Cohen and Lewis 1978; Fisher 1978; İnalcık 1973, 131). Imports were paid for with goods either produced in the towns under Ottoman rule or traded through them from elsewhere, particularly Europe. Ottoman expansion brought control of the region's transit route, and the sultans sought to defend and exploit them. A combination of customs and guardpost (*derbent*) emerged at strategic points on the land routes, especially the mountain passes, and caravanserias were refurbished. Rhodes was taken from the Knights of St John (1522) to secure the sea route from Istanbul to Egypt, while the fall of Chios (1566) completed Ottoman domination over the Aegean. The conquest of Cyprus (1570) rounded off Muslim control of the entire eastern Mediterranean, though it did not eliminate the pestilential activities of Christian privateers and pirates from further west.

Considerable debate has taken place over the effects of Portuguese activities in the Indian Ocean (Fig. 8.2) on the flow of goods along the Middle East's transit routes (Braudel 1972, 543–70; Lane 1940). There seems little doubt that, following a considerable rise in the fifteenth century, the amounts of spices, drugs and dyestuffs reaching the Levant coast and being shipped by European merchants did decline in the early sixteenth century (Magalhães-Godinho 1969, 713–80). This was not due to the size of Portuguese purchases in India and beyond; nor was it due to their ability to take over the complex trading systems of the Indian Ocean (Tibbetts 1971). They had neither the intent nor the means of doing so. In some ways, the intrusion of the Portuguese into the region's trade after 1497, when Vasco da Gama reached Calcutta, was simply that of another group of merchants who, in their commercial activities, behaved very like all the others in the region. That is, they fitted in with the established traditions of 'peddling' (pp. 74–5); they bought and sold comparatively small quantities of goods, or exchanged bullion and base metals (copper particularly) for goods to trade, while travelling from port to port (Steensgaard 1974, 28, 154). Much more important to the transit trade of the Middle East in the sixteenth century was the disruption of shipping on its way to the Red Sea ports (Fig. 8.2), for this had effects on the amount of goods carried by the caravans through western Arabia. The Portuguese disrupted trade by their attempts to blockade the Bāb al-Mandab (from 1502), their seizure and destruction of Muslim ships, and their abortive attacks on Aden and Jiddah (the last in 1541) (Hess 1970; Sergeant 1963). Threats to the gateway of Mecca, only a day's journey in the interior (Cortesão 1944, 11), were particularly worrying to the Muslim authorities. Portuguese motives were more those of the crusader than the merchant, with the desire and the oppportunity to strike at the very heart of Islam. The Mamlūks of Egypt and then the Ottomans countered the threats and collaborated to some extent with other Muslim states. However, the size of the forces committed suggests, despite the overcoming of considerable difficulties in deploying them, that affairs in the Indian Ocean itself were really only of peripheral concern.

Disruption of the Red Sea and dependent routes was offset by greater reliance upon those leading from the Gulf. As part of their strategic design and out of concern for Indian trade, the Portuguese were content to keep these open. They did, however, seek to exploit, even to develop, them through their control of Hormuz, which they acquired in stages between 1507 and 1543 (Magalhães-Godinho 1969, 764–72). This barren island at the mouth of the Gulf had become a major entrepôt during the İl-Khān period in Persia (p. 178), but reached its zenith as a somewhat precarious, cosmopolitan trading community of around 40,000 people under Portuguese rule (Steensgaard 1974, 193–205). The Ottoman conquest of Mesopotamia (complete in 1538) revived the ancient river routes and perhaps weakened trade along the routes across Persia, already disrupted by Ottoman-Safavid conflict. As a consequence, Aleppo emerged as the main commercial centre for European merchants, a position which it retained well into the seventeenth century (Davis 1967, 36). It completely eclipsed Bursa, which, as the leading city of the growing Ottoman Empire before the capture of Istanbul, had rivalled it during the late fourteenth and fifteenth centuries (İnalcik 1960). Cairo, by contrast, could be described as in a dilapidated state even by the beginning of the fifteenth century. By mid-century two-thirds of Alexandria was in ruins (Lopez, Miskimin and Udovitch 1970).

Use of the Red Sea route revived during the second half of the sixteenth century, following a shift in Portuguese strategy, but it never recovered its former importance (Magalhães-Godinho 1969, 773–8). In the end, Portuguese action did not destroy the transit trade of the Middle East. Around 1600 Europe still obtained about 60 per cent of its pepper and 52 per cent of its other spices and drugs through the Levant ports (Steensgaard 1974, 168 Table 12). The decline of this lucrative trade came in the seventeenth and eighteenth centuries, when conditions in both the Ottoman Empire and the Indian Ocean were very different from those prevailing in the sixteenth century.

### 'The Sick Man of Europe'[7]

The historiography of the Ottoman Empire in the seventeenth and eighteenth centuries has been dominated by themes of decay and decline (Hourani 1957; Lewis 1958). Viewed with a long-term economic and political perspective, Europe was certainly in the ascendant. The Ottoman Empire ceased to grow after the acquisition of Crete (1669) and Podolia (1672), and from the end of the seventeenth century it began to lose territory in Europe, as it had done earlier in the century on the Persian frontier. None the less, it is arguable whether the balance of military advantage inevitably lay with Europe, at least before Bonaparte's invasion of Egypt (1798) (Parry 1975). Coincident with territorial losses were formidable internal problems. The Ottoman solutions, voluntary and involuntary, transformed the centralised organisation of the classical period (pp.

190, 195) into something approaching a loose federation of largely autonomous principalities of varying size and significance, frequently at war with each other over the control of territory and revenues. It is remarkable that the Empire held together at all (Owen 1976). Two of the most important pointers to internal decay are references to the abandonment of villages and the 'deplorable condition' of agriculture (Pinkerton 1802, 27, quoted Smith 1975). Settlement desertion is alleged to have increased during the seventeenth and eighteenth centuries, while decline in farming is seen as an associated phenomenon. The widespread distribution in the eighteenth century of pastoral nomads, with their propensity to raid (e.g. Beaujour 1829, 129–30; Eton 1798, 254–66; Southgate 1840, vol. 2, 198–9; Thévenot 1727, vol. 2, 701–10; Volney 1786, vol. 2, 127, 135, 302–4, 372–9), seems to confirm the picture, since the nomads' presence can be explained as both the cause and the effect of decay in settled agriculture (Hourani 1957). Against the apparent evidence for decline, however, has to be set that for an expansion in cash cropping, particularly in areas accessible from the Mediterranean. The most reliable is the increase in cotton-growing in Syria–Palestine towards the end of the seventeenth century (Owen 1981, 7 Table 2), followed by an enlargement of silk production in Syria, Mt Lebanon, Cyprus and parts of Anatolia after 1725 (Davis 1967, 143–60). It is also clear that attempts were made to curb the nomads (Cohen 1973, 104–7).

At least part of the taxes and other dues were paid in cash during the seventeenth and eighteenth centuries. This implies marketing, as does the collection of revenues in kind. Although periodic markets presumably continued in the countryside (p. 196), much of the selling must have taken place in towns. In any case, the towns — especially the larger ones — were dependent upon the rural 'surplus', not only for provisions but also for industrial raw materials. Towns continued to flourish during the period. Rank-size hierarchies may have evened up in the lower levels, while primacy weakened in response to greater decentralisation of political power (Erder and Faroqhi 1980). Existing regional centres such as Aleppo and Damascus expanded in population and built-up area, though at a slower rate than in the sixteenth century, while Cairo revived and developed (Abu-Lughod 1971, 56–79; Mantran 1962; Raymond 1979). A few new centres emerged on the twin bases of political and commercial control. Notable examples are Smyrna (İzmir) and Acre. Smyrna began to grow in the early seventeenth century, but really took off after the devastating earthquake of 1688, to become the chief port of the eastern Mediterranean during the eighteenth century (Davis 1967, Ülker 1974). Acre was deliberately promoted as a port after 1746 by the successive rulers of Palestine, Dāḥir al-ʾUmar al-Zaydāni (c.1690–1775) and Cezzâr Ahmed Paşa (c.1736–1805) (Cohen 1973, 128–37). Small satellite towns grew in number and population to mediate provincial power and collect the rural 'surplus' for sending to regional centres. Appearance and organisation remained much as in earlier periods of Muslim dominance, while wealthy individuals continued to add to the suite of public buildings (Cohen 1973, 128–57; Rafeq 1966, 95–6, 121–2, 150–1, 309; Raymond 1979; Ülker 1974, 37, 50–3), as well as restoring old ones. A significant introduction was the *kahve hane* (coffee-house), 'where idle people, strangers and others who were not of the first rank assembled and passed their time'; some provided music and story-telling (Pococke 1745, vol. 2, 122; Russell 1734, vol. 1, 146–8). Although coffee-drinking had spread rapidly across the Empire in the second half of the sixteenth century (p. 193), coffee-houses seem to have proliferated in the towns during the eighteenth century (Rafeq 1966, 186).

By about 1800 the urban population of the Ottoman Empire was proportionately high. Issawi (1969) has estimated that towns of 10,000 or more inhabitants contained about 10 per cent of the Egyptian population of about 3.8 million, 20 per cent of Syria–Palestine's 1.0–1.5 million people and 15 per cent of Mesopotamia's 1.0–1.5 million. We may guess that the proportion in Anatolia was not less than 20 per cent. These figures compare with 23.9 per cent for the population of England and Wales living in towns of 10,000 or more people in 1801, and 16 per cent for that of France in settlements of more than 2,000 people in 1806 (Law 1967; Pounds 1974, 319). Istanbul, including the suburbs and Üsküdar, was still the largest city, although there had been a decline from the peak in the late seventeenth century. Such high proportions for the urban population can be explained in various ways. Immigration must have been considerable, mainly because urban mortality rates were normally high and made worse by frequent outbreaks of plague (Panzac 1973; Russell 1734, vol. 2, 336–8). Most migrants must have come from the countryside and were drawn in by various advantages of urban living. Food supplies are likely to have been relatively more certain than in the countryside and charity was more accessible. As the preferred residences of the elite, the towns offered a wide variety of work opportunities (Owen 1981, 24–5).

Largely due to the work of Raymond (1973–74), we know most about the socio-economic structure of Cairo. Around 1800, about 5 per cent of the population of 250–300,000 were soldiers and administrators, and another 2 per cent landed proprietors and members of the 'ulamā. At the end of the eighteenth century many of them were living outside the walled area of the city, a marked contrast with the situation 250 years before. The demands of the elite for goods and services gave employment to perhaps 70–80,000 people, about 27 per cent of the urban population. About 30 per cent of the economically active population of Cairo were master craftsmen, working for the most part in units of three or four employees and organised into more than 74 separate guilds. A further 10 or 11 per cent of the active population were retailers and merchants (Raymond 1973–74, vol. 1, 203–6). In terms of numbers of employees and establishments, the most important industry was textiles, including dyeing (Raymond 1973–74, vol. 1, 229–31). This seems to have been typical of towns right across the Empire, though the larger towns in particular were associated with various specialities: mohair cloth at Ankara, silks and cottons at Aleppo, muslins in Mosul, damask at Damascus and linens in Cairo and various towns in the Nile Delta (Owen 1981, 45). While food processing, metal-working, tanning and pottery-making were also widespread, some manufacturing industries had a more restricted distribution, dictated in large measure by the availability of raw materials. Soap boiling, for example, was located in areas of Mediterranean climate where the olive flourished (Gibb and Bowen 1950, 298, n.2).

Trading systems of varying scale and complexity continued to centre on the region's towns. At the lowest level, farmers brought their produce for sale. In small towns, raw materials and non-perishable foodstuffs such as cereals and dried fruits were bulked for sending on to higher order centres and the ports (Chevallier 1970; Raymond 1973–74, vol. 1, 244–89; Raymond 1979). Short-haul caravans were organised as need arose (Chevallier 1970; Tavernier 1684, 45–50). A similar sort of trade was sustained by the nomads' need to exchange livestock and livestock products for foodstuffs, weapons and other goods which they did not themselves produce on any scale. Inter-regional trade reflected both continued governmental concern for the provisioning of the major cities, including Mecca and Medina, and also local specialisations in cultivation and manufacturing (Fig. 8.5) (Alexandrescu-Dersca 1957; Raymond 1973–74, vol.. 1, 289–305; Shaw 1964, 125–30, 134–6). After recovering in the sixteenth century, the Ottoman Empire's transit trade in spices virtually disappeared during the early seventeenth century. A hundred years later, Middle Eastern markets themselves were being supplied through the western Mediterranean. This secular change was largely the result of the successes of the European East India Companies, based on national monopoly markets and internalised protection costs (Rousseau 1809, 117–22; Steensgaard 1974, 142–55). Down to 1725 considerable quantities of raw silk crossed Ottoman territory from north-western Persia *en route* to Europe and local manufacturers. The flow was interrupted from time to time by

Figure 8.5: Patterns of Trade in the Ottoman Empire in the Late Eighteenth Century

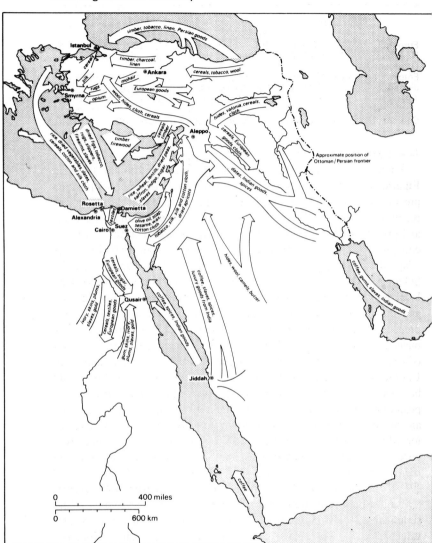

Turko–Persian wars, but it dwindled to an uncertain trickle following Russian and Turkish occupation of the main producing areas (1726–36) (p. 206). The search for alternative sources encouraged the expansion of silk production in Ottoman lands, as noted above (p. 199) (Davis 1967,

Table 8.1: External Trade of Egypt, 1783

| | Imports Paras | % | Exports Paras | % | Total Paras | % |
|---|---|---|---|---|---|---|
| Europe | 124,453,719 | 15 | 111,144,334 | 14 | 235,590,053 | 14 |
| N. Africa | 21,705,617 | 3 | 41,140,00 | 5 | 62,845,617 | 4 |
| Ottoman Empire | 305,791,869 | 36 | 431,275,125 | 36 | 737,066,994 | 46 |
| Jiddah | 382,500,000 | 46 | 191,250,000 | 25 | 573,750,000 | 36 |
| Total | 834,451,205 | 100 | 774,809,459 | 100 | 1,609,260,664 | 100 |

Source: Trécourt 1942, quoted Raymond (1973–74), vol. 1, 193.

141–3; Steensgaard 1974, 34–5, 186–7).

English, French and Dutch merchants entered the trade of the Ottoman Empire during the sixteenth century. They rapidly usurped the position previously enjoyed by Italians. Their textiles were cheaper; they had more ready money; and their great ships were less at risk from the elements and predators (Braudel 1972, 606–42). The newcomers were fortunate enough to gain considerable trading and group privileges (*Imitiyāzāt* or 'Capitulations'), beginning with the French in 1536 and 1569, when the Sultan was particularly anxious for allies. Even so, friendly relations with local notables were essential; they controlled the hinterlands of the ports and the flow of exportable commodities (Veinstein 1975). Like their predecessors, the French and their rivals were interested in obtaining raw materials (cotton, mohair, silk and valonia) and semi-manufactured goods (silk and cotton yarn), though a contraband trade in cereals still flourished. Until the opening decades of the eighteenth century, cotton and silk could be obtained virtually nowhere else in the quantities required by European industry. After 1700, however, alternative sources were opened up and the Middle East's importance began to slip (Davis 1967, 27). In return for Middle Eastern products, the Europeans imported manufactured goods, especially woollen cloth, and colonial produce, including the region's former specialities, sugar and coffee (Davis 1967, 41; Raymond 1973–74, vol. 1, 156–7, 173–84; Shaw 1962, 171–2; Ülker 1974, 76–138).

While the Levant trade was of crucial importance to Europe and supplied the Ottoman territories with desirable commodities, its total value was much lower than the internal trade of the Empire. The commercial duties paid by Europeans amounted to less than £100,000 per annum around 1750, when the total income of the Porte was more than £1 million (Davis 1967, 40). At the same date, Egypt exported an estimated £100,000 worth of goods to France, compared with £500–800,000 to Syria. In 1776 the province's trade with Europe was estimated at 13 million francs, compared with 67 million francs for trade with the rest of the Empire (Owen 1981, 52). Figures for 1783 (Table 8.1) indicate that the intra-Empire sector of Egypt's trade was valued at more than three times that with Europe which, in turn, was reckoned at less than half the value of that still carried on with Jiddah, at that time the principal port on the Red Sea.

## THE PERSIAN EAST

Scholarly work on Persia during the period c.1250–1800 has tended to emphasise either political and diplomatic history or artistic activity. Research on society and economy, and on anything which could be called historical geography in the modern sense, has been neglected. Consequently, this historico-geographical review of Persian landscape development is much briefer than that on the Ottoman Empire.

### Political Background, c.1250–1800 (Fig. 8.2)

Eastward expansion during the sixteenth century brought the Ottoman state into conflict with the new Safavid polity in Persia. The Safavid family first came to prominence three or four hundred years earlier (Savory 1980). They were the leaders of a Sufi religious movement (the *Ṣafiviyya*) based in the town of Ardabīl in Azerbaijan. Adherents increased during the thirteenth and fourteenth centuries, not only throughout Azerbaijan but also in eastern Anatolia and northen Syria. Türkmen recruits were particularly numerous, and they were to provide the movement with its fighting power. Safaviyya teaching acquired a definite Shī'ite character towards the end of the fourteenth century, and its propagandists began to proclaim that the leader was the representative on earth of the Twelfth Imam in direct descent from 'Ali, the Prophet's son-in-law, who had been occluded in 873–74. Towards the middle of the fifteenth century, adherents adopted the distinctive scarlet headgear which earned them the nickname of 'redheads' (*kizilbash*) from the Sunni Ottomans, who were becoming increasingly concerned about the proselytisation, and hence the subversion, of the Türkmen tribes of Anatolia.

Up to this point, eastern Anatolia, Azerbaijan and the rest of Persia had been subject to the Mongol Īl-Khāns who had succeeded Hulagu (c.1217–65), grandson of Çingis Han. By the end of the fourteenth century, the original unity of the Īl-Khānate had been replaced by a mosaic of warring principalities. The whirlwind success of

Timūr imposed a new unity on Persia, but it did not last beyond his death (1405) (Petrushevsky 1968). In the north-west a struggle resumed between confederations of Türkmen known respectively as the *Kara-Koyünlü* ('Those of the Black Sheep') and the *Ak-Koyünlü* ('Those of the White Sheep'). The balance gradually tipped in favour of the latter, but the Ak-Koyünlü began to disintegrate politically towards the end of the fifteenth century. Disintegration provided the Safavids with their opportunity.

After several false starts and the virtual elimination of the Safavid family itself, the Safavids were able to defeat the Ak-Koyünlü in 1501, take over their capital of Tabrīz and dominate Azerbaijan. One of the first acts of the victor, Shāh Ismā'il I (1501–24), was to declare the 'Twelver' form of *Shī'ism* to be the state religion, despite the predominance of Sunni Muslims in the newly acquired territory. A conversion campaign was launched. It was sufficiently successful to define the Safavid state for the Ottomans, to provide a dynamic for wars of expansion, and then to hold the new state together through crises of defeat and disaffection. Within ten years of capturing Tabrīz, Ismā'il was master of Persia and Lower Mesopotamia. Successful proselytisation in Anatolia, the encouragement of revolt and the support of rival candidates to the Ottoman throne eventually brought war between the two Empires. The Ottomans' command of logistics, which allowed them to use guns, destroyed Ismā'il's tribal levies at Çaldirān, north-west of Khvoy, in August 1514. Substantial territorial losses came in the 1550s and 1580s (Fig. 8.2). The Türkmen became restive, while the Üzbecks of central Asia made repeated onslaughts on the eastern provinces. Yet the Safavid state held together. Its fortunes were restored by Shāh 'Abbās I (1588–1629). He had contained the Üzbecks by 1602, and went on to reconquer Azerbaijan (1603–07) and to retake Baghdad (1624), while returning good order to the provinces. The bases of 'Abbās' success were the personal qualities of authority and courage, together with civil and military reorganisation. Like the Ottomans long before, 'Abbās created a professional army loyal to himself. It was recruited from among the descendants of the thousands of Caucasian and Armenian prisoners taken by his grandfather, Tahmāsp I (1524–36), between 1540 and 1553. Known collectively as *ghulāmān-i khāṣṣa-yi sharīfa* ('slaves of the royal household'), they were armed with the firearms despised by earlier generations of kizilbash troops. Creation of the new force diminished the power of the Türkmen tribes by removing military dependence on them. On the civil front, 'Abbās began a process of transferring provincial governorships from kizilibash chiefs, who had hitherto been predominant and fairly autonomous, to *ghulāms*. The main concern of the new governors was to please their royal master by extracting as much revenue as possible from their jurisdictions. Both the civil and the military changes proved their value in the short term, but, in the long run and under weaker royal authority, they proved disastrous. Although they fought well in the early days, ghulām troops seemed to lack the fighting spirit of the tribesmen and, like the Janissaries, gradually eased themselves into the relative comfort of court and garrison life. At the same time, lacking any personal interest in the provinces which they administered, ghulām governors became extortionate and corrupt. The taxes levied to support a luxurious court and to pay for endless campaigning became so burdensome as to sap the economic vitality of the countryside.

'Abbās I was succeeded, in the main, by weak and inexperienced tyrants who had little grasp on administration. A revolt by the Ghalzay Afghans of Kandahar in 1709 marked the beginning of the end for the Safavids. For much of the eighteenth century Persia was little more than a set of city and tribal polities of fluctuating fortune, size and number (Abrahamian 1974). In 1719 and 1720 the Afghans raided across the Dasht-e-Lūt. They penetrated further still in 1722 and defeated the Persian royal army near the village of Gulnabād, east of Esfahān. Following the capture of the Safavid capital, the Afghans ruled much of central and southern Persia for seven years, though Safavid claimants survived in various northern districts. Russia and the Ottoman Empire took advantage of Persian weakness in the 1720s to occupy territory in the west and north. Further encroachment was stemmed by Nādir Khān Afshār who, after sponsoring a brief revival of Safavid rule (1729–36), made himself Shāh. He reunited the country, subdued Kandahar, and marched into Mughal India. His capture of Delhi (March 1739) yielded booty valued, it is claimed, at 700 million rupees or £70–80 million (Lockhart 1938, 156). Nādir Shāh next campaigned successfully against the central Asian khanates of Bokhara and Khiva (1740). Love of war, though, led to neglect of internal affairs. Provincial reconstruction was ignored, and the Empire was held together by fear and violence from Mashhad, in the north-east. In the end, Nādir Shāh's blind savagery brought about his assassination (1747). A power struggle followed. Karīm Khān Zand emerged victorious, and the southern and western territories under his direct control knew relative peace and prosperity (1757–79). His death provoked yet another round of internal conflict. This resulted in the dominance of rulers from the Qājār tribe; the dynasty ruled Persia from 1794 to 1925. Although the Qājārs managed to maintain the unity of the country, they lost substantial territories in the north to Russia, notably in the early nineteenth century (Treaties of Gulistan, 1825, and Turkmanchai, 1828), while Persia became the scene of a long diplomatic and commercial contest between the Russians and the British.

## Towns

After the initial destruction caused by the Mongols, the stabilisation of their regime in Persia revived the prosperity of many towns, though population levels were rarely as high as in previous centuries. Transit trade revived and manufacturing increased (Petrushevsky 1968). The disintegration of the Īl-Khān state in the fourteenth century led to the promotion of several provincial centres before the Safavids restored unity; the process of atomisation resumed when their control ended. Each ruler wanted to locate his court in a town close to his territorial base.

The Safavids first took over the Il-Khān centre of Tabrīz (1501), but Ottoman pressure in the north-west and the capture of the city forced Shāh Ismā'il to move to Qazvīn (1514). Esfahān became the capital in the 1590s, under Shāh 'Abbās I, and remained so until the fall of the dynasty. The triumph of Nādir Afshār led to the promotion of Mashhad from 1737, while Karīm Khān preferred Shīrāz. The success of the Qājārs resulted in the choice of Tehrān as the capital (1796), a role it retained under the Pahlavis. A sizeable court, together with the separate entourages of its various principals and a large portion of the army, swelled the urban population of the capital. For example, the Qājār court and royal bodyguard together have been estimated as forming about 6 per cent of the total population of Tehrān (about 60,000) in the late eighteenth century (Brown 1965, 41). As elsewhere in the Middle East, the demands for goods and services generated by these relatively wealthy people attracted others to the capital. Manufacturing and trade were encouraged thereby (Firoozi 1974).

As well as leading to an increase in population and economic activity, the acquisition of capital status often led to significant morphological development. Most rulers sought to embellish their chosen city, and their courtiers and followers built accommodation for their own needs. The Qājārs, for example, added the citadel and various palaces to Tehrān (Brown 1965, 43–5), while Karīm Khān's chief legacy to Shīrāz was a complex which combined bazaar, mosque, baths and government offices (Lerner 1976, 205–72). But nothing rivalled the new royal suburb established by Shāh 'Abbās I 1.5 km south of Esfahān in the opening decades of the seventeenth century (Fig. 8.6). Its bazaar, tree-lined ceremonial avenue, meydan (a square or esplanade use for polo and mounted archery

Figure 8.6: Developments Ordered by Shāh 'Abbās I at Esfahān

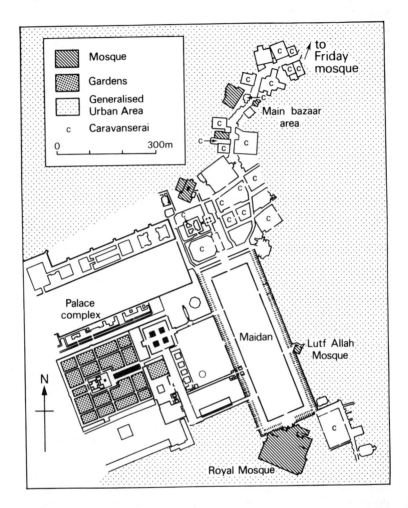

contests) and garden palaces continued earlier Persian traditions, but the
scale and beauty of the ensemble made Esfahān the wonder of the age,
at least to impressionable Europeans (Beaudouin and Pope 1939, 1391–410;
Chardin 1735, vol. 2; *Iranian Studies* 1974). The fall of the Safavids and
the relocation of the court brought about the withering of Esfahān. Population fell from Jean Chardin's estimate of 600–700,000 in the second half
of the seventeenth century (the same as London, he thought) to perhaps
100,000 by the end of the following century, though John Malcolm

(1769–1833) thought that it perhaps doubled again in the first 15 years of the nineteenth century (Malcolm 1815, vol. 2, 519–20). By the middle of the nineteenth century the population may have been down to about 50,000. Athough some public buildings had been repaired by that date and a few new ones added, the great bazaar was in decay. Squatters had moved into the dilapidated pavilions off the ceremonial avenue, which had been stripped of its trees. European visitors thought that between a half and two-thirds of the city was in ruins, while the reduced population of Persia's largest city had refocused itself in, and just to the north-west of, the pre-Safavid city (Brown 1965, 47–53).

Shīrāz suffered a similar fate after the fall of the Zand dynasty. Its population declined from an estimated 40,000 to perhaps 20,000 (Kinnier 1813, 64; Morier 1818, 111; Ouseley 1819, vol. 1, 26). In 1811 James Morier (1780–1849) reckoned that about a third of the buildings were in ruins, including the tomb of Hāfez (c.1325–89/90), whose verses had made his native city the most celebrated in Persia (Morier 1818, 111; Waring 1807, 40). A few years later, everything in Shīrāz appeared to Robert Ker Porter (1777–1849) to be neglected: 'the bazaars and maidans falling into ruins: the streets choked with dirt, and mouldering heaps of unrepaired houses' (Porter 1821, vol. 1, 693). In this case, though, at least some of the decay was attributable to the destruction ordered by the first Qājār ruler, Aqā Muḥammad (1794–97), and a number of earthquakes (Lerner 1976).

Persia's entire urban population in the late eighteenth century was probably 10–12 per cent of the total of about 5 million.[8] The economically active people were engaged in the same types of activity as their brothers in the Ottoman Empire — services and administration, trade and manufacturing — and lived in a similar built environment with characteristic Islamic features. As in the Ottoman Empire, the popularity of coffee-drinking brought a new amenity (*khanah* or coffee-house) to the urban scene during the seventeenth century, and supplemented the baths and the traditional wrestling clubs (*zur khanah*) as social centres (Abrahamian 1974). Many towns were famous for particular kinds of manufactured goods. For example, in the late eighteenth–early nineteenth century, Yazd had a reputation for shawls made from a mixture of silk and Kirmanian wool, while Kashān was known for its carpets and Shīrāz for weapons (Hambly 1964). These commodities entered the circuits of Persia's internal trade through the wholesale merchants (*tajjārs*) who constituted, at most, an estimated 1 per cent of the urban population (Floor 1976). Some items, of course, entered international trade, in which merchants from neighbouring states, as well as the Shāh's subjects, were active and often resident in Persian towns.

## Trade

Local and international trade was promoted under the early Safavids. Unification extended the internal market, while strong local governments improved the safety of the routeways, a contrast with parts of the Ottoman Empire on which Europeans often commented (Steensgaard 1974, 68–74). New caravanserais were built on the principal routes, perhaps especially those connecting the interior with the Gulf ports and particularly Gombroon (Bandar 'Abbās) (Newberie 1905, 462, 465, 466; Siroux 1974). These took on a new significance in the early seventeenth century. At that time north European merchants began their long-drawn-out attempts to divert the flows of Persian trade towards the Gulf ports. Their objectives were mixed. In part there was a desire to tap the 'fabled wealth' of Persia (Searight 1969, 35). More pressing, though, was the need to lubricate Europe's trade with India and the Far East without shipping large quantities of specie. The English attempted to achieve this end from 1615 by disposing of their woollen cloths. Little market existed in tropical India, but there seemed to be a potential in Persia with its cold winters. From 1622 the Dutch, by contrast, sought to ship spices and drugs direct from their factories in the East Indies (Steensgaard 1974, 374–6). Neither was particularly successful. The Persian market was probably too small in the seventeenth century to absorb large quantities of European goods and, in any case, there was stiff competition from commodities moved inland from the eastern Mediterranean and the Black Sea. Spices could be obtained more cheaply by alternative land routes from the East (Steensgaard 1974, 377–86). But the whole picture was made more complicated still by the decision of Shāh 'Abbās I to use a monopoly over the supply of silk (affected in 1619) as a diplomatic weapon. 'Abbās aimed to secure European armaments and military advisors for his wars, while depriving the Ottomans of both an important raw material for their own industry and the valuable revenues earned from transit dues. For their part, the English and Dutch grasped at the opportunity to secure a monopoly on the supply of Persian silk to Europe, where it was much in demand in wealthy circles. In the event, they were not successful. Considerable amounts of silk continued to find their way on to the overland routes (pp. 199, 202–3) and neither the Dutch nor the English East India Companies had the resources to acquire the amounts being offered to them at the prices demanded, especially since they were expected to pay all transport costs and duties on the two-month journey from Qazvīn, where the silk arrived from Gilan, to Gombroon (Ferrier 1973; Steensgaard 1974, 377). The Shāh's monopoly lapsed with the death of 'Abbās (1629), and by 1650 the Dutch and English had given up all attempts to divert the silk trade. Even in their best years, they had never been able to obtain more than 18 per cent of the 70–100,000 kg exported annually from Persia (Steensgaard 1974, 387–96).

The European merchants had more success in achieving a southward diversion of trade in the eighteenth century (Hambly 1964; Perry 1979, 246–71; Ricks 1973 and 1974). In some ways this was simply the result of being able to channel and expand the flows of goods which had entered the Gulf trade for centuries, both the coasting trade in foodstuffs and raw materials (which tied the coastal areas together in a single maritime system) and also the transit trade with India. Indian requirements, especially for markets, were crucial. During the eighteenth century European ships acted as carriers for 'country' merchants — Arabs, Persians and Indians — who dominated the trade and also used local vessels for trading. They imported Indian chintzes, muslins and other manufactured goods, as well as sugar and coffee. These were increasingly paid for by exporting specie and commodities such as dried fruit, wines, nuts, opium and gums rather than the local manufactures which had been important in earlier centuries. Trading on their own account, European merchants sought to exchange their woollen cloth and other manufactures for silk, carpets and particularly the soft goats' hair produced in villages near Kermān and known as Kirmanian wool. There were ups and downs, of course, largely determined by political conditions in Persia but also by the spread effects of wars in Europe. Persian failure to re-establish heremony over the Gulf following the Omani seizure of Bahrain (1717) initiated a struggle for maritime supremacy between various coastal communities which endangered both local and European shipping. Qājār concentration on the interior exacerbated the situation. Eventually, the rulers of Muscat were forced to withdraw from the Gulf and turned their attention to building a maritime empire in East Africa. The continued danger to their ships brought British military intervention (1805–06, 1809–10, 1819–20), which finally stabilised the shifting mosaic of small polities on the Arabian side of the Gulf and created the outlines of the present-day political fabric.

## The Countryside

Developments in external trade could be expected to have had effects on land use and agricultural production in Persia. Rising demand for silk in the seventeenth century is likely to have stimulated clearance of the jungle vegetation in Gilan, especially near Rasht and along the Safid Rūd, in the effort to expand production. Increased mulberry planting is likely to have taken place in the neighbouring provinces of Mazandaran and Azerbaijan.

War with Russia and Turkey, resulting in foreign occupation of the silk-producing provinces, disrupted output in the 1720s and 1730s (p. 206). Cotton was grown in many districts, but Russian demand in the 1790s probably increased the areas devoted to it in the Caspian lowlands and Khurasan (Issawi 1971, 249; Perry 1979, 249–52; Ricks 1973 and 1974). Tobacco was introduced in the early seventeenth century. A ban on imports imposed by Shāh 'Abbās I stimulated local cultivation in southern Persia. By 1630 tobacco was also being grown in Gilan. Before the end of the eighteenth century it was widespread, and distinctive local qualities were recognised (Issawi 1971, 247). The overall impression, then, is one of increased commercial orientation in agriculture, at least in areas with access to export outlets. It appears somewhat paradoxical, therefore, that the European image of a smiling countryside during Safavid times was replaced by one of deserted villages and abandoned land at the beginning of the nineteenth century. John Malcolm believed that no more than one seventh of the land of Persia was cultivated (Malcolm 1830, quoted Issawi 1971, 265), while a great under-use of resources is implied by estimates that between a third and a half of the population was nomadic (Abrahamian 1974; Issawi 1971, 265), presumably following in the main a life of transhumant nomadism. The picture of general decay is so similar to the one often painted for the Ottoman Empire that it is important to examine its validity. This is the opening theme of the next chapter.

## Notes

1. 'Gazi': one who fights for Islam.
2. I prefer to keep 'livings' as the translation of the Turkish *diriklır* rather than 'fiefs' with its misleading connotations. The remainder of the revenue went direct to the Sultan.
3. For example, Kayseri 3.1 per cent 1500–83; Karaman 3 per cent 1523–87. Calculated from Jennings (1976).
4. The average rate per urban household in Palestine in 1596–97 was 93.5 *akca* compared with an average of 245 akça for village households (Hütteroth and Abdulfattah 1977, 97).

5. Aleppo from 4 to > 9 ha; Cairo from about 22 to 38 ha (Raymond 1979).

6. In the previous century, Taqi al-Dīn Ahmad al-Maqrīzī (1364–1442) reckoned that the average man of middle rank would spend almost 20 per cent of his income on spices (Wiet 1962).

7. This famous remark is attributed to Tsar Nicholas I when in conversation with the British ambassador, Sir Hamilton Seymour, in 1853 (ffrench Blake 1971, 6).

8. Derived from Abrahamian's estimate (1974) of the total population in 1812. His estimate of the urban population is < 10 per cent. Cf. 12.6 per cent (648,700) calculated from various contemporary estimates collated by Hambly (1964).

# 9 LANDSCAPE EVOLUTION TO c.AD 1800: A REVIEW

## Assessment

The previous chapter suggested that both the Persian and Turkish Empires entered a phase of decline which lasted for much of the eighteenth century. It was manifest most clearly in political relations with states in northern and western Europe, but it embraced the economies of the two Empires and their landscapes. Towns were in decay, the land was frequently under-used and many villages were deserted. The present chapter attempts, first, to test the validity of this image of decay and decadence and then to try to explain it.

Much of the evidence currently available comes from comparatively late European sources, many of them written by travellers. They came to the region in growing numbers during the late eighteenth and early nineteenth centuries, initially diverted from the course of the Grand Tour by the Revolutionary and Napoleonic Wars (1792–1815), but always attracted by the associations of the region with biblical history and its reputation as 'the gorgeous East' (Milton, *Paradise Lost*, 2.3). The building of railways to the southern shores of France and the institution of steam-packet services on the Mediterranean Sea facilitated their progress in the 1840s and later (Chevallier 1968). The objectivity of the accounts which travellers produce has been questioned (e.g. Ernst and Merrens 1978). As Ahmed (1978, 11, 16) has observed, travel itself is often an exploration of the traveller's own myths and legends. These were quite complicated for European — and American — visitors to the region. On the one hand, the travellers expected, if not the world of the *Arabian Nights*, then at least that of Byron's poems and Chateaubriand's romances. The mundane, faded reality which they found was often a big disappointment. On the other hand, some compensation was found in the 'luxuriance of ruin' (Burton and Tyrwhitt-Drake, 1871, vol. 1, 3) and the apparent emptiness of the plains. Both provoked appropriately melancholy feelings and stimulated thoughts of death and decay (Daniel 1966; Jullian 1977; *Syria, The Holy Land, Asia Minor etc.* 1837–38). When they turned from fantasy and introspection to look at actual landscapes and real activity, the travellers did not make their observations in a systematic way. They were often casual observers and acquired information almost incidentally to the main business of travelling. Accordingly, travel accounts are suspect. For example, lack of familiarity with local conditions could lead to an interpretation of a cereal-growing area as desolate, if it was crossed after harvest. Nomads would have been the only people visible, not because the land was uncultivated but for the simple reason that the stubble provided grazing for their animals. Similarly, the absence of villages may have been more apparent than real, for they would have been obscured from view by the dusty haze characteristic of the plains in summer.

The travellers' reports may have been erroneous, but they were not entirely wrong. The existence of abandoned arable land and of deserted villages in many parts of the region does seem to be confirmed in various ways for the eighteenth and early nineteenth centuries. Sixteenth-century Turkish defters yield settlement names and numbers. Comparison of these with similar data from later sources suggests decline, at least in some parts of the Ottoman Empire, and this has been assigned to the seventeenth and eighteenth centuries (e.g. Hütteroth and Abdulfattah 1977). Both topographical and archaeological surveys carried out in the late nineteenth or early twentieth century revealed numerous standing ruins in many areas, though the dates of desertion and their explanation were not clear (Butler 1920; Palestine Exploration Fund 1881–88). Against this evidence must be set that for the expansion of cash cropping during the course of the eighteenth century. Even the average European tourist in the early nineteenth century was warned that he would find some areas in 'the highest state of cultivation' (*Constantinople* . . ., 1838, vol. 1, 72; vol. 3, 28). There is thus a paradox, but it can be resolved.

Although the precise area of well-cultivated and populous land cannot be computed, much of it seems to have lain close about the towns, where arboriculture was visually striking, or to have been located in districts readily accessible to European commerce. Some contemporary observers noted that, even in these territories, however, there was what we would now recognise as a considerable under-use of land (e.g. Bowring 1840, 9; Niebuhr 1792, 292; *Syria, The Holy Land, Asia Minor etc.*, vol. 1, 6 and 18; Volney 1788, 98–113). Year-round use by pastoral nomads, if it can be established, might be confirmatory evidence, for their mode of living involves an extensive, rather than an intensive, use of resources. By contrast, any increase in agricultural production, in conditions of low productivity, required the allocation of additional land. This meant in turn the deployment of more labour. The needs could have been met by concentrating population in areas favoured by precipitation, as well as access to the ports. Less favoured areas might be expected to have lost population, although a few large villages could have been created on commercially orientated estates by amalgamating populations and abandoning less conveniently placed settlements. This could be the explanation for desertion, but European visitors would not know that. The underuse of land would have been much greater in areas not favourably located with respect to large towns and the coasts, even if physical conditions looked conducive to farming. Parts of interior Persia were probably in the worst condition of all by c.1800 (p. 212), perhaps chiefly because many basins and valleys are almost shut off from the Gulf by rugged mountains, and transport costs were accordingly high.

Contemporary western observers adopted an additional perspective, closely allied to the widely accepted explanation for what they regarded as a generally sorry state of affairs in the Middle East. It was that the widespread depopulation and under-use of land apparent in the late eighteenth and early nineteenth centuries constituted the latest stage in a protracted and progressive deterioration which had followed the decline of Roman power. Such a view had two principal foundations. One was the respect for classical antiquity inculcated by an education based on Greek and Latin authors and apparently confirmed by the quality of ancient works of art and architecture. Classical antiquity was a 'golden age', a time of prosperity and large population, of rational government and efficient exploitation of resources. The second support for the view that deterioration had been continuous from at least the Muslim conquests, was anti-Islamic polemic. It developed during the Middle Ages, but was sustained long after the 'Age of Faith'. The Muslim world came to be portrayed as the very antithesis of the western Christian world of enlightenment (Ahmed 1978, 16; Daniel 1966). Muslim society was represented as authoritarian, bigoted, unprogressive and inhumane; its well-springs were dark and evil (Asad 1973). The people were indolent, sensual and cruel. Little effort was made to change this view until Edward Lane's *Manners and Customs of the Modern Egyptians* (1836) attempted to describe a local society sympathetically in its own terms (Ahmed 1978, 111–19). As long as this prejudiced view prevailed — and it survives still — any notion of good coming out of an Islamic political regime was inconceivable. An inexorable downward spiral was initiated, possibly at the time of Muslim conquest.

The evidence to support the view that decline began with the Muslim Arabs is fragmentary in time and space, as well as indirect and highly suspect. Only one district has been studied in detail over a long period of time, and that is the Diyālā plain (Fig. 9.1) (Adams 1965). In terms of both numbers and size of built-up areas, settlement peaked there in early Sasanian times (AD 225–400), that is, what the western world would

Figure 9.1: Population and Settlement Change in the lower
Diyālā Valley, c.4000 BC–AD 1957

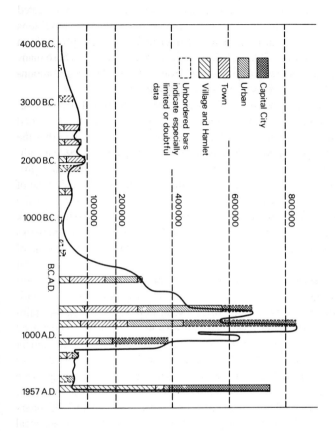

Source: Adams (1965), Table 25.

regard as the late classical period. Cultivated land probably reached the
maximum extent possible at the same time. There were fluctuations in
both settlements and cultivated area thereafter, but the trend — initiated
before the Arab conquests — was broadly downwards, precipitately so
in the ninth and tenth centuries and again in the twelfth and thirteenth
centuries. Extremely low levels of settlement numbers and cultivated area
were soon reached. These were sustained, with slight perturbations, for
over 600 years. While the Diyālā experience is very special, since it comes
from a district where the cultivation of large areas and the support of

great numbers of settlements were dependent upon elaborate but vulnerable irrigation systems, it may be more widely representative. Previous chapters have suggested that, on a regional scale, deterioration was not uniform everywhere in space or time. There were ups and downs. But the tax yields imply that everywhere the broad trend was downward throughout the Middle Ages. Deterioration may have been halted, or even reversed, for a while during the sixteenth century in both the Safavid and Ottoman Empires. However, evidence from both sub-regions suggests a resumption of the downward spiral in the seventeenth century, followed probably by local stabilisation and even some re-advance in the subsequent one. As discussed above, contrary trends could take place in different parts of the region. For example, retreat was characteristic in the steppe of Syria, while advances were taking place not far away in cotton- and silk-producing areas to the west. Much depended on access to markets and local political conditions.

There can be little doubt that one trend was progressive and nearly universal from almost the earliest times: the loss of woodland (Thirgood 1981). Populations may have declined, settlements may have been deserted and cultivated land abandoned, but it is doubtful whether forest would have regenerated, particularly on any scale, though Willcox (1974) believed he had evidence for such a possibility from eastern Anatolia. Scrub would colonise abandoned farmland, especially cultivated terraces on hillsides distant from inhabited villages, but the current climatic regime would probably militate against the rapid growth of mature trees. The loss of soil from steep, unterraced hill-slopes would work in the same direction. If, for these reasons, timber should be regarded as effectively a non-renewable source in the Middle East, the demands of even a reduced and declining population for kindling, fuel and construction materials would almost certainly reduce the area of surviving woodland. The results would be cumulative. Economic distance would provide some protection, but even that would be breached when herders chose to burn woodland to thin out the trees and encourage plants which could be eaten by their stock. Often only careful management and direct human intervention can create conditions for afforestation. They were to come later than the periods under consideration now.

## Explanations

Contraction of cultivation and the desertion of settlements can frequently be attributed to specific events. Periods of low floods must have been

extremely serious in the Nile Valley, though the available data on settlement numbers do not allow precise correlation. Reoccupation was comparatively easy in times of high flood. By contrast, in Lower Mesopotamia,
high floods and shifts in river channels were devastating, and it took years
of work to reverse the effects. The cutting of four major canals between
the Tigris and the Euphrates in the vicinity of Baghdad, soon after the
Muslim conquest, raised the water table and probably led to rapid salination (Gibson 1974; Le Strange 1895). Earthquakes must have wrecked
many villages, as well as the towns whose fortunes often appear in the
chronicles and histories (Ambraseys 1978; Poirier 1980). Restoration may
not have been too difficult, but there are likely to have been cases of complete abandonment and slow recovery. Warfare brought destruction, as
in the 'Great War' between the Roman and Persian Empires in the early
seventh century AD, or when the Mongols swept through the region in
the thirteenth century, and when the Safavids adopted a scorched-earth
policy on their north-western frontiers in the sixteenth century. The
pandemic plagues of the sixth and fourteenth centuries brought depopulation; Russell (1968) estimated that 20–25 per cent of the population died
in the epidemic of AD 451–44, and perhaps as many as one third of the
5–9 million people of Egypt and Syria may have died in the Black Death
(1347–51).[1] The effects on settlement survival and land use, however, are
not likely to have been very long-term or permanent. In fact, none of
the specific events mentioned above could, of itself, produce long-term
abandonment of arable land and settlement desertion. For that to
happen, a succession of short-term disasters, with mutually reinforcing
effects, would be necessary. Despite the dangers in monocausal explanation, scholars have sought for a major cause to explain the phenomenon
of long-term and widespread decay and deterioration. They have seen
it in processes acting on a long time-scale and with a fundamental bearing on land use. The favourite contenders — not mutually exclusive —
are secular changes in climate and, as Huntington (1911, 249) described
it, 'human greed, misgovernment and folly'. Each will be examined in
turn.

## Secular Change in Climate

A previous generation of scholars was clear on the role of climate: it
had exercised a controlling influence on most human activity. Several
individuals were convinced, however, that desiccation had been the dominant trend in the Middle East from the Roman period onwards. They

could cite considerable evidence in their favour. For example, the distribution of ruined settlements in the steppes of Syria suggested that population had been higher there in the past and that the carrying capacity of the land had subsequently declined. Apparently well-used routes had gone out of use. In both cases the crucial element was water, and that appeared to be missing (Huntington 1911, 262–3, 268–78). Yet there was evidence for its greater abundance in the past. Settlement sites and artificial terrace sytems in the Negev, for instance, clearly indicated cultivation and that, in turn, presupposed damper conditions. Several stream channels were known to be spanned by massive Roman bridges, though in modern times they carried little discharge, even in such extremely wet winters as that of 1904–05 in Syria–Palestine. Similarly, ancient cisterns were found to be dry in northern Egypt, while spring houses (fountains) and large public baths were found in Syria and Jordan, where there were no running springs in modern times (Huntington 1911, 288–95). Thus, desiccation was a plausible hypothesis, and it could be supported by evidence from the Sistan and Tarim basins (Chappell 1970). Recently, Sperber (1974) has published evidence from the *Talmud* and the *Midrah* dating from the late second and third centuries AD which shows an acute awareness of the importance of precipitation, and may even indicate 'a protracted period of unusually low rainfall in Palestine'.

Perhaps the major protagonist of this view was Ellsworth Huntington. He recognised, however, that the diachronic distribution of dating evidence for the remains themselves suggested successive cycles of occupation, desertion and reoccupation over time. He therefore proposed that, while desiccation may have been progressive, it did not follow a smooth curve. The trend was punctuated by periods in which water availability increased. 'Pulsatory desiccation' was suggested as a hypothesis (Chappell 1970).

There are, of course, alternative explanations for the phenomena used to support the case for desiccation. Population estimates and carrying capacity may have been exaggerated, while economic forces could explain the abandonment of land which would be regarded, in any case, as marginal for cultivation under any climatic regime approximating that of the present. A damper climate is not the only explanation for the evidence of cultivation in the Negev. There had to be an incentive to settle the pockets of good soil. This may have been political and strategic. Water conservation, however, was the key to successful occupation (Evenari *et al.* 1961; Kedar 1957; Mayerson 1956; Woolley and Lawrence 1915, 13). Bridges might conceivably have been built simply to ease the crossing of a deep gully for laden pack animals, rather than to ensure a dry passage. As for springs, the flow could have been diverted or stopped by earthquakes and neglect of maintenance, as well as by lack of rain. In general, as Tchalenko observed (1953, 66–9), the protagonists of desiccation have taken little account of local geological, pedological and topographical relationships, which influence both sub-surface and surface availability of water. Physiological drought is more commonplace in the Middle East than many commentators realise.

Support has grown in recent years for climatic change of an opposite kind, namely an increase in dampness. The case is largely based on an interpretation of the so-called Younger Fill, which is widespread in the region and beyond (Vita-Finzi 1969a and b). This is a distinctive, buff-coloured, well-bedded valley fill, containing pottery and, at various places, standing structures as well. These artefacts have suggested that the fill was laid down after about AD 400, though terminal dates have ranged as late as AD 1800. Both the widespread distribution of the Younger Fill and its apparently synchronous nature suggested that a universal agency must be responsible. Examination of various possibilities produced the conclusion that it was due, fundamentally, to a significant shift towards damper climatic conditions. Either precipitation increased overall, or else its distribution changed over the year. Whichever it might have been, run-off would increase, resulting in greater erosion and higher transportation capacities for streams, though clearly transport efficiency was not sufficient to remove all the load, and aggradation took place.

This hypothesis appears to fit in well with, and to be confirmed by, the accepted view of the climatic history of northern Europe (Bintliff 1982). A warmer, drier period down to AD 400 was succeeded by colder and damper conditions which lasted to 700–800. Warming was renewed in the following centuries and culminated in the 'Little Optimum', around 1200–1300. Colder, damper conditions then set in — the 'Little Ice Age' — and continued through into the nineteenth century. Independent evidence for a similar sequence in the Middle East is sparse. Coarse-fraction analysis of twelve cores from the northern bed of the Gulf suggests that a humid phase of climate began about 2000 BC (Diester-Haas 1973). Records of extremely cold winters are known from the eighth to the thirteenth century AD, when the Bosporus and the Nile reportedly froze (Erinç 1978; Lamb 1977, vol. 2, 148), and again in the first half of the seventeenth century. Some places may have experienced wetter conditions than usual, like Egypt in the 1290s. Unfortunately, the Middle East's copious literary sources have not been systematically searched for information on weather conditions, despite their evident potential for recording extreme events.

Various possible developments can be suggested as likely consequences of damper, cooler conditions in the region. Erosion rates on some Mediterranean watersheds are reported to be as much as 3–4 m³/ha per annum (Glesinger 1960). Increased precipitation would have raised them, thereby quickly reducing the supportive capacities of sloping land, while allowing for the occasional bumper harvest which wetter conditions generally produce in the dry-farming districts. Damper conditions, however, would encourage more locust swarms. Greater and more rapid run-off would produce serious seasonal flooding in many valleys and perhaps discourage cultivation there. Frost episodes would destroy crops, while a general lowering of the temperature would reduce the growing season in mountainous and plateau areas, leading in a few years to the abandonment of the higher, more marginal land.

Unfortunately, there are various problems with the hypothesis of greater wetness. First, zonal and temporal changes in climate may not be symmetrical (Lamb 1977), so that north European events cannot be used to predict those in the Middle East. Second, several recent studies have indicated that the crucial Younger Fill accumulated at a variety of dates in different locations, and even the same location may have experienced successive phases of aggradation (Wagstaff 1981). Third, explanation of the fill's physical characteristics does not necessarily require the hypothesis that flow and precipitation regimes were different from those prevailing at the present (Wagstaff 1981). These difficulties have encouraged the formulation of alternative explanations to that of climatic change to account for increased sedimentation. The most obvious depends upon the undoubted relationship between the clearance of vegetation and the opening up of land for cultivation, on the one hand, and an increase in erosion and stream load, on the other. Activity is likely to have been greater at some times than others. In some locations, a phase of increased erosion and sedimentation clearly corresponded with the Hellenistic and Roman periods. While this hypothesis accepts that erosion and sedimentation were closely connected in time, another possibility has been advanced recently (Forbes, Koster and Foxhall n.d.). The construction of hill-slope terraces, whether primarily as a conservation measure or simply to make use of sloping land under conditions of population pressure, would reduce the effectiveness of erosion, however caused. Should maintenance be neglected and the terrace walls begin to collapse, then large quantities of eroded material accumulated behind them would be washed into streams. The streams would be unable to carry the increased load for any distance and sedimentation would result. The trigger to the whole process might be loss of labour, rather than increased precipitation. Loss of labour, of course, could have developed for a variety of reasons, but now that discussion has moved away from climatic change we come to a consideration of 'human greed, misgovernment and folly'.

## 'Human Greed, Misgovernment and Folly'

Shortage of labour seems to be a recurring theme in the agricultural history of the region. Three periods stand out as being particularly important to our considerations. They are the late Roman period, the second half of the fourteenth century, and the seventeenth and eighteenth centuries. Examination of their characteristics will provide a way of entering the thorny maze of possible human causation.

In the Roman Empire shortages of labour were already noticeable when Septimius Severus attained the purple in AD 193, but they increased thereafter (Boak 1955; Leake 1824, 68). The evidence is varied. The army experienced difficulties in recruiting, despite its privileged status. Edicts tried to bind the peasants to the land, while forcing them to take over responsibility for neighbouring agri deserti. The rather sudden appearance of water mills, which were labour-saving devices, adds to the argument. By the early fourth century, the labour force in the lower Meander Valley was about half that recommended by the agricultural writers, Cato and Columella, though there had been no significant change in farming methods (Jones 1953a).

Various explanations for this degree of apparent labour shortage have been offered. Successive plagues and wars must have taken their toll, but reports of peasants voluntarily giving up their land, even in second-century Egypt, and of a marked and widespead increase in brigandage (Peters 1970, 577) point to heavy taxation as the major cause. Contemporaries were more sure that this was the case than were later historians (e.g. Lewis and Reinhold 1966, 434). Taxation mounted for various reasons. A large army had to be paid for, and both men and material were rapidly consumed in a series of civil wars, as well as in countering barbarian incursions (such as those of the Goths and Huns into the Eastern Empire during the third century) and in the vastly expensive wars against Parthia (AD 114–17; 161–64; 195–98; 216–17). Macrinius (217–18) actually paid 200 million sestertii to buy off the Parthians. Elaboration of the eastern defences under Septimius Severus (193–211) and Diocletian (284–305) must have been expensive, as was the refurbishing or construction of town walls. Public buildings were still given to various towns and, like defence, were paid for out of imperial revenues. Civil servants increased in number, especially as a result of Diocletian's reforms.[2]

Meanwhile, monetary inflation mounted during the late second century, and continued to be high until well into the fifth century, when it was finally 'squeezed out' of the system (Jones 1953b; Whittaker 1980).

The main burden of taxation fell on the peasants. They were also the chief victims of extraordinary levies. It can be argued, therefore, that as the land tax increased, so farms which were already economically marginal, perhaps because of physical conditions, must have been forced out of business (Jones 1966b, 308–9). Tenants and labourers simply abandoned the land. Some became brigands; in the fourth century others entered monasteries. Perhaps the majority found their way to the towns. Those people who remained on the land had to bear an even heavier burden. It was in these circumstances that neglected terraces would collapse.

To an extent, we have simply noted the symptoms of economic depression. There are signs that it was over by the middle of the fifth century. Small-villa estates suddenly emerged on the Belus massif in association with plantation-type olive cultivation (Tchalenko 1953). The large number of contemporary milestones from western Asia Minor indicates that the roads were being repaired and used once more. Public buildings in some of the 'twenty towns of Asia' were repaired and extended. Churches were built, colonnaded streets laid out and some residential quarters added (Foss 1972 and 1976). Yet the degree of recovery can be exaggerated. It was found principally in the administrative centres of the new provinces and it may be significant that some of the earliest signs come from Antioch-on-Orontes, military capital of the East (Downey 1961, 451–5, 500–3). Other towns may have stagnated. Even towns as large as Ephesus still contained large tracts of unoccupied and ruined buildings, ready quarries for building material. It is possible to envisage, therefore, that the economic crisis of the third and fourth centuries AD took the countryside to a low threshold, marked by reduced use of land, lowered production and depopulation. An upward spiral of recovery may have been difficult to initiate, let alone sustain. Except locally, little seems to have happened before wars with Sasanian Persia renewed the universal burden of taxation and produced devastation in some areas.

Harsh and arbitrary taxation has been used to explain the apparent shortages of labour which are evinced from time to time following the Arab conquests (Ashtor 1976a, 67–8, 159, 316). Taxes were needed to maintain an exceptionally luxurious court and to pay for troops. People fled because short-term military and political gain was won at the cost of long-term security and prosperity, by distant and alien lords whose 'fiefs' were allocated simply for their revenues and acquired through support for the powerful. Often the lords ignored their responsibilities for co-ordinating the farmers' efforts and ensuring that disputes were settled fairly. Repairs and maintenance essential to the functioning of irrigation systems were neglected. Supportive capacities were reduced. The effects were localised and discontinuous in time and space. Of wider and more lasting significance, however, were epidemics, and especially the Black Death (1347–51). These reduced the entire population of large areas (p. 219). Since draft animals, as well as people, were affected by bubonic plague, the surviving population would have found great difficulty in maintaining cultivation at the old levels. It is impossible to say whether the cultivated area in the western parts of the region really contracted more in the fourteenth than in the third or fourth century; there are just too many uncertainties. Whatever the scale, high mortality rates among young children and women would have limited replacement, while fertility is likely to have been reduced in response to continuing socio-economic difficulties. Reoccupation of the countryside and a corresponding re-advance in the cultivated area would have been retarded.

The spread of arbitrary taxation, partly associated with the use of tax-farming methods to raise revenue, has been blamed for settlement desertion (implying loss of labour) and abandonment of arable land (a consequence) in the Ottoman, Safavid and Qājār Empires during the seventeenth and eighteenth centuries. More evidence is available to examine the proposition from the Ottoman provinces. Although tithe and head-tax were the basic sources of revenue to the Empire, special exactions (*avariz*) were required from time to time to deal with emergencies. Rare at first, they became increasingly common once offices were opened to purchase. The development is a complex one to trace, but a start might be made with inflation. Price inflation appears to have been 'imported' into the Ottoman system through rising European demand for food and raw materials in the second half of the sixteenth century (Barkan 1975). Access to abundant American silver allowed European merchants to bid prices upwards. But the situation must have been worsened by a decade of poor harvests (1566–76) in central Anatolia, which must be traced back ultimately to the deterioration in the weather associated with the Little Ice Age (Griswold 1977). One response to rising prices was debasement of the currency, but this simply added to the mounting financial chaos.

During the same period, the expenses of the state increased enormously (Lewis 1958; Owen 1981, 12–14; Shaw 1976, 169–219). War was being fought on two fronts, in Europe and against Persia (1576–1639), while developments in warfare itself forced the Sultan to employ more full-time troops armed with muskets. The 'living-holders' of the classic cavalry army not only found themselves becoming redundant. They were also unable to meet their obligations from the revenues assigned to them. As titles came up for renewal, many 'livings' were not reallocated. Their revenues were diverted to the direct use of the state. Pressure from those who had benefited from price inflation — merchants and some courtiers — encouraged the government to let land to the highest bidder. In the same circumstances, the farming of tax-collection in the provinces became attractive. Both offered money in advance, the possibility of steady increases, and no state expenditure on collection. On the other hand, tax-farming is open to abuse. The collectors represent the government, but they are not under its immediate supervision. In the Ottoman Empire, control of major tax-farms was identical with provincial governorships. Exercise of their functions often required tax-farmers to use force, and this encouraged the development of private armies.

Local financial and military control promoted the emergence of autonomous polities within the Empire, a development which was further encouraged by the practice of alloting particular tax-farms for periods of several years, or even a lifetime. Frequent changes of personnel, as well as strife between the controllers of the various tax-farms, allowed the central government to play off one against another in a classic application of the divide-and-rule principle. The purchase of office meant the contraction of large loans which had to be repaid from the revenues collected. It was this necessity which converted the collection of extraordinary taxes into an oppressive and arbitrary routine. Abuse was exacerbated by the tax-farmers' abilities to acquire properties on their own account through purchase and foreclosure on loans to indebted peasants and sipahis. Direct interference with agricultural routine and land use followed in the search for increased revenues. Such a fusion of tax collection, indebtedness and direct control over land gave enormous power. While this could lead to increased production, as we have seen in the previous chapter (pp. 191–3, 199, 212), it must be conceded that the thoughtless and capricious use of concentrated power, backed by the sword, would produce the oppression which contemporaries saw as the principal cause of underproduction, land abandonment and depopulation.

Sir Paul Rycaut (1629–1700), one of the shrewdest western observers of the Ottoman Empire in which he lived for some 13 years, wrote as if he thought the system had been deliberately devised to subdue the spirit of the Sultan's subjects in order to make them 'more patiently suffer all kinds of injustice and violence that can be offered them, without thoughts or motion of Rebellion' (Rycaut 1686, 143). The *philosophe* historian,

Volney (1757–1830), who travelled in Egypt and Syria in 1782–85 and learnt Arabic, believed that governmental abuses ultimately derived from the licensing of unbridled authority by the Qur'ān (Volney 1788, vol. 2). Muslim observers would not have agreed, and Volney's anti-religious prejudices are clear throughout the rest of his account. One of the best-known Muslim commentators was Haci Halifa (1608–57), who worked in the Ottoman *Hazine-i Âmire Dairesi* (Treasury). He pointed out that the ruin of the countryside was well known, and that in his twelve years of travelling he had seen many deserted villages. The basic cause, Haci Halifa thought, was that the Ottoman Empire had moved into the decline phase of the cycle of growth, stasis and decline through which all states passed during their history. However, the immediate reason for the run-down of the countryside was extortion occasioned by the sale of offices. This had lowered the standards of loyalty and integrity required in the public service and produced the injustices so roundly condemned in the Qur'ān itself (Lewis 1962). In fact, one might suppose that the moral power of the Şeriat, together with enlightened self-interest, would probably act as powerful constraints on reckless self-seeking. All three of the commentators cited seem to imply that good government could repair the damage, though they would have disagreed on what its nature might be; Haci Halifa clearly thought a revival of the old Ottoman style of military dictatorship was necessary.

Another strand in the web of 'human greed, misgovernment and folly' remains to be unravelled. This is variously described as 'bedouinisation' and 'nomadisation', and it is a favourite element in most western explanations for the abandonment of cultivation in the Middle East, which was allegedly and mistakenly thought to be progressive during the whole of the Muslim era (Ashtor 1976a). One consequence of the early Muslim Arab successes was certainly an outward movement of pastoral nomads from Arabia, not only into Syria and Mesopotamia, but also into Egypt and the interior of Persia. Numbers are impossible to estimate. Although some of the nomads appear to have settled remarkably quickly and become cultivators, others continued to migrate with their flocks and herds, oblivious of the rights and needs of the cultivators whose livelihoods they at least disturbed and must frequently have destroyed. Despite the problems of distance and discontinuous occupation characteristic of the region, as well as the nature of 'desert power', a concerned government could perhaps have curtailed nomadic activity. The Umaiyads, however, chose to do nothing.

The first half of the tenth century brought a fresh influx of nomadic Arabs into Upper Mesopotamia and Syria (Ashtor 1976a, 158–9), possibly occasioned by the effects of drought on the steppe-grazing of Arabia. In combination with the effects of Byzantine attacks, the consequences — vividly described by Ibn Ḥawḳal, who came from the area — were neglected irrigation systems, more abandoned arable land and an increase in the number of deserted villages (Kramers and Wiet 1964, 1, 203–23). The spread of Türkmen tribes across the northern parts of the region some 100 years later similarly extended the nomadic domain at the expense of settled life in the slow, but progressive fashion already described for Asia Minor (p. 185).

When western travellers began to visit the region in any number, which was during the eighteenth century, they were forcibly struck by the ubiquitous presence of nomads — bedouin in Egypt and Syria, Türkmens in Asia Minor and a variety of different groups in Persia. Although Europeans always found nomads picturesque and later saw them as embodiments of the 'noble savage', they also regarded them as the principal causes of the perceived devastation of the region's agriculture. Travellers were told of the nomads' rapacity — extorting 'protection money', raiding villages and attacking caravans. It was not only the Muslim world which was shocked when the news broke that the Pilgrimage caravan had been virtually annihilated on its return journey to Damascus in September 1757 (Rafeq 1966, 213–22; Volney 1788, vol. 2, 87).

Expansion of the nomads' domain seems to have been real enough. Two examples may be cited. From the late eleventh century onwards, central Anatolia was controlled by Türkmens who made their winter quarters in the hills between the damp basins across which they ranged in summer (pp. 185, 191). Some tribes moved into lowland Cilicia (the Çukorova) during the seventeenth century and were joined by other groups during the next century, some apparently from as far away as western Anatolia. Patterns of transhumance were either developed or consolidated, so that summer pastures in the Taurus Mountains and on the nearer edge of the plateau were linked with winter grazings in the milder but swampy plains (Gould 1973, 25–6). On the steppes of Syria, the number of pastoral nomads may have increased during the seventeenth century, but it is from the mid-eighteenth century onwards that we have evidence to suggest that they were pressing into the cultivated areas across the desert margins of the south and east (Hourani 1957). Volney (1788, vol. 2, 6, 10, 275–6) was clear about the devastation which they wrought.

The increased nomadic pressure in eighteenth-century Syria is associated with two developments in Arabia itself. One was an obscure series of changes within the tribal pattern of the peninsula. This involved the northwards movement of the Shammar and 'Anaza confederations and the destruction of the old pattern of alliances on which the Ottomans had relied for the security of their desert borders (Bodman 1963, 8–13; Hourani 1957). The other development was the emergence of a Muslim fundamentalist movement in the small oasis towns and villages of Najd, in central Arabia. Its inspiration came from the teaching of Muḥammad ibn 'Abd al-Wahhāb (1703–92), but military command was under the house of Sā'ūd. In their efforts to spread a purified Islam, the Wahhābis pressurised the neighbouring tribes. They captured the Holy Cities of Mecca and Medina (1803, 1805), pushed into Lower Mesopotamia (Karbalā sacked 1802; Najaf attacked 1807) and even threatened Damascus (1810). The Saudi emirate was finally destroyed in its heartland by an Egyptian army (1818). Power passed to a bedouin state centred on Hā'il (Hitti 1964, 740–1; Holt 1966, 149–55).

While politics and religion encouraged pastoral nomads to spread out from their traditional dirās, the fundamental stimulus may well have been provided by shortage of grazing, following droughts. These were noted in Syria for 1733–34, 1747–48 and 1756. They may have been as extensive as those reported from Upper Mesopotamia in the 1770s (Alexandrescu-Dersca 1957; Rafeq 1966, 114, 176, 215). The nomads would have been forced to move their animals or lose their livelihood. They would follow lines of least resistance so that relatively stable and active governments were at least able to reduce their ravages, while in the west, forts and upland retarded their progress (Cohen 1973, 104–5). Lower Mesopotamia, however, lay wide open to nomadic incursion. There were few natural obstacles to in-movement from the west; cultivation had already contracted into scattered enclaves along the main rivers, leaving much land free as potential grazing; and Ottoman control was precarious because of distance from Istanbul and military pressure from Persia (Adams 1965, 107–11; Adams and Nissen 1972, 71–83). The Nile Delta was similarly exposed to nomadic infiltration. Bedouin tribes were reported to be in control of large areas there during the eighteenth century and, given his circumscribed position, there was little that the governor in Cairo could do about it, even if he had wanted (Shaw 1964).

## Discussion

The previous pages have shown that there is abundant evidence from the late Roman period onwards for the deleterious effects of human actions in bringing about the abandonment of cultivation and the desertion of settlements. Economic necessity may often have been the mainspring, driven by the costs of waging war, but sheer greed and short-sighted opportunism were, no doubt, powerful forces. Questions arise, however, as to how far 'human greed, misgovernment and folly' together form a sufficient and necessary explanation for the generally run-down appearance of Middle Eastern landscapes around 1800, and whether this was an accumulative result of long-term processes or the product of relatively recent activity. An answer must depend upon two considerations. The first is whether human actions, immediate or short-term in their incidence, could have long-term effects. Where elaborate irrigation systems were necessary to the cultivation of large areas and the support of high levels of population, any disruption of the system could have potentially long-term consequences. This was the case with the canal systems of Lower Mesopotamia and, on a humbler scale, with the qanāts of Persia. Both needed continuous maintenance, while large-scale canal systems required the exercise of power over large areas. Restoration was always difficult to organise and slow to implement. In the dry-farming districts, by contrast, the lightening of the tax burden and the improvement of security might have been sufficient, other things being equal, to bring about an extension of cultivation within a season or two. Alternating phases of expansion and contraction can be envisaged.

At this point, the second consideration comes into play. It is the 'impossibility of distinguishing between the work of man and that of nature, because human modes of expression are so variable and because adjacent regions are so diverse' (Huntington 1911, 261–2). The debate over the origins of the Younger Fill is a case in point. It is clear, though, that damper conditions in the period AD 400–700 and from 1300 onwards would have made it easier to extend cultivation on the drier margins of the region: on the Syrian steppe, in Upper Mesopotamia, central Anatolia and the north-western parts of Persia. There would be no need for pastoral nomadism to increase merely to ensure subsistence. The fact that pastoral nomadism did extend its domain, while cultivation retreated, suggests either that the dominant tendency in climate was towards greater aridity or that the human factor was decisive. The impact of climatic shifts would be felt most clearly in the arid fringes of the dry-farming areas. As Volney (1788, vol. 2, 344) observed, a year or two of drought there would be sufficient to force people out. Some would seek security in the towns; others would become pastoral nomads. They would probably not be willing to return to cultivation without some incentive in the form of market demand for their products and lower taxes. In these fringe areas, then, the role of the towns would be crucial, both as the centres and mediators of demand, as well as the seats of government authority. It could be a

Figure 9.2: Model of Man-Environment Interaction in the Middle East Over Time

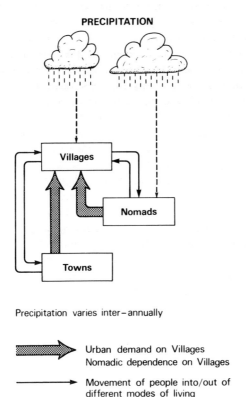

PRECIPITATION

Villages

Nomads

Towns

Precipitation varies inter–annually

Urban demand on Villages
Nomadic dependence on Villages

Movement of people into/out of
different modes of living

role both positive and negative. On the positive side, it would encourage cultivation in the neighbouring villages. On the negative side, the greed and oppression of landlords and government could force the abandonment of these villages (English 1966). Much would depend upon the character and attitudes of the urban elite. Those in turn would be affected, to some extent, by the ability of the farmers to meet demand. Weather conditions would obviously be an important constraint.

Such heuristic considerations allow the production of a model which integrates the interdependent relationships between town and country, cultivating villages and pastoral nomads, with the constraints imposed by precipitation conditions, whether inter-annual variations or a secular shift (Fig. 9.2). Shifts in either precipitation or urban population affect the rest of the system, bringing about movements of people and changes in both the style and the location of their way of life. Movements in and out of village-based cultivation are envisaged. Although primarily descriptive of the dry-farming margins, where a cycle of developments can be most easily envisaged, the model is plainly transferable to other parts of the region where relatively high levels of precipitation and low inter-annual variability meant that cultivation was less precarious. At the same time, the model does not underplay the significance of man's own decision-making in the face of abundant precipitation or drought, rapacious or benevolent officials. The model accommodates not only the contraction of cultivation, but also the expansion which was characteristic of the nineteenth century.

## Notes

1. Dols (1977, 197–8) gives 4–8 million as the population of Egypt. Poliak (1938) gives 1.2 million as the population of Syria.

2. A minimal number for the 32 provinces of the three eastern *dioceses* under the new system can be calculated at 4,135, on the basis of each provincial governor having a staff of 1,000 and each diocesan *vicar* one of 300 (Peters 1970, 607).

# 10 RECOVERY AND MODERNISATION

## Introduction

Nights are rarely, if ever, absolutely black. So 'desolation' was not uniformly characteristic of the region's landscapes around 1800, despite the romantic emphasis often bestowed by western travellers. Land was cultivated. Accessibility to the sea had induced a degree of cash cropping orientated towards export. Not all villages had been abandoned. Domes and minarets still floated over the urban skyline; religious buildings were repaired from time to time and new ones were erected. Towns retained their traditional functions, even behind crumbling façades. Moreover, the early nineteenth century saw not a deepening of the gloom, but rather the beginnings of a bright transformation initiated perhaps by a growing export trade, but certainly assisted by the re-establishment of central authority and expanding local populations. This chapter seeks to outline the landscape developments of the early revival period and to indicate how impulses given in the nineteenth century have continued to reverberate down to the late twentieth. The emphasis throughout is on the nineteenth century.

During the period, the Ottoman and Qājār states in the Middle East remained substantially intact, though territory was lost at the fringes, notably in the Caucasus, and western political intervention increased (Avery 1967; Kirk 1964; Shaw 1977). British involvement in the Gulf and south Arabia crystallised political developments there. Egypt became virtually independent, though, like Cyprus (acquired by Britain in 1878), owing tribute to the Sultan. Later it passed under British control (1882–1922). Also towards the end of the century, the stirrings of cultural and political nationalism laid the foundations for the independent Arab states which emerged from the collapse of the Ottoman Empire after the First World War. First, they had to pass through a phase either of tutelage under British and French mandates or of direct colonial rule under Britain. Israel emerged from the Palestine mandate in 1948. In central Arabia, a new Wahhābi state began to arise from 1902. British and Russian intervention, including partial occupation (1941–45), kept Persia/Iran divided and weak until, perhaps, as late as the 1950s. A Turkish republic emerged in Asia Minor between 1919 and 1923.

These events form the dramatic background to the mundane

233

geographical changes outlined in the following pages. Attention is given first to the countryside and then to the towns. Population growth, sedentarisation of the pastoral nomads and early industrialisation are taken up as the chapter proceeds. It begins, however, with a consideration of foreign trade and the development of communications.

## The Countryside

### Trade and Communications

Nineteenth-century European merchants continued the interests of their predecessors in the raw materials and foodstuffs of the Middle East. Export figures are both incomplete and suspect, but they indicate that the volumes of primary products leaving the region increased during the nineteenth century. The best-documented case is the export of long-staple cotton from Egypt. This rose from an annual average of 0.12 million *kantars* in 1821–25,* shortly after cultivation began, to an annual average of 6.72 million kantars in 1908–12 and 9.10 million kantars in 1939 (Mabro 1974, 14; Owen 1969). The fragmentary evidence from Syria and Lebanon reveals increases in the exports of raw silk (Owen 1981, 81–2, 154–8, 249–51), while even the poorer data from Persia are sufficient to show a general rise in the volume of exports (Issawi 1971, 130–4). Although the basic cause was undoubtedly an expansion in the world economy created by the industrialisation of north-western Europe and the US, political crises and shortages were important boosters to Middle Eastern trade since they often forced prices to levels sufficient to overcome the region's high transport costs. Thus, Egypt and coastal districts of Anatolia supplied cereals to British forces in the Mediterranean during the Revolutionary and Napoleonic Wars (1793–1815) and capitalised on post-war food shortages in Europe, while the Crimean War (1853–56), through interrupting grain supplies from Russia, encouraged their export from Syria too. The European 'Cotton Famine' created by the American Civil War (1861–65) not only raised cotton exports from Egypt to 2.0 million kantars in 1865, but also pushed those through Smyrna from 5–7,000 bales in 1860 to over 80,000 bales in 1865 (Owen 1981, 245). Even Persia was affected (Issawi 1971, 245). Anatolian, Lebanese and Persian silk exports were encouraged by a decline in French cocoon production in the late 1840s and 1850s caused by the loss of silkworms from muscardine disease. The booms passed, of course, but exports did not automatically sink to the original base levels. During the ninteenth century the Middle East

---

* 1 kantar = 44.9 kg.

became part of the integrated world-wide economic system which replaced structures of partial, imperfectly linked systems, and its exports responded relatively quickly to external demand. Movements in foreign demand were mediated to potential producing areas of the Middle East in a variety of ways.

The capacities and freight costs of ships were fundamental. Ships carried most of the region's exports at the beginning of the nineteenth century and remained largely unrivalled until the completion of the Trans-Caspian Railway in 1885 and the line from Vienna to Istanbul in 1888. The chief impact of the new, long-distance railways was on Persia, for they were instrumental in making Russia the country's dominant trading partner by the First World War. At sea, shipping rates fell steadily in the first half of the nineteenth century. Steamships were introduced to the region, first on the Mediterranean and Black Seas (1836), then on the Gulf (1838) and finally on the Caspian Sea (1846). Regular services were established in the 1840s on the Mediterranean, but not until the early 1860s elsewhere (Chevallier 1968). Although they carried little cargo at first, steamships assisted trade by improving the circulation of merchants, news and bullion. By the 1850s, however, they were carrying more cargo than sailing vessels in the Mediterranean and were making rapid advances in other seaways. Their great advantages were that they overcame the constraints of seasonal winds, particularly important in the Red Sea and the Indian Ocean (*Parliamentary Papers 1831–32*, 245), and they were generally faster than sailing ships. Reliability and speed not only reduced storage costs and the waste of perishable materials, but also mediated the demands of the world market more rapidly than before, thereby making it possible for land use to adapt within, literally, one or two growing seasons. The cutting of the Suez Canal (1859–69), which reduced the sailing distance from the head of the Gulf to Cape St Vincent by some 10,500 km, brought Lower Mesopotamia and southern Persia into direct contact with European markets and had a profound influence on the land-use patterns of the former area. Increased shipping and larger quantities of cargo demanded improvements to the major ports. These were gradually supplied with breakwaters, quays, open and closed basins, and warehouses — facilities which quite transformed the waterfronts, as well as increasing efficiency.

The response of the port hinterlands to the transmission of information about world demand and price levels was largely mediated by traditional caravans down to the 1850s, though surprisingly few improvements were made to the roads. For example, Pamuk (1978, 42–3) has shown that the most rapid expansion of exports from Smyrna took place before the construction of railways in its hinterlands. None the less, high transport costs effectively limited the range of traditional land carriage for bulky, low-value items like cereals to about 75 km, that is, much the same distance as in the early fourth century AD (Jones 1966b, 311–12; Kurmuş 1974; Quataert 1977). The main exception was provided by river navigation in Egypt, which was encouraged by Muḥammad 'Ali (1805–49) and enhanced by the re-excavation of the Maḥmudiya Canal (1819) to provide access from the western branch of the Nile direct to Alexandria (Owen 1981, 66).

Marketing systems probably had a dendritic structure (Kelley 1976; Vance 1970). Village 'surpluses' appear to have been collected in bulking centres (small towns) by local intermediaries before being consigned to European principals who, for the most part, remained in the ports (Chevallier 1970; Issawi 1970; Lafont and Rabino 1910). Intermediate centres (large towns) may also have been important in some marketing structures. The efficient operation of the networks was increased by the favourable trading conditions exacted from host governments through, for example, the Treaty of Turkmanchai between Persia and Russia (1828) and the Anglo-Turkish Commercial Convention (1838) (pp. 7–8, 263). Local monopolies were ended. Duties on goods transported and exported by foreigners were relaxed or abolished and in such a way as to discriminate against indigenous exporters. Consular systems were extended and commercial tribunals established (Hershlag 1964, 144; Issawi 1980, 75–88; Owen 1981, 88–91).

The inland reach of foreign marketing systems was increased from the 1850s. Steam navigation was established on the Tigris in 1855. This not only reduced the round trip from Baghdad to Basra from three or four weeks to about 6 days, but it also drew Mosul and its umland into the trading system of the Gulf (Owen 1981, 182–3). The Kārūn was opened to steamers as far as the rapids at Ahvāz in 1888 and, with a short porterage, subsequently up to Shushtar. The first railway was built in 1853, from Alexandria to Cairo. By the First World War Egypt had 4,600 km of track and a particularly dense network of standard and narrow-gauge lines in the Delta. The first railways in Anatolia were built for similar commercial purposes in the hinterland of Smyrna and the south-western corner of lowland Cilicia. Imperial and strategic considerations were more important in the building of the Anatolian Railway across the interior, and military requirements remained a prominent reason for the addition of another 3,042 km of track under the early Republic (1923–38). Similar motives influenced the construction of the Hijāz Railway (1900–08) through the western fringes of Syria and Arabia. Commercial reasons explain the penetration of the cereal-growing Hauran (1894), the tying together of the principal inland towns of Syria and the links through the fringing mountains to the coast. Although plans were made to construct railways in Persia from the 1850s onwards, British and Russian rivalry effectively prevented the construction of all but three short lines before the First World War. It was left to Reza Shāh (1921–41) to span the country with the Trans-Iranian Railway (1927–38), though its principal extensions to Mashhad and Tabrīz were not finally completed until 1956 and 1959 respectively (Fig. 10.1).

The effects of the railways can be illustrated from Anatolia. By the end of the 1860s, the Smyrna–Kasaba line had reportedly captured about 90 per cent of the caravan traffic moving down the Gediz Valley, while the Smyrna–Aydın line had taken over about 50 per cent of that in the Menderes Valleys (*Parliamentary Papers 1867–68*, 231). This was the consequence of competitive freight rates, and of the speed, regularity and security of the services offered. In addition, the caravan networks were reorientated towards agencies established in towns just off the railway line and fed to the nearest railway station (Kurmus 1974, 97–8). Finally, the 'pull' of Smyrna was extended to the edge of the central plateau and, with the extensions of the lines to Alasehir in the north (1872) and Civril in the south (1888), well beyond it. Here, at the end of the nineteenth century, the rival companies encountered competition from the Anatolian Railway, which had also forced reductions in freight rates on parallel caravan routes and reshaped much of the system into its own feeder network (Quataert 1977). However, the constraints of time-related costs for caravan carriage prevented the Anatolian Railway's influence from penetrating central and eastern districts much beyond 100 km of the railheads. The agriculture of large areas of central and eastern Anatolia was barely affected by western commercialisation, and such export outlets as it required still lay in the ports of the nearer Mediterranean and Black Seas, rather than on the Straits (Quataert 1977). Central and eastern Persia were even more remote and unaffected since they lacked all access to modern means of transport, while low capacity and expensive caravans remained viable until after the First World War. Arabia was almost completely untouched until even later; it produced little of commercial interest.

The constraint of high, time-related freight rates was finally reduced with the construction of roads for wheeled, and ultimately motor vehicles. An early and economically significant development was the Beirut–Daṃascus carriage road (1859–63), with its southward extensions into Mt Lebanon (from 1867). These developments, followed by the Damascus Railway, helped to promote Beirut into the most important port in the

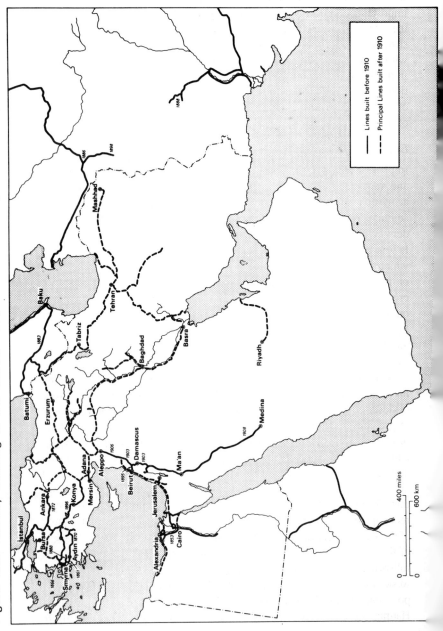

Figure 10.1: Railway Building in the Middle East, AD 1853–c.1910, with Subsequent Extensions

eastern Mediterranean by the end of the nineteenth century. However, with this exception, significant development of arterial roads and, more important, of road networks was delayed until the 1890s and even until the 1950s. In the 1890s, for example, British interests promoted road construction in Egypt (Owen 1981, 222), while the Russians linked Tehrān with the Caspian Sea at Enzeli (Bandar Pahlavi) (Issawi 1971, 157, 200). In the 1950s indigenous governments sought to promote national integration, as well as economic development, by often ambitious programmes of road-building in which graded tracks linked villages to the new systems of all-weather highways.

*Extension of Cultivation*

Improved access to the coast, together with increasing exports of food and raw materials, might indicate that greater-than-ever success was achieved in extracting 'surplus' from the countryside during the nineteenth century. While the proposition cannot be dismissed entirely, independent evidence suggests that increased productivity was more important. It was achieved not by significant improvements in farming methods, which scarcely began to change in most districts before the middle years of the twentieth century, but by extensions to the cultivated area and increases in the amount of land devoted to cash crops. This section will outline developments first in the major dry-farming sub-regions of Greater Syria and Anatolia and then in the principal areas of irrigation, Egypt and Lower Mesopotamia. A brief discussion of Persia will separate the two sets of descriptions.

Much of the potential arable land east of the Aleppo–Damascus road may have lain uncultivated about 1800 (Lewis 1955), though the shells of deserted villages indicated that some of it at least had been worked in previous generations. About half of the villages in the fertile district of Hauran were reported either to be deserted completely or to be occupied only occasionally by a shifting population. Reoccupation and cultivation were first noticed by western observers during the Egyptian occupation (1831–40) (Fig. 10.2). Consul Werry, for example, believed that 170 previously abandoned villages in the vicinity of Aleppo were reoccupied then (Polk 1962), though if Eton (1798, 267) is right, some of them may not have been deserted for very long. British consular reports of the 1840s and 1850s suggest that something of a reversion occurred following the withdrawal of the Egyptian army, which was forced by the western Powers. Only 35–40 per cent of the villages reoccupied near Aleppo remained inhabited by 1845 (Polk 1962), while of 100 villages in Hauran, 75 were deserted in 1842–43 and more than 50 still in 1849 (Ma'oz

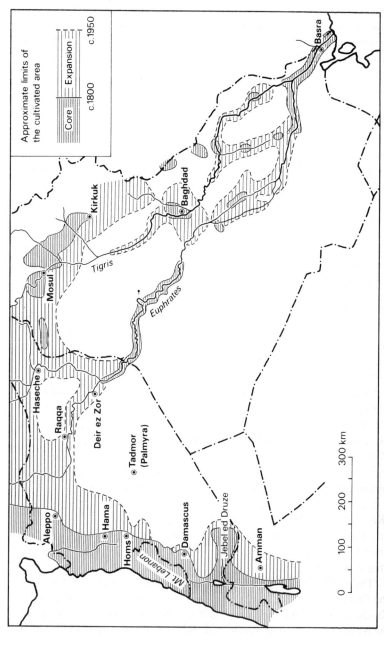

Figure 10.2: Expansion of the Cultivated Area in the Syrian Steppe and Mesopotamia after c.AD 1840

Sources: Lewis (1955) and Wirth (1962).

1968, 164–5). None the less, the advance was resumed. By 1860 the frontier of cultivation had reached about half-way from Aleppo to the Euphrates. Around 1900 it had touched the river and was edging downstream towards Raqqa (Lewis 1955). Drought in the years 1868–74 may have retarded the advance, and some emigration was reported from villages north of Damascus (Huntington 1911, 352–4). The early 1890s and much of the 1900s, however, received precipitation significantly greater than the mean (Huntington 1911, 297–8; *World Weather Records*, vol. 79, 413), and this must have greatly assisted the colonisation process by ensuring successful harvests. Most of the advance was achieved, after all, by dry-farming methods. Irrigation played a comparatively minor part in Syrian agriculture until after the Second World War. The country entered the 'big dam' era in 1973 with the completion of a 60-m-high dam at Tabqa on the Euphrates, holding up a lake some 80 km long (Beaumont, Blake and Wagstaff 1976, 358–61; Zuckermann 1971). Cereals were the major crops grown in the nineteenth century, generally on a biennial rotation, but fluctuations in world prices encouraged the growth of cotton from time to time as, for example, between Aleppo and the coast in the 1880s and again in the 1900s.

The process of colonisation, which was replicated east of the Jordan (Lewis 1955), seems to have followed a fairly standard pattern. A group of families, frequently of refugee origin or from an existing 'congested' mountain community, moved into an area of the plains. Sometimes they were sponsored by enterprising landlords, including the Sultan, and often they were encouraged by immunities from taxation and conscription for a period. The group may have established a completely new village, but the chances are that they reoccupied a deserted one. Wells would be cleared and, in some cases, old subterranean irrigation canals would also be refurbished (Lewis 1949). Communal ownership of the land (*mush'a*) (Owen 1981, 256–9) may have been established down to, and perhaps in spite of, the Ottoman Land Code of 1858, and with it the periodic reallocation of properties was introduced. A circle of ploughed land, with a diameter of perhaps 12 km, was soon established around the village (Kelman 1908, 175). Once that apparently critical limit had been reached, daughter settlements might be founded, if there was still sufficient unoccupied land left and provided the initial community continued to increase in population.

The frontier of cultivation crossed the Euphrates during the 1920s when the French mandate authorities began to pacify the area and refugees — Kurds, Assyrian and Armenian Christians — streamed southwards from the repression of the new Turkish Republic. More than one million ha had been brought into cultivation in the Syrian Jezira and perhaps as many as 2,000 villages established by 1951 (Lewis 1955). This involved the re-settlement of the Khābūr Valley, whose agriculture had enjoyed such a high reputation in the early Middle Ages. Enterprising merchants from western Syria were frequently involved, and early use was made of machinery to compensate for labour shortages (Warriner 1962, 91–2).

Within the boundaries of modern Lebanon, cereal cultivation expanded in the Beq'a Valley down to the 1880s, but the most notable changes in land use arose from the rise and fall of the silk industry. After a phase of stagnation or decline in the late eighteenth century, silk production began to increase again during the Egyptian occupation. It was supported by a considerable expansion in the area devoted to mulberry plantations, achieved in part by the conversion of arable land previously used for other crops and partly through colonisation. New villages were established. Demands for both fuel and land depleted the surviving remnants of the historic forests. A struggle over land developed in the 1850s between the largely Druze chiefs and their Maronite Christian tenants. It culminated in community violence (1858–60) and massacre (Owen 1981, 155–67). Silk production peaked in the 1900s, but falling prices and profits forced the industry to contract even before the First World War virtually destroyed it. The post-war revival was short-lived. Lebanese silk production was eliminated in the 1920s through a combination of the industry's own in-efficiencies and poor-quality products, on the one hand, and the cheapness of better-quality silk from China and Japan, on the other hand, as well as shifts in women's fashions and the emergence of the first artificial fibres (Owen 1981, 249–53). In an effort to adjust, the virtual monoculture of the late nineteenth century was gradually replaced by a diversity of tree crops in which apples were prominent (Jones 1963), though much of the more marginal land began to go out of cultivation as families left the villages, either to settle in the growing coastal towns (chiefly Beirut) or to emigrate.

As C.G. Smith (1975) has observed, the notion that Palestine was transformed solely by Jewish colonisation is a myth. The first moder-ately successful Jewish agricultural settlements were not established until the 1880s, while the extension of cultivation was apparent more than 40 years earlier in the plain of Esdraelon and southwards from Jerusalem (Owen 1981, 79). Further expansion took place in the 1850s and late 1870s, and by 1886 it could be claimed, for example, that 'almost every acre' of the plain of Esdraelon was in the 'highest state of cultivation' (Oliphant 1886, 58). Two aspects of the general expansion were significant. One was the southward advance into areas of high inter-annual variabilities in precipitation (> 30 per cent). Figures from Jerusalem suggest that this must have been helped by precipitation levels in the 1850s and 1860s which were sporadically above the mean for 1846–1954 of 560 mm and significantly so throughout the 1880s (Rosenan 1955). The other development was the spread of orange-growing in the coastal plain, especially in the vicinity of Jaffa. It began after the Crimean War and was largely organised by Greeks involved in coastal shipping (Schölch 1981). Originally exported to Egypt and Asia Minor, oranges were shipped to Europe in increasing quantities after 1875. This development was encouraged by the discovery that a local variety was sufficiently large and thick-skinned to be able to stand a long sea voyage without losing quality. Exports grew, and around 1910 some 3,000 ha of oranges were being cultivated near Jaffa (Owen 1981, 178, 265). In most districts the extension of cultivation was initiated from existing Arab villages, often located in the hills. They established daughter settlements in neighbouring, but uncultivated areas (Amiran 1953). From the 1880s they were joined by growing numbers of Jews (pp. 255–6). Within 15 years of the foundation of the state of Israel (1948), the cultivated area within the cease-fire lines and the *de facto* frontiers had increased by over 400 per cent, chiefly through the careful use of irrigation water. Shortages during the First World War forced Jewish communities to retreat from a reliance on citrus, cereals or vines and adopt forms of mixed farming.

In Anatolia, sudden surges in world commodity prices brought erratic and episodic increases in the exports of one or more of six agricultural products: cereals, cotton, dried fruits, opium, silk and tobacco. Before the 1890s these items came from a limited number of naturally favoured coastal districts, where export-orientated agriculture was already well established in the eighteenth century (p. 199). Perhaps the widest diversity of cash crops was found in the valleys of the Gediz, Büyük and Küçük Menderes, tributary to Smyrna. The price rises for cotton in the 1860s coincided with a fall in transport costs brought about by the extension of the railways, with the result that the amount of cotton grown increased and uncultivated land was brought under the plough. The amount of cotton produced subsequently fluctuated, but land remained in cultivation and from the 1860s there was a marked change-over from long fallows of six to ten years to short ones of one or two years' duration (Kurmuş 1974, 35, 150–1). The neighbouring district centred on Bursa was already a major silk-producing area in 1800. Production increased during the 1850s, and this was supported by planting much of the wide zone between the town and the Sea of Marmara with mulberries. By the end of the nineteenth century, agriculture in the area was considered to be

'advanced' by European observers (Cuinet 1894, vol. 4, 54–64). Traditional cotton-growing in lowland Cilicia (the Çukorova) was boosted during the 'Cotton Famine'. Though production fell back when American
cotton returned to the market, the district continued to respond to market
influences (Quataert 1980). Cultivation expanded rapidly in the 1860s
and 1870s, particularly to the south and west of Adana, where a considerable amount of drainage was undertaken. In 1868–69 alone 78 new
villages were established (Gould 1973, 145–6). By 1875 cultivation was
expanding across the upper plain as well (Gould 1973, 193–6). In the opening decade of the twentieth century, the amount of land under cultivation
in the Çukorova was increasing by about 5 per cent per annum (Quataert
1981). On the eve of the First World War, agriculture in the area was among
the most commercially orientated in the Ottoman Empire, well advanced
in the use of machinery and, because of chronic labour shortages, drawing in 50–100,000 migrant workers, some allegedly from as far afield
as Bitlis, Harput (Elâzığ) and even Mosul (Kemal 1963; Pamuk 1982, 46).

Although the interior of Anatolia had long supplied a limited range
of high-value commodities, notably Angora wool, it was really opened
up to commercial forces with the settling of refugees in the 1870s and
then the construction of the Anatolian Railway in the 1890s. Somewhat
surprisingly, in view of contemporary developments in North America,
the response came mainly from cereals. These normally constituted 50–75
per cent of all freight carried by the railway during the 1890s and covered
as much as 80 per cent or more of the cropped land in 1909–10 (Quataert
1977). While the statistical data alone do not reveal whether increased
exports resulted from a diversion of the normal harvest or from an extension in the arable area, it is clear from evidence about the process
of settlement development that significant advances in the amount of land
cultivated were involved.

The evidence comes from the series of interconnecting basins which
form the Great Konya basin in the south-central part of Anatolia and the
furthest limit of the Anatolian Railway (Konya reached 1895) (Hütteroth
1968 and 1974). Down to the 1870s there were comparatively few permanent settlements in the area. These were mainly along the mountain edges
of the basins. The hills and mountains between them were dotted with
the winter quarters (*kışla*) of nomads, some of them containing permanent structures and inhabited by sizeable seasonal populations. They were
linked to summer grazing stations (yaylas) out on the damp and exposed
basin floors. During the 1870s the basins began to be studded with the
agricultural settlements of refugees (*muhacır*) arranged generally in zones
roughly parallel with the fringing mountains. Individual villages of refugee

## Figure 10.3: Model of Settlement Patterns Development in Central Anatolia after c.AD 1870

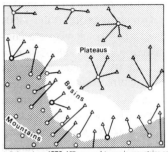

I. Before about 1870. Villages and towns in mountains and at mountain edge (circles) have yaylas (triangles) in the basins; Kişla villages of nomads and semi-nomads on the plateaus have numerous yaylas roundabout. (Yaylas in the mountains have not been shown in this map)

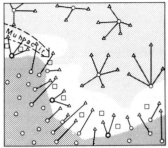

II. (a) Since 1870-1880. New villages of Muhacir (squares) develop in between indigenous yaylas (triangles) in the basins. The zone of agriculture and permanent settlement advances into the basins.

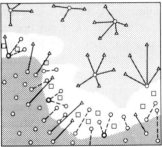

II. (b) 1870-1880. Yaylas in the basins begin to develop into çiftliks, hamlets and finally into independant villages. Connections to parent villages are gradually disrupted.

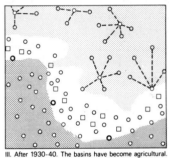

III. After 1930-40. The basins have become agricultural. The yaylas of Kişla villages on the plateaus become independant, too. Nearly all of them have developed into permanent settlements, but are still registered as quarters (mahalle)'of their parent villages.

Sources: Hütteroth (1968 and 1974).

origin are recognisable now by preserved or partially decayed grid-iron plans, as well as a pattern of long, strip fields surrounding them and evidently laid out at the time of foundation as a way of sharing the land equally between all families in the community. The encroachment on their traditional grazing land forced the indigenous pastoralists to make more use of their yaylas in order to maintain title to the land, and this began to involve some cultivation (Fig. 10.3). The long run of precipitation figures for Istanbul (Erinç and Bener 1961) hint that the expansion of agriculture was probably assisted by damper-than-average conditions during the late 1870s and much of the 1880s. The cultivated areas around the former yayla and refugee settlements increased during the 1890s, despite a succession of four poor or bad harvests (Quataert 1977) which, in the absence of precipitation data from either the sub-region itself or from Istanbul, were probably produced by drought, since the centre of Anatolia is notorious for its inter-annual variability in precipitation (Erinç 1950; Tümertekin 1955). Expansion at that time was probably stimulated by the building of the Anatolian Railway.

The extension of the cultivated area continued in Anatolia after the War of Liberation (1919–22). The process culminated after the Second World War with large increases in the cultivated area of republican Turkey, especially during the early 1950s when annual increases exceeded 5 per cent in all but one of the seven years, 1949–56 (*Ziraî İstatistik Özetleri 1941–62*). They were achieved, chiefly in central Anatolia, by ploughing up still more of the former grazing land and reducing fallow periods. Encouragement came from government pricing policy, but the increases were made possible by, on the one hand, favourable precipitation conditions and, on the other hand, by a rise in the number of tractors. Unfortunately, the expansion of cultivation involved ploughing up land of marginal fertility and yields began to fall — disastrously when precipitation oscillated unfavourably in the late 1950s. Much surviving woodland was cleared in the hills and more accessible parts of the mountains, partly for construction timber, partly to provide more farmland, but mainly for domestic fuel to supply an increasing population. The removal of vegetation and the opening up of former grazing land increased run-off and advanced erosion in the upper parts of river catchments, while creating disastrous floods in some low-lying areas. One of the solutions attempted was a programme of dam construction aimed principally at regulating stream flows, but also at providing water for irrigation.

Irrigation has been of considerable historical importance in Persia, but a significant proportion of its agriculture has been dependent upon direct precipitation (Hershlag 1964, 134–5, 194–207). The sub-region's mountain frame, long distances and consequently high transport costs restricted the expansion of agricultural land and confined commercially orientated farming to a limited range of relatively high-value products. A rapid rise in exports during the years 1834–64 suggests that this was a period in which cultivation was extended. Silk production increased in the 1850s in response to French difficulties (p. 234), but was itself hit by the same problems in the following decade. Its fortunes varied thereafter, with government attempts made in the twentieth century to foster its revival. Since about 80 per cent of the production came from Gilan by 1910 (Lafont and Rabino 1910, quoted Gilbar 1978), this is where an expansion in mulberry-planting must have taken place. However, the district was also an important producer of rice, output of which increased by nearly three times between 1865 and 1913, partly to meet domestic demand and compensate for a decline in silk production, but partly for sale to Russia (Gilbar 1978). Cotton also became important, mainly in response to Russian industrialisation in the late nineteenth century, though some stimulus was given by the earlier 'Cotton Famine' (Gilbar 1978 and 1979). Accordingly, the jungular vegetation of Gilan was dotted with an increasing number of agricultural clearings by the end of the century (Curzon 1892, vol. 1, 361), and its unique pattern of rather dispersed settlements had probably emerged. Tobacco production, which was already widespread in Persia by about 1800 (p. 212), increased in the 1870s and 1880s, mainly in response to demand from the Ottoman Empire, where Persian pipe tobaccos were much appreciated (Issawi 1971, 247–52). Opium was first exported in appreciable quantities in 1853 from the area of Esfahān. As demand rose, first for export to China and then also to Europe, so poppy cultivation spread northwards and eastwards into areas where physical conditions were not really suitable. High prices forced the abandonment of cereal cultivation in some districts. This was probably a factor in the severity of the 1871–72 famine, though the basic cause was five years of low precipitation in several parts of the country (Gilbar 1976 and 1978; St John 1876). Adverse precipitation conditions, coupled with the world recession, halted the expansion of cultivation in the 1870s, but it seems to have resumed as commodity prices recovered after 1900. As elsewhere in the Middle East, the nineteenth century advances seem to have been made in piecemeal fashion as local people responded to world demand. It is likely that abandoned villages were re-occupied and that much of the extension in cultivated area was achieved by cleaning out and repairing ancient qanāts (English 1966). In the late 1950s, however, government began to take a hand. A number of dams were built to control flooding and extend cultivation. Attempts were made to revivify the agriculture of Khuzestan, which had been so successful in ancient and early medieval times (pp. 148–50, 179). The device chosen was that of the high-powered, capitalistic agrobusiness. Four companies were established and took up 68,000 of the 100,000 ha made available for irrigation by the Pahlavi Dam (1972), but the results were disappointing and the scheme was clearly in trouble by 1976 (Beaumont, Blake and Wagstaff 1976, 467–9; Weinbaum 1977).

In the neighbouring plains of the Tigris and Euphrates, the extension of cultivation in the ninteenth century was achieved largely by traditional means and was mainly directed to supply wheat and barley (Hasan 1958 and 1970). The elaborate irrigation systems of antiquity had long since collapsed (pp. 179, 191, 216–18), and water was supplied by blocking 'natural' channels with temporary dams of reed matting and brushwood to divert water into canals or to flood the land directly (Fernea 1970, 39–41). Various Ottoman governors in the 1850s and 1860s attempted to improve the system by organising the clearance of old canals and cutting some new ones, but without much success. Except in the traditional date-growing belt along the Shatt al'Arab, land use was both extensive and shifting, while output per unit area was low. The system had developed these characteristics for two principal reasons. The first was to accommodate nomadism and cope with salinity, while maintaining fertility by long fallowing (Fernea 1970, 38–47). The second was a response to the vagaries of water distributaries. By the early nineteenth century, the Euphrates flowed mainly down an easterly channel. This was the result of avulsion, encouraged by the silting-up produced by small-scale, piecemeal irrigation systems. During the 1860s, however, the western channel began to receive more of the flow and to reassert its earlier importance to local irrigation. The process was assisted in various ways. To prevent the flooding of Baghdad, the Saqlawiyya escape channel, through which excess water had flowed from the Euphrates to the Tigris, was blocked. In consequence, the eastern channel of the Euphrates became overloaded and it overflowed into the western branch, gradually cutting a new channel for itself; thus the lower sections of the eastern channel were deprived of water (Ionides 1937, 67, 74–5). In 1869 the Daghara Canal was blocked in preparation for a military attack by Ottoman forces on the tribes. Later still, a new canal was cut from the eastern branch of the Euphrates to irrigate land to the south of Baghdad. People moved to where irrigation water could be secured, principally westwards, and violent struggles for land and water resulted. During the late nineteenth century, barley became more important than wheat, in some measure because of shifts in world prices but also in response to creeping salination.

The major changes of the 1920s and 1930s in Iraq were heralded by an attempt to regulate the flow of the Euphrates and again feed water to the eastern branch. This involved the construction of a barrage near Hindīyah (1891). Unfortunately, reduced flow in the eastern channel and the normal process of constructing temporary dams simply increased the silting. When the barrage broke in 1903, water reverted to the western channel and about 240 km of the eastern branch became dry. Within three years many of the settlements along it had been abandoned (Cadoux 1906). Efforts at restoration were made with the reconstruction of the barrage in 1911–14 and its repair in 1925. After the First World War the new authorities in the country began to implement an integrated plan for flood control and irrigation development which had originally been presented to the Ottoman authorities in 1905. Another barrage was constructed on the Euphrates, this time at Habbānīyah (1934–39), to protect Lower Mesopotamia from flooding by diverting flood water into a large depression where it would be stored for summer irrigation. A permanent weir was erected on the Diyālā (1927–28) to feed six irrigation canals. While ancient canals were subjected to piecemeal restoration and numerous regulators were constructed, a striking development was the increased use of motor pumps, from 143 with 1,500 hp in the whole of Iraq in 1921, to 2,500 with 90,000 hp in 1939; most of the increase took place in central areas (Burns 1951; Lebon 1955). Though cultivation spread further and further away from the main water channels, greater and more frequent applications of moderately saline water increased soil salinity, and the problem grew to major proportions.

The Tigris was relatively neglected at this stage as a potential source of irrigation water. Its flow was considered too unpredictable and too prone to sudden rises for it to be harnessed successfully. The first barrage — at al-Kūt — was not erected until 1943. It fed water to the Dujaylah Canal and thus allowed the settling of some previously deserted sections of the plain. Another barrage was added at Sāmarrā (1956) so that water could be diverted into the Tharthār depression to reduce flooding and allow further extensions of irrigation. Flood control and irrigation provision have continued and by 1952, 5,322,000 ha (including fallow) were cultivated throughout Iraq, compared with perhaps 1,470,000 ha in 1913 (*Middle East and North Africa* 1960).

Unlike Iraq, Egypt seems to have retained a good deal of its historic irrigation system down to the beginning of the nineteenth century (Hamdan 1961; Richards 1978). However, the immense increase in exports of long-staple cotton noted at the beginning of the chapter (p. 234) was achieved, and partly accompanied, by radical changes in the agricultural system of the country. Cotton is a summer crop, unlike the previous staples of Egyptian agriculture, and its areal expansion was accompanied on peasant farms by change to a system of biennial rotation involving maize, to feed the family, and *bersim* (clover), to feed livestock. More important still was the increase in the cultivated area and the gradual transformation of the irrigation system from one dependent upon the seasonal flooding of basins to one consisting of canals and distributaries providing water all the year round.

The biggest increase in the cultivated area (Table 10.1) took place in

Table 10.1: Areas Cultivated and Cropped in Egypt, 1821–1939

| Year | Area cultivated (thousand *feddan*) | Areal increase per annum (%) | Area cropped (thousand *feddan*) |
|---|---|---|---|
| 1821 | 2,032 | | |
| 1835 | 3,500 | 5.2 | |
| 1840 | 3,856 | 2.0 | |
| 1852 | 4,160 | 1.1 | |
| 1862 | 4,053 | −0.3 | |
| 1877 | 4,742 | 1.1 | |
| 1882 | 4,758 | 0.02 | |
| 1890 | 4,941 | 0.2 | |
| 1897 | 5,043 | 1.4 | 6,764 |
| 1902 | 5,335 | 1.2 | 7,429 |
| 1907 | 5,403 | 0.2 | 7,662 |
| 1912 | 5,285 | −0.04 | 7,681 |
| 1917 | 5,269 | −0.07 | 7,677 |
| 1920 | 5,305 | 0.2 | 7,807 |
| 1925 | 5,420 | 0.4 | 8,123 |
| 1930 | 5,549 | 0.5 | 8,634 |
| 1935 | 5,229 | −1.1 | 8.054 |
| 1939 | 5,338 | 0.5 | 8,522 |

Note: 1 *feddan* = 4,200.8 m².
Source: Mabro 1974, 9, 14.

the 1820s and 1830s. It was largely achieved by reclamation in the Delta and the use of conventional irrigation canals to distribute the low flow of summer, ponded up by simple weirs. Sāqiyas were employed to water the fields. Extensive use was made of forced labour down to the gradual abolition of the *corvée* system in the 1880s. The efficiency of the Delta system was improved in 1843–61 by the construction of barrages where the natural distributaries diverged below Cairo. Further improvements were made in 1890. The barrages raised the water level to feed three arterial canals, which in turn supplied an elaborate network of subordinate canals and channels. This allowed greater use of 'flow' instead of 'lift' irrigation. As reclamation spread, new estate villages (*izbas*) emerged, consisting of tenant houses grouped around a store and the house of the landowner or his overseer (Owen 1981, 146–8).

A second phase of expansion in the cultivated area took place in the 1870s. It followed construction of expensive feeder canals alongside the sides of the valley from take-off points higher up river. Very often, though, they supplied insufficient water in summer to affect a change of land-use patterns. None the less, the introduction of perennial irrigation encouraged the expansion of Egypt's cropped area by allowing the possibility of growing rather more than two crops in the year. In 1897, the first year for which there are reliable figures, the cropped area was actually 35 per cent greater than the area cultivated. By 1912 it was 45 per cent greater and 60 per cent in 1939.

The final transformation of the rural landscapes of Egypt began with the construction at the first solid constriction in the river channel, at the First Cataract, of the Aswān Dam. That was in 1902, but increasing demand for water required the raising of the dam in 1912 and again in 1933. The course of the Nile was rapidly transformed from a series of seasonal swamps created by the northward movement of the flood into a dry valley, lined with hundreds of canals, taking water from an embanked and controlled river (Hamdan 1961). Basin irrigation survived only in the remotest reaches of Upper Egypt. But not all was gain. Protein intake fell as the area under beans declined. Bilharzia spread in the population from long hours spent working in water throughout the year. Soil fertility began to decline since shorter fallow meant a reduction in the beneficial effects of heating and baking the soil, a process which aerated it, broke up colloids and encouraged the growth of nitrifying bacteria (Richards 1978). Water-logging developed as a result of the raised water table and, together with the ending of natural flushing, encouraged the spread of salination. The land-use problems became really noticeable around the end of the century, when a striking fall in cotton yields took place, especially in the apparently more fertile central areas of the Delta. They were eventually dealt with by applying artificial fertilisers and by constructing a drainage system carefully interdigitated with the irrigation network.

The era of major change was rounded off by the construction of the Aswān High Dam (1960–68) (Benedick 1979; Field 1973; Holz 1968; Little 1965; Nour el Din 1968; Owen 1964). It is 111 m high and holds back a lake 500 km long. Whatever its unfortunate hydrological and ecological effects (George 1972; Kassas 1972; Schalie 1974), it allowed the reclamation of more arable land, particularly in 1962–66 when 555,000 feddans (233,100 ha) were recovered, mainly in the Valley. Some 850,000 feddans (357,000 ha) were converted to perennial irrigation, a net gain to multi-cropping of 490,000 feddans (205,800 ha). Meanwhile, land use in Upper Egypt has changed from the cotton which became dominant in the nineteenth century, to the more suitable sugar-cane — previously cultivated in the Middle Ages. In parts of the Delta wheat has given way to rice, and winter maize to the more productive summer maize (Mabro 1974, 83–106). Paralleling these changes has been interest in reclamation away from the Delta, notably in Tahrir ('Liberation') province west of the Delta, and in the string of oases in the Western Desert, where artesian water has been tapped at depths of 3,600 m (Beaumont, Blake and

Wagstaff 1976, 477–9). The incentive is the need to feed a rapidly expanding population.

*Population Increase* (Fig. 10.4)

So far, the expansion of the cultivated area and changes in cropping patterns have been discussed as if they were simple responses to the state of the world market. The chronology of developments in central Anatolia, however, has shown that this was not the only factor (pp. 244–6). Other elements in the causal nexus were important. They include population growth and various measures taken by government. There was a complex interplay between them all and the different components have been separated simply to ease discussion.

Population increase is of particular consequence, since both Hicks (1969) and Boserup (1970) have suggested that mounting population pressure is a potent influence on land-use change. Population statistics for the Middle East in the nineteenth century have to be drawn from a variety of sources which are less than accurate. Those for the Ottoman provinces are derived from official censuses taken to assist the administrative and military purposes of the state, though they were rarely published in detail (Karpat 1978). No such political arithmetic was used in Persia, and the first attempt at a census in modern times was not made until 1956. Population estimates are therefore less reliable for that sub-region.

Graphing of the available data, such as they are, facilitates comparisons and the picking out of trends (Fig. 10.4) (Gerber 1979; Owen 1981, 217, 224; Shaw and Shaw 1977, 117). Population generally mounted throughout the nineteenth and twentieth centuries, except where it was locally cut back by a combination of war losses, community massacre, and emigration (as in the case of Anatolia in 1913–22), or decimated by famine (Persia 1869–72, 1895–1905) and epidemic (as in Persia with cholera in 1867–72). Rates of increase clearly rose in the second half of the nineteenth century, though the precise timing varies across the different sub-regions. The increases exceeded 1.0 per cent per annum in Egypt and Anatolia throughout the nineteenth century, climbing to more than 2 per cent per annum at the end of the century in the former and from the first quarter of the twentieth century in the latter. In Iraq and Greater Syria (including Palestine but excluding Lebanon) a rate of 1 per cent per annum was not exceeded until the middle of the nineteenth century. The total population of the Ottoman provinces was about 12.7 million in 1850 and about 32.4 million by 1900. Gilbar (1976 and 1979) has postulated an annual rate of increase for Persia in the 1850s of just over 0.5 per cent — a relatively

Figure 10.4: Population Change in Different Sub-regions of the Middle East, AD 1820–1970

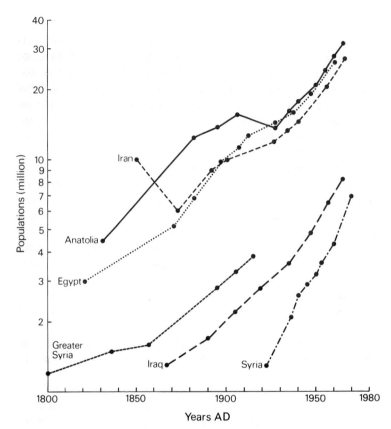

low rate due to the effects of famine and cholera, and it is doubtful if growth exceeded 1 per cent until the first quarter of the twentieth century. The resulting curve would fit well with a sluggish advance in the cultivated area of Persia and a slow, but highly differentiated spatial pattern of commercialisation in agriculture (pp. 246–7). In the rest of the Middle East, population growth may have lagged behind land-use change.

The increases in population can be attributed to falls in the death rate and to immigration, especially to the Ottoman Empire. Falling death rates were achieved by changes in the rhythm of disease. While plague had been a frequent visitor to the region from at least the fifth century AD, it gradually disappeared; for example, there was no outbreak in

Syria–Palestine after c.1830 (Gerber 1979). Why this should have happened is rather mysterious. It might be related to the receding of serious famine (except in Persia), and thus of the movement of contaminated grain, but the reduction in the number and frequency of direct contacts with India may possibly have been significant. The diffusion of other contagious diseases, however, was restricted by the establishment of quarantine stations. Meanwhile, the incidence of malaria was reduced through draining the breeding grounds of the *Anopholes* capable of propagating it and by extending cultivation. Sedentarisation of the nomads (pp. 258–60) may have reduced mortality rates for both women and children, while stronger governments gradually reduced feuding and local wars, with their shameful toll of male deaths.

Although the loss of territory in the Caucasus provoked a movement of refugees southwards into Persia, the major stream of immigration flowed to the Ottoman provinces. The Ottoman Empire had a long tradition of receiving individuals and groups seeking economic and social betterment, as well as security. The hospitality of its central provinces was severely strained, however, with the progressive loss of territory around the periphery of the Black Sea and in the Balkans. Particularly large numbers of refugees (*muhacılır*) came after the Crimean War (1853–56) and the Turko-Russian War of 1877–78, since both resulted in the persecution of Muslim Tartars and Cerkes (Circassians) living in the Crimea and Caucasus respectively, and their transportation to Siberia and territories further east. Over 1 million Tartars are reported to have reached the Empire between 1854 and 1870. Many of them were settled in southern and central Anatolia, as well as Syria. Something like 100,000 Cerkes were settled in Syria in 1855–60 and 600,000 in Anatolia. More than 1 million refugees arrived between 1876 and 1895, of whom about a quarter were Cerkes and the rest a mixture of Tartars and Muslims from Slav and Greek territories (Gould 1973, 63; Shaw and Shaw 1977, 115–17). The last major wave of refugees came in 1922, when about 500,000 Muslims left Greece and Bulgaria in exchange for the 1.5 million Orthodox Christians who went to Greece (Pentzopoulos 1962). Under the terms of the *Muhacırın Kanunnamesi* (Refugee Code) of 1857, refugees were given state land and were exempted from taxation and conscription for up to twelve years. Although many refugees died from disease and others reverted to nomadism, this device was effective in bringing land into cultivation, as in central Anatolia (pp. 244–6), and in containing some of the indigenous nomadic groups, notably in Syria. But there was a dark side. Some of the Muslim refugees began to take vengeance on the non-Muslim populations near whom they were settled. This fostered the growth of religious intolerance which, on the whole, had been absent from the Ottoman domains. It may have contributed to the bitterness of the 'Armenian Question' when this emerged towards the end of the century, resulting in forced migration for the Christians and massacres by both sides — though the situation was also provoked by the terrorist tactics of Armenian nationalists instigated and supported by Russia (Shaw and Shaw 1977, 200–5, 314–17). Subsequently, Muslim refugees were able to occupy villages and lands deserted by both Armenian and Greek Christians.

Despite the smaller numbers involved, the immigration of Jews to Palestine is much better known than that of Tartars and Çerkes. It is better known, too, than the settling of other groups in Palestine. For example, a small group of American millenarians made largely unsuccessful attempts to settle in the Promised Land during the 1850s and 1860s (Kark 1983), while from 1867 members of the *Tempelgesellschaft* (Association of Templars), an offshoot of the German pietist movement, began to establish the seven villages which they occupied on the eve of the First World War (Carmel 1975). Jewish immigration is important, however, since it led to the foundation of a state and made distinctive contributions to landscape development (Beaumont, Blake and Wagstaff 1976, 404–25; Gilbert 1974).

Pious Jews had drifted back to Palestine in small numbers ever since the Diaspora of the first century AD, but the first waves of immigrants came in the period 1883–1903, largely as a result of pogroms in Russia. The newcomers founded 20 settlements. These resembled the cereal-growing villages of central Europe, but they were not an economic success until rescued by Baron Edmond de Rothschild and reorganised along other, more profitable lines such as viticulture. The Ottoman authorities were generally hostile to these developments because the immigrants maintained their status as foreign nationals, despite the ownership of land. Arab opposition began to emerge about 1886, when the first attacks on Jewish settlements took place. Another 40,000 Jews were added in 1904–14 to the original immigrant total of about 25,000. They included a number of young idealists inspired by the Zionist vision of restoring Jewish self-esteem through a return to the land, farm work and the creation of a just society. The first *kibbutz*, a village community run on collectivist lines, was founded in 1910. Most of the other early settlements, however, did not have socialist ideals. They were more economically orientated. Most were found in the coastal plain near Jaffa, but a few were established in eastern Galilee. The Balfour Declaration (1917) and the securing of the Palestine mandate by Britain encouraged further Jewish immigration.

About 30,000 arrved in 1919–31. The first *moshov*, in which the rigours of collectivist life were relaxed, was established in 1921. By the end of the next decade, there were 175,000 Jews in Palestine, equivalent to about 17 per cent of the total population. They were settled not only in the coastal plain and Galilee, but also in the Hula basin and the plain of Esdraelon.

The rise of Nazism in the 1930s brought more immigrants and raised the Jewish population to about 457,000, or 20 per cent of the population. It is no wonder that the Arabs felt threatened and turned upon the newcomers. About 100 new settlements were established. Many of them were located for political reasons in an attempt to disperse the Jewish presence and to consolidate their hold so that, in the face of proposals from successive commissions of enquiry to divide Palestine into Jewish and Arab sectors, they would be able to sustain the Zionist claim to the whole of the land promised to Abraham's seed (Genesis 12:7; 15:1–21; 17:1–9). Despite British restrictions on further immigration, imposed to allay Arab fears and to prevent population exceeding what was believed to be the carrying capacity of the country, 110,000 Jews succeeded in entering Palestine between 1940 and May 1948. About 20 *moshavim* and 60 *kibbutizm* were established. They were clearly located to establish the Jewish presence and to provide the territorial foundation for an independent state. This appeared, almost fully formed as an organisation, in May 1948. The Law of Return, enacted by the Israeli parliament in 1950, provided that 'every Jew shall be entitled to come to Israel as an immigrant'. About 600,000 had arrived from Europe alone by 1970, and more than 20,000 from India. Israel's 'war of independence', however, meant the flight of 700–800,000 Arabs (about half the Arab population) and the virtual end of the historic Jewish communities in the Arab countries of the Middle East and north Africa.

Although more than 80 per cent of Israel's population is urban (Blake 1972), the Jewish agricultural settlements have had a considerable effect on the extension of cultivation and the appearance of the countryside (p. 243). Their success is technologically based, but it is perhaps exaggerated by the comparatively small scale of the country and the effectiveness of Israeli propaganda.

## Government Measures

After a phase of near-parity with European armies, Ottoman and Qājār forces began to show organisational and technical weaknesses in the late eighteenth century. Recognition of the problem led to modernisation along European lines. Large sums of money were required, especially as the rapid increase in the sophistication of weapons over the second half of the nineteenth century meant imports rather than local manufacture. The Ottoman and Qājār responses were much the same as those of the autonomous ruler of Egypt, Muḥammad 'Ali.

An important element in their programmes was an attempt to raise the taxable capacity of the population through the promotion of agriculture. A variety of measures was adopted. The Ottoman government, for example, tried to promote cotton cultivation during the 1860s by making seed available. At the end of the century, it sought to improve the quality of silk produced by importing eggs from Japan. It established an Agricultural Bank (1888) which developed an extensive branch network designed to make credit available to farmers (Quataert 1975). Finally, the Ottoman government tried to reassert the state's historic rights over the land, while at the same time seeking to give security of title to the cultivators. These were the aims of the *Arazi Kanunnamesi* (Land Code) of 1858, amended in 1867 (Fisher 1919; İnalcik 1955; Tute 1927). It misfired. The reasons were the rising value of land occasioned by increasing commercialisation of agriculture, the established power structure of the rural areas, and peasant mistrust of the authorities. In consequence, much land was registered in the name of locally powerful individuals, whose ability to dictate land use was thereby enhanced.

Much more effective in encouraging the expansion of cultivation were measures not directly associated with agriculture. They included the effective centralisation of government and the improvement of provincial administration. In addition, security was improved in two ways. First, powerful local rulers were destroyed, as in 1837–47 when Ottoman forces pacified Turkish Kurdistan, and in 1865–67 when they established central authority in the Çukorova (Gould 1973, 89–110). Second, operations were mounted against nomadic tribes which finally broke their political power and curtailed their depradations (pp. 258–60).

Despite these measures, the three major governments of the region were unable to raise all the income they needed from local resources. From the middle years of the nineteenth century they embarked on a programme of foreign borrowing. Much of the consequent investment was unproductive. It created mounting debt, bankruptcy and ultimately foreign financial control; in the case of Egypt, bankruptcy led to British occupation (1882–1922). British influence in Egypt, first as financial controller then as protector, encouraged agricultural development both through state assistance to irrigation and by insisting, to the detriment of industrial growth, that the natural advantages of the country in raising revenue lay with cash cropping. In territories still under direct Ottoman rule, the Ottoman Debt Administration (from 1881) exercised considerable control over agriculture and land use through its control over the revenues of several provinces and its monopoly on the silk tithe (Blaisdell 1929). Its offshoot, the *Régie Cointéressée de Tabacs de l'Empire Ottoman* (1883), promoted tobacco cultivation as did a similar, but less successful, body operating in Iran (1890–92) (Issawi 1971, 247–51). Since independence, the various national governments have played a much more direct role in agricultural development, but, like their predecessors, they have tended to under-invest in it when compared with the amount sunk in infrastructure and expended on promoting industry. In addition, they and their forebears encouraged the sedentarisation of pastoral nomads.

## Sedentarisation of Pastoral Nomads

Amongst the most striking differences between the countryside of today and that of the early nineteenth century are the virtual absence of pastoral nomads and the contraction of grazing land. Since about 1840 governments have sought to sedentarise the nomads with greater vigour and persistence than at any earlier time. Without realising the ecological basis for the nomads' seasonal migrations, governments have sought two objectives. First, they have tried to remove the challenge to their authority represented by autonomous communities used to bearing arms and easily deployed from daily life into warfare. Second, governments have attempted to increase security for the cultivator by removing the disruption caused, on the one hand, by the nomad's delight in raiding — whether to collect more animals or to ensure the payment of 'protection money' — and, on the other hand, by his disregard for the standing crops of military inferiors. A third objective has sometimes been evinced, most notably in Arabia under the Sā'ūd family. That is, the elimination of a way of life which has been seen as incompatible with man's overriding religious obligations. To this can be linked an expressed concern for the nomads' welfare, especially following the pauperisation of many camel herders with the development of motor transport.

Force has been met by force. Governments sent their armies against the pastoral nomads. In Syria, for example, military expeditions were mounted against the bedouin during the 1840s. They were rarely successful at the time, since insufficient troops were deployed, while the government lacked any firm purpose. The nomads simply withdrew into the desert, and they matched the soldiers in arms and fighting ability. More successful expeditions were mounted in the 1860s and 1870s, when the soldiers possessed superior firepower. Real headway began to be made in containing the nomads with the fortification of the desert frontier, beginning at Deir ez-Zor (1862) and extending, first, to the north-west and, then, after the building of the Hijāz Railway, towards the south (Ma'oz 1968, 49–50, 134–5). In the 1920s Reza Shāh used his forces to curtail or prevent nomadic migrations. Nomads were transferred to distant territories. They were forced to settle in villages built specifically for them. Everywhere since the First World War, the availability of the aeroplane has meant that pastoral nomads have been unable to outrange government forces or simply 'disappear'.

As well as being compelled to settle, pastoral nomads have given up their migrations for other, sometimes more pressing reasons. Many nomads have lacked any ideological commitment to wandering or to nomadism which, by transcending economic necessity and advantage, might have held them to their ancient ways (Bates 1972). Farming was often one of their activities, and they have been prepared to concentrate on it rather than herding when threatened with loss of land which they regarded as theirs or when enticed by government distributions. Opportunities for paid work in construction or in manufacturing have proved attractive and the consequent disengagement of many active men has had profound effects. Herds have been reduced to sizes which can be handled by women, children and elderly men, and the temptation has been to give up bothering and live on wages. Absence of the men meant that decisions could not be taken as to when and where to move. Smaller herd size, in any case, reduced the need to migrate very far and, if the animals were given up, there was no need to move at all. Skills have been lost and, when a whole tribe has given up its livestock, it is very difficult to rebuild the herds (Birks 1978). Loss of livestock through bad weather or disease has also been important in causing sedentarisation by reducing herd size below that necessary to support a family (pp. 65–8). While demand for camels probably increased in the late nineteenth century so that more goods could be carried, the advent of the motor vehicle in the 1920s gradually destroyed much of the long-established market for camel-breeders. Finally, the extension of cultivation meant the loss of grazing and the reduction of fallows. Difficulties arose over access to what grazing remained and rents became high, while the expansion of such crops as cotton meant little edible residue. In consequence, pastoral nomadism has become difficult to pursue.

Just how many people still follow this way of life is hard to determine. All the estimates, however, point to a progressive decline in the second half of the nineteenth century which continued into the following one. For example, there may have been 3.0–3.5 million pastoral nomads in Persia c.1850 (30–33 per cent of the total population), but these had fallen to perhaps 2.0–2.5 million by the end of the century (Gilbar 1976).

In Iraq, there may have been about 450,000 in 1867 (35 per cent of the total population), but 433,000 by 1890 (25 per cent) and 250,000 (5 per cent) in 1947 (Hasan 1958).

The results of sedentarisation were striking. On the one hand, those people who turned to farming created their own villages or reoccupied abandoned ones, where they were free from seasonal hunger. In Adana province, 19 per cent of the villages were created by former nomads and in some districts the proportion rose to 93 per cent (Gould 1973, 176–81). 'Bedouin agriculture' in the western Beersheba basin was carried out from perhaps ten temporarily inhabited hamlets of widely spaced houses, replicating nomadic encampments, during the 1940s. The number increased to 73 by 1953 and they had become permanent (Amiran 1953; Amiran and Ben-Arieh 1963; Marx 1966). Those nomads who attempted to struggle on with their old way of life often found that their animals died simply because they were not adapted to the particular local environment and circumscribed range. Diseases spread more rapidly among the people themselves, largely through inadequate shifts of camp site and consequent lack of hygiene. As already indicated, migrations were restricted. In the mountains, the main problems were created by the agricultural cycle of the villages whose territory had to be crossed and with which 'way leaves' had to be negotiated. In the steppe of Syria and Arabia, movement became centred upon secure water points, sometimes supplied by the government. In all areas, overgrazing resulted as the nomads strove to satisfy the rising demands of towns for urban animal products.

## Towns and Industry

Published information about urban population in the nineteenth century is sparse and unreliable. Analyses of growth have been limited in scope and the results are difficult to compare directly because of different starting assumptions and statistical definitions. None the less, Issawi (1969) attempted the task. He concluded that growth rates were generally low. This is not confirmed by detailed investigations. Use of Gerber's (1979) estimates for Beirut gives an annual rate of increase for 1831–1905 of 0.9 per cent, but it was possibly as high as 13.3 per cent in 1831–36, when Beirut was being transformed from a 'third-rate Arab town to a flourishing commercial city' (Moore 1835, quoted Issawi 1977). These compare with an annual rate of 1.8 per cent which can be calculated for Baghdad during 1837–1900, with a peak of 2.4 per cent in 1853–77 (Jones 1969; Mantran 1962). Damascus may have grown at a similar overall rate in 1836–1915, but at 2.4 per cent in 1853–77. Aleppo grew at the still higher rate of 2.1 per cent from 1837 to 1905, but even that was probably exceeded in 1892–1905 (Gerber 1979), and Sykes (1915, 298) was struck by how quickly the gardens which had once surrounded the city had been transformed into new living quarters. These figures compare with rates of less than 1 per cent for towns in England and Wales during the second half of the eighteenth century, when industrialisation was getting underway, and a peak of 1.8 per cent per annum in 1811–21 (Lawton 1978).

Overall, the increases of population in Middle Eastern towns are perhaps to be explained, like those in the late twentieth century, by rates of natural increase higher than in the countryside (Clarke 1972). In the Ottoman provinces, however, rural-urban migration was probably important. Gilbar (1976) claimed that, by contrast, there was no evidence for substantial immigration to Persian towns in the mid-nineteenth century.

Published estimates show that, if settlements of 2,000 or more population are accepted as towns, then the urban population may have exceeded 20 per cent in Syria–Palestine, Anatolia and even Mesopotamia for much of the nineteenth century, and was steadily increasing (Issawi 1969). Persia perhaps lagged; in the 1850s, 8 or 9 per cent of the population may have lived in towns of 10,000 or more and by 1900 this may have more than doubled (Bharier 1972; Gilbar 1976). Across the region, growing urbanisation was partly the result of the population increases in a few large centres, perhaps the ports in particular (Issawi 1969).

Evidence from Syria–Palestine, Egypt and Anatolia, however, demonstrates also that small towns proliferated, and a similar development has been postulated for Persia (Baer 1968; Bharier 1972; Erder and Faroqhi 1980; Gerber 1979). Most of the new towns grew from established communities, but a few were either completely new foundations or else arose from shifts out of older, less convenient sites. Their origins will be discussed below. Whether it is appropriate to think of 'over-urbanisation' (Ibrahim 1975) in the nineteenth century is debatable, but it was certainly incipient in Egypt, where more than 11 per cent of the population were already living in towns of 20,000 or more inhabitants as early as 1846 (Baer 1968).

As population grew, built-up areas expanded, though seldom by more than a 0.5–0.75 km radius during the nineteenth century (Wirth 1966 and 1968). The direction of growth was often dictated by the roads approaching the town or, in the cases where there was one, by the location of the railway station. New towns and new suburbs were characterised by wide, straight streets capable of taking the wheeled vehicles which reappeared in the region during the nineteenth century. The streets intersected at right angles or radiated from public gardens. The most comprehensive new developments were perhaps in the Ismāʿīliyah suburb of Cairo, laid out on Housmann-type lines in 1867–69 (Abu-Lughod 1971, 104–13). Gas street-lighting began to appear in the larger towns during the 1860s. European-style hotels were built, and occasionally theatres. Government offices, in styles reminiscent of public architecture in France and Germany, dominated townscapes perhaps even more than the new mosques and bazaars. In the old-established towns of the region, the historic core was completely encased by the new suburbs and gradually became, in areal terms, a relatively minor part of the urban sprawl. It is not clear, however, whether at this stage shanty-towns developed in wedges of undeveloped land between the centre and the new suburbs. The fortifications sometimes disappeared. They were certainly neglected. Large old houses were subdivided and left to decay. Streets became congested as wheeled vehicles attempted to negotiate their awkward corners. The result was the cutting of the first wide avenues through the old quarters, as in Cairo under Muḥammad ʿAli and in Baghdad during the governorship of Midhat Paşa (1869–72). While the historic bazaars usually survived and were even rebuilt, the distribution of activities within them frequently changed over time (Schweizer 1972).

Urban functions remained much as in earlier centuries, but it was their expansion which provided additional work and thus lay behind urban growth and increasing urbanisation. Administrative functions were enlarged by reforming and interventionist governments. Lower-order administrative centres were created. The existing towns remained nuclei for collecting and forwarding agricultural produce, but as activity increased, so new bulking centres emerged. Hinterlands of the major towns grew and became more integrated as commercial agriculture spread and transport improved (Beckett 1966; Ehlers 1975; Kurmuş 1974). The towns, in addition, continued to be centres for the distribution of foreign products. Unlike the region's primary exports, which required bulking, foreign products were capable of being broken down into very small consignments and were thus able to penetrate far inland. Processing and manufacturing continued. A high degree of spatial differentiation within individual towns was apparent where several separate jobs were involved, as in carpet-making (Costello 1973). This might extend out to neighbouring villages and involve a putting-out system (p. 265).

During the nineteenth century, Middle Eastern industry began to change. There were various stages. At the beginning of the century, western manufacturers, especially in textiles, affected a considerable penetration of the region's markets. Exports to the Ottoman provinces, including Egypt, from Britain alone rose from an annual average of £500,000 in the mid-1820s to £1.5 million in 1836–39 and £2.5 million in 1845–49 (Owen 1981, 93). It was estimated in 1842 that enough British cloth was being imported to supply 4 yd (3.7 m) to every person in the region (*Parliamentary Papers 1844*, 125).

Penetration was assisted by the favourable duties on foreign imports and the abolition of internal duties secured by the Treaty of Turkmanchai between Persia and Russia (1828) and especially the Anglo–Turkish Commercial Convention (1838) and the Anglo–Persian Commercial Treaty (1841) (Issawi 1966, 38–40; Issawi 1971, 72–3). Their terms were extended to other nations and, though modified, they remained in force until at least the end of the century. The consequences for local industry are unclear; European visitors were often struck by the quantity of European goods displayed in the bazaars (Gilbar 1979; Issawi 1971, 72–3). Foreign commentators on the Ottoman Empire, including British consuls, were impressed by reports, or possibly their own observations, of closed workshops. They concluded that local artisan industry, especially textiles, was in decline (e.g. Ubicini 1856, 341–4; Urquhart 1833, 141–4). This conclusion is not unequivocally supported by successive estimates of the number of workshops producing in various towns during the course of the nineteenth century. It is clear that workshops did survive, though numbers appear to have fluctuated according to the local circumstances of trade. The resilience of craft producers can be explained partly by the protective effects of distance and high transport costs so that, for example, the remoter areas of eastern Anatolia were less affected by European competition than coastal areas, while the whole of Persia may have escaped better than the Ottoman Empire. Owen (1981, 93–5) has suggested, however, that factors other than distance have to be taken into account. The expansion of the local market through population increase is one. Another is the difficulty which European textile manufacturers experienced in copying Middle Eastern patterns and styles, and there is some evidence from late nineteenth-century Syria of indigenous manipulation of local fashion (Owen 1981, 261–3). For their part, Middle Eastern artisans could readily copy European fabrics and finish the cloth according to local taste. Finally, Middle Eastern textile manufacturers could use cheap, high-quality but imported yarn to make their cloths.

Although artisan production survived, the lack of tariff protection against foreign competition reduced the scale of development in modern, western-style industry with its characteristics of relatively large factories and inanimate power (Issawi 1971, 76; Sarç 1941). This was reinforced by the subordination of manufacturers to local merchants. The restrictive practices of the guilds, even though their power was in decline in the

Ottoman Empire at least from the 1850s, worked in the same direction. The loss of export markets was also important. There was a chronic short-age of personnel familiar with the use of modern machinery, especially in maintenance and engineering. A further disadvantage was often the expense of fuel, much of which had to be imported. Hard coal was mined in the Ereğli district on the Black Sea from 1829 and, despite vicissitudes, the Anatolian coal and lignite mines were able to supply most of the penin-sula's needs of about 1 million t per annum on the eve of the First World War (Issawi 1980, 290–1). This was exceptional. Elsewhere coal produc-tion was limited to a few primitive pits in the Elburz Mountains and in Lebanon. These were mainly geared to domestic consumption (Bowring 1840, 20; Issawi 1971, 282 n.4). In any case, the Middle East's coal reserves are not very great (0.27 per cent of the world's total). The region's power problems only came near to a solution with the discovery in the late nineteenth century of how to generate and transmit electricity.

Despite the difficulties, some success was achieved in establishing modern manufacturing industries, partly by foreign capitalists but also by some local entrepreneurs. Success rested upon one, or some com-bination, of three factors. First, there was the need to process various agricultural products at source and prior to export. Second, cheap raw materials and — a mixed blessing — cheap labour offered the possibility of developing simple transformation industries, where a degree of indirect protection was offered by relatively high transport costs to the region. Third, some specialist luxury goods, notably carpets, were in high demand abroad and could still compete in foreign markets.

Processing became particularly well developed and widespread for cotton and silk. Cotton gins and presses proliferated in such major pro-ducing areas as the Nile Delta, the Aegean valleys of Anatolia, and the Çukorova. Silk-reeling and -spinning using modern machinery developed in and around Bursa from 1838 and throughout Mt Lebanon. The 1850s were a period of particular expansion. Lebanon contained about 57 small factories by 1867, and about 100 by 1880. Although the Lebanese plants were often founded by local entrepreneurs who were able to guarantee supplies of cocoons and labour, they were subservient to French capital and French commercial interest. Consequently, they survived while prices for silk remained high and so long as they had a more or less certain share of the French market, but they were decimated at the end of the century when the basic conditions changed (Owen 1981, 157–9, 251–3).

A census of manufacturing taken in 1915 for the province of Istanbul and the towns of Smyrna (İzmir), Magnesia (Manisa), Uşak, Bursa, Karamüsel and Panderma (Bandırma) showed that 70.3 per cent of the plants covered were in food processing and 11.9 per cent in consumer goods, chiefly textiles. Most plants were small, hardly more than workshops (Issawi 1980, 310; Ökcün 1970). The picture was probably much the same in other parts of the region, though the units may have been smaller. Steam-powered corn mills became widespread in the Ottoman provinces during the nineteenth century, while powered olive pressing and soap-making developed in the Mediterranean fringes. Sugar refining expanded and adopted some modern methods in both Egypt and Persia. By the outbreak of the First World War, there were at least 5 modern cotton spinning mills in Smyrna and 2 in Adana, while woollen mills in Smyrna, Uşak, Panderma and Bursa were producing coarser yarns and cloths for the lower end of the market neglected by foreign importers (Kurmuş 1974, 210–14). Attempts to establish other types of consumer industries during the nineteenth century were generally short-lived, both in the Ottoman Empire and in Persia. Altogether, though, there was sufficient use of, and demand for, modern machinery to stimulate the beginnings of an engineering industry in the Ottoman Empire. For example, in the Smyrna district there were 6 engineering firms in the mid-1880s, some of them capable of building steam engines (Kurmuş 1974, 210–14).

The development of succcessful export-orientated consumer industries was virtually confined to carpets. In Persia, production was promoted and co-ordinated by foreign firms once expansion began in the 1870s under the stimulus of outside demand. British entrepreneurs, for example, controlled clusters of carpet-making villages around Sultanabad (Arāk) and in Azerbaijan, but made carpets on their own account with local labour in Hamadan and Tabrīz (Housego 1973; Issawi 1971, 301–5). Capitalist transformation was even more marked in the hinterland of Smyrna. British merchants first entered the carpet business there in the 1860s and six concerns dominated the industry by the 1880s. Until that date, the industry was organised on a putting-out basis. Woollen yarn was bought from peasants, distributed to dyers (chiefly in Smyrna), and then the coloured yarn was given out to village carpet-makers by brokers, with instructions as to size and quantity. Completed carpets were subsequently collected from village headmen. A challenge to the British monopoly in the early twentieth century modified the whole system. Spinning and dyeing became concentrated in modern establishments. Agencies with salaried staff were established in 14 towns to organise the distribution of yarn and the collection of carpets. The final stage of development, however, was to concentrate the looms, still hand-operated, into a few factories (Fig. 10.5). These developments restored much of the British monopoly (Kurmuş 1974, 179–86).

Figure 10.5: Operations of 'Oriental Carpet Manufacturers Ltd', c.1913

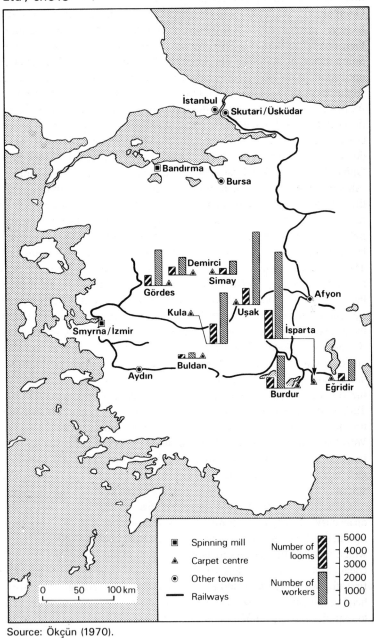

Source: Ökçün (1970).

Industrial development even in the nineteenth century was not left entirely to private entrepreneurs. The politico-military concerns of both the Ottoman and Qājār states led them, like Muḥammad 'Ali, to establish factories to supply munitions, uniforms and other equipment needed by their remodelled armies. Most were located in or near the capital cities. Some of the plants established by the Ottomans in the 1840s survived until another boost was given to state-backed industrialisation under Abdulhamit (1876–1909), and a few are still in production. Most failed. As in Persia, where a more tentative programme of state industrialisation was launched in the 1850s and 1860s, skilled manpower was scarce, costs were very high, poor communications created bottlenecks in production and there were several disasters at individual plants (Clark 1974; Gilbar 1979; Hershlag 1964, 134–54; Issawi 1971, 292–7). Muḥammad 'Ali's programme was much more ambitious. Instead of supplying just the politico-military complex, he aimed at import substitution over a wide range of products. At the peak of this programme in the 1830s, 30 textile mills, for example, were at work in different parts of the country, powered by human and animal muscle as well as seven or eight steam engines. The experiment gradually ran down during the next decade; machinery became dilapidated, motive power was lost in the cattle murrain of 1843; the chief market contracted when the army was reduced in size; and the extension of the Anglo–Turkish Commercial Convention to Egypt removed the protection necessary to nascent industries of the type being promoted (Owen 1981, 69–76). The revival of experiments in import substitution and of attempts to establish basic heavy industry, notably iron and steel, had to wait until the twentieth century. The first moves were made with the removal of disadvantagous trading agreements by Iran (1928) and Turkey (1929) and the raising of protective tariffs by Egypt. Other countries followed suit as they secured political independence after the Second World War. Industrialisation was subsequently aided in the case of Iran, Iraq and Saudi Arabia by revenues from the exploitation of oil and natural gas, as well as from the later use of these directly as fuel and then feed stocks.

Oil seepages were known in the region in ancient times and commented on as curiosities by many travellers in subsequent periods. They became significant during the world-wide search for exploitable petroleum sources which followed the successful exploitation of oil in Pennsylvania in 1859 (Beaumont, Blake and Wagstaff 1976, 268–94; Issawi 1971, 311–34). The first discoveries in the Middle East were actually made in 1869, at the head of the Gulf of Suez, while under Midhat Paşa attempts were made to exploit seepages in Mesopotamia. Oil in commercial quantities was

Table 10.2: Middle East Oil Production as a Percentage of World Output

| 1913 | 0.6 |
|------|------|
| 1920 | 1.9 |
| 1930 | 3.4 |
| 1946 | 9.1 |
| 1950 | 16.9 |
| 1960 | 23.8 |
| 1970 | 31.8 |
| 1978 | 36.0 |
| 1979 | 35.2 |
| 1980 | 32.0 |
| 1981 | 29.1 |

Sources: Beaumont, Blake and Wagstaff (1976) and UN (1983b).

struck, almost as the explorers were about to give up, in 1909 at Majd-i Sulaiman in Khuzestan. A pipeline was constructed to Abādān and the first shipment of oil was made in 1912. This propelled Iran to the position of the largest producer in the region, a rank it retained until the 1950s. The pipeline, as well as the storage facilities, large refinery and company town built at Abādān, foreshadowed industrial landscapes elsewhere in the oil-producing areas. Meanwhile, intense interest developed in the oil shows along the rest of the geosyncline stretching from the Gulf through Mesopotamia, along the western edge of the Zagros Mountains. With it came extreme political rivalry. Control of petroleum resources grew vitally important as the use of oil developed from lighting to fuels capable of raising steam and powering the new internal combustion engines (Hoffman 1927, 67–72; Mejcher 1970; Williams 1968). As the Middle East's contribution to world oil supplies increased (Table 10.2), the way was prepared for a reassertion of the perennial importance of the Gulf as one of the world's commercial arteries. World politicians began to reassess the geopolitical significance of the Middle East. It is a process which has continued.

# BIBLIOGRAPHY

Abrahamian, E. (1974) 'Oriental despotism: the case of Qajar Iran', *International Journal of Middle East Studies* 5: 3–31

Abu-Lughod, J.L. (1971) *Cairo: 1001 Years of the City Victorious*, Princeton

Adams, R.M. (1958) 'Survey of ancient watercourses and settlements in central Iraq', *Sumer* 14: 101–4

—— (1962) 'Agriculture and urban life in early southwestern Iran', *Science* 136: 209–22

—— (1965) *Land Behind Baghdad. A History of Settlement on the Diyālā Plains*, Chicago and London

—— (1972) 'Patterns of urbanisation in early southern Mesopotamia' in P.J. Ucko, R. Tringham and G.W. Dimbleby (eds.), *Man, Settlement and Urbanism*, London, 735–49

—— (1974) 'Historic patterns of Mesopotamiam Irrigation agriculture' in T.W. Downing and M. Gibson (eds.), *Irrigation's Impact on Society*, Anthropological Papers of the University of Arizona 25, Tucson, 1–6

—— and H.J. Nissen (1972) *The Uruk Countryside. The Natural Setting of Urban Societies*, Chicago and London

Ahmed, L. (1978) *Edward W. Lane, A Study of his Life and Works and of British Ideas of the Middle East in the Nineteenth Century*, London and New York

Ahrweiler, H. (1962) 'L'Asie Mineure et les invasions arabes (VIIe–IXe siècles)', *Revue Historique* 227: 1–32

Albright, W.F. (1963) *The Archaeology of Palestine*, Harmondsworth

Alexandrescu-Dersca, M.M. (1957) 'Contribution à l'étude de l'approvisionnement en blé de Constantinople au XVIIIe siècle', *Studia et Acta Orientalia* 1: 13–37

Allan, W. (1972) 'The influence of ecology and agriculture on non-urban settlement' in P.J. Ucko, R. Tringham and G.W. Dimbleby (eds.), *Man, Settlement and Urbanism*, London, 211–26

Ambraseys, N.N. (1978) 'Studies in historical seismicity and tectonics' in W.C. Brice (ed.), *The Environmental History of the Near and Middle East since the Last Ice Age*, London, New York and San Francisco, 185–210

Amiran, D.H.K. (1953) 'The pattern of settlement in Palestine', *Israel Exploration Journal* 3: 65–78, 192–209, 250–60

—— and Ben Arieh, Y. (1963) 'Sedentarisation of bedouin in Israel', *Israel Exploration Journal* 13: 161–81

Ammianus Marcellinus, *The Histories*, transl. J.C. Rolfe, Loeb Classical Library, London and Cambridge, Mass. 1935

Asad, T. (1973) 'Two European images of non-European rule' in T. Asad (ed.), *Anthropology and the Colonial Encounter*, London, 103–18

Ashtor, E. (1976a) *A Social and Economic History of the Near East in the Middle Ages*, London

—— (1976b) 'The Venetian cotton trade in Syria in the later Middle Ages', *Studi Medievali* 17: 675–715. Reprinted in E. Ashtor, *The Medieval Near East: Social and Economic History*, London 1978

Avery, P. (1967) *Modern Iran*, 2nd edn, London

Aymard, M. (1966) *Venise, Raguse et le Commerce du Blé pendant la seconde moitié du XVIe siècle*, Paris

Ayrout, H.H. (1963) *The Egyptian Peasant*, Boston

Bacon, E.E. (1946) 'A preliminary attempt to determine the culture areas of Asia', *Southwestern Journal of Anthropology* 2: 117–32

—— (1953) 'Problems of delimiting the culture areas of Asia', *Memoirs of the Society for American Archaeology* 9: 17–23

—— (1954) 'Types of pastoral nomadism in central and southwestern Asia', *Southwestern Journal of Anthropology* 10: 44–68

Baer, G. (1964) *Population and Society in the Arab East*, London

—— (1968) 'Urbanization in Egypt, 1820–1907' in W.R. Polk and R.L. Chambers (eds.), *Beginnings of Modernization in the Middle East*, Chicago and London, 155–69

Baines, J. and J. Málek (1980) *Atlas of Ancient Egypt*, London

Bakhīt, M.A.S. (1972) 'The Ottoman province of Damascus in the sixteenth century' (unpublished PhD thesis, University of London)

Barkan, Ö.L. (1946–50, 1952, 1954) 'Les déportations comme méthode de peuplement et de colonisation dans l'Empire Ottoman', *Revue de la Faculté des Sciences Economiques de l'Université d'Istanbul* 11: 524–69; 13:56–79; 15:209–39

—— (1957) 'Essai sur les données statistique des registres de recensement dans l'Empire Ottoman aux XVe et XVI siècles', *Journal of the Economic and Social History of the Orient* 1: 9–36

—— (1970) 'Research on the Ottoman fiscal surveys' in M. Cook (ed.), *Studies in the Economic History of the Middle East*, London, 162–71

—— (1975) 'The price revolution of the sixteenth century: a turning point in the economic history of the Near East', *International Journal of Middle East Studies* 6: 3–28

Barker, E. (1948) *The Politics of Aristotle*, Oxford

Barnett, R.D. (1958) 'Early shipping in the Near East', *Antiquity* 32: 220–30

Barth, F. (1962) 'Nomadism in the mountains and plateau areas of south-west Asia' in *The Problems of the Arid Zone*, Arid Zone Research 18, Paris 341–55

—— (1968) *Nomads of South Persia: The Sasseri Tribe of the Khamseh Confederacy*, Boston

—— (1973) 'A general perspective on nomad-sedentary relations in the Middle East' in C. Nelson (ed.), *The Desert and the Sown: Nomads in the Wider Society*, Berkeley, 11–21

Bass, G.F. (ed.) (1972) *A History of Seafaring*, London

Bates, D.G. (1972) 'Differential access to pasture in a nomadic society: the Yörük of southeastern Turkey', *Journal of Asian and African Studies* 7: 48–59

Beaudouin, E.E. and A.U. Pope (1939) 'City plans' in A.U. Pope (ed.), *A Survey of Persian Art*, vol. 3, 1391–1410, Oxford

Beaujour, L.A. Felix de (1829) *Voyage militaire dans l'Empire Othoman*, Paris

Beaumont, P. (1968) 'Qanāts in the Varamin Plain', *Transactions of the Institute of British Geographers* 45: 169–80

—— (1981) 'Water resources and their management in the Middle East' in J.I. Clarke and H. Bowen-Jones (eds.), *Change and Development in the Middle East: Essays in Honour of W.B. Fisher*, London, 40–72

——, G.H. Blake and J.M. Wagstaff (1976) *The Middle East: A Geographical Study*, London

Beckett, P.H.T. (1953) 'Qanāts around Kerman', *Royal Central Asian Journal* 40: 47–57

—— (1966) 'The city of Kirman, Iran', *Erdkunde* 20: 119–25

Beek, M.A. (1962) *Atlas of Mesopotamia*, London

Belgrave, C.D. (1934) 'Pearl diving in Bahrain', *Journal of the Royal Central Asian Society* 21: 450-2

Bell, B. (1970) 'The oldest records of the Nile floods', *Geographical Journal* 136: 569-73

—— (1971) 'The Dark Ages in ancient history. 1. The first Dark Ages in Egypt', *American Journal of Archaeology* 75: 1-26

—— (1975) 'Climate and the history of Egypt: the Middle Kingdom', *American Journal of Archaeology* 79: 223-69

Benedick, R.E. (1979) 'The High Dam and the transformation of the Nile', *Middle East Journal* 33: 119-45

Benet, F. (1963) 'The ideology of Islamic urbanisation', *International Journal of Comparative Sociology* 4: 211-26

Bent, J.T. (1891) 'The Yourouks of Asia Minor', *Journal of the Anthropological Institute of Great Britain and Ireland* 20: 267-76

Benvenisti, M. (1970) *The Crusaders in the Holy Land*, Jerusalem

Bernard, P. (1967) 'Aï Khanum on the Oxus: a Hellenistic city in central Asia', *Proceedings of the British Academy* 53: 71-95

Beug, H.-J. (1967) 'Contributions to the postglacial vegetational history of northern Turkey' in H.J. Cushing and H.E. Wright (eds.), *Quaternary Palaeoecology*, London and New Haven, 349-56

Bharier, J. (1972) 'The growth of towns and villages in Iran, 1900-66', *Middle East Studies* 8: 51-62

Bianquis, T. (1980) 'Une crise frumentaire dans l'Egypte fatimide', *Journal of the Economic and Social History of the Orient* 23: 67-101

Bintliff, J.L. (1982) 'Climatic change, archaeology and Quaternary science in the eastern Mediterranean region' in A.F. Harding (ed.), *Climatic Change in Later Prehistory*, Edinburgh, 143-61

Birks, J.S. (1978) 'Development or decline of pastoralists: the Banī Qitab of the Sultanate of Oman', *Arabian Studies* 4: 7-20

Blaisdell, D.C. (1929) *European Financial Control in the Ottoman Empire*, New York

Blake, G.H. (1972) 'Israel: immigration and dispersal of population' in J.I. Clarke and W.B. Fisher (eds.), *Populations of the Middle East and North Africa: A Geographical Approach*, London, 182-201

Blouet, B.W. (1972) 'Factors influencing the evolution of settlement patterns' in P.J. Ucko, R. Tringham and G.W. Dimbleby (eds.), *Man, Settlement and Urbanism*, London, 3-15

Boak, A.E.R. (1955) *Manpower Shortage and the Fall of the Roman Empire in the West*, Ann Arbor and London

Boardman, J. (1977) 'The olive in the Mediterranean: its culture and use' in J. Hutchinson, G. Clark, E.M. Jope and R. Riley (eds.), *The Early History of Agriculture*, Oxford, 187-96

—— (1980) *The Greeks Overseas: Their Early Colonies and Trade*, new edn, London

Bodman, H.L. (1963) *Political Factions in Aleppo, 1760-1826*, Chapel Hill

Bohrer, V.L. (1972) 'On the relation of harvest methods to early agriculture in the Near East', *Economic Geography* 26: 145-55

Bökönyi, S. (1976) 'Development of early stock rearing in the Near East', *Nature* 264: 19-23

Boserup, E. (1970) *The Conditions of Agricultural Growth. The Economics of Agrarian Change under Population Pressure*, London

Bottema, S. (1975-77) 'A pollen diagram from the Syrian Anti-Lebanon', *Paleorient* 3: 259-68

Bowen, R.L. (1951) 'The pearl fisheries of the Persian Gulf', *Middle East Journal* 5: 161-80

—— and F.P. Albright (eds.) (1958) *Archaeological Discoveries in South Arabia*, Baltimore

Bowring, J. (1840) *Report on the Commercial Statistics of Syria*, London

Boyle, J.A. (1970) 'The last barbarian invaders: the impact of the Mongol conquests upon the East and West', *Memoirs and Proceedings of the Manchester Literary and Philosophical Society* 112: 1–15. Reprinted in J.A. Boyle, *The Mongol World Enterprise, 1206–1370*, London 1977

Braidwood, R.J. and L.S. Braidwood (1950) 'Jarmo: a village of early farmers in Iraq', *Antiquity* 24: 189–95

Braudel, F. (1972) *The Mediterranean and the Mediterranean World in the Age of Philip II*, London

Breasted, J.H. (1927) *Ancient Records of Egypt*, Chicago

Briant, P. (1979) 'L'élevage ovin dans l'Empire Achéménide VIe–IVe siècles avant nôtre ère', *Journal of the Economic and Social History of the Orient* 22: 136–61

—— (1982) *Etat et Pasteurs au Moyen-Orient ancien*, Cambridge and Paris

Brice, W.C. (1955) 'The history of forestry in Turkey', *Oram Fakültesi Dergisi, Istanbul Üniversitesi* 5: 19–42

—— (1966) *South-west Asia*, London

—— (1978a) 'The desiccation of Anatolia' in W.C. Brice (ed.), *The Environmental History of the Near and Middle East since the Last Ice Age*, London, New York and San Francisco, 141–8

—— (1978b) 'Conclusion' in W.C. Brice (ed.), *The Environmental History of the Near and Middle East since the Last Ice Age*, London, New York and San Francisco, 351–6

Brichambaut, G. Perrin de and C.C. Wallén (1963) *A Study of Agroclimatology in Semi-arid and Arid Zones of the Near East*, World Meteorological Organisation, Technical Note no. 56, Geneva

Brooks, E.W. (1898) 'The Arabs in Asia Minor, from Arabic sources', *Journal of Hellenic Studies* 18: 182–208

Brothwell, D.R. and P. Brothwell (1969) *Food in Antiquity: A Survey of the Diet of Early Peoples*, London

Broughton, T.R.S. (1938) 'Roman Asia Minor' in T. Frank (ed.), *An Economic Survey of Rome*, vol. 4, Baltimore, 499–918

Brown, J.A. (1965) 'Geographical study of the evolution of the cities of Teheran and Isfahan' (unpublished PhD thesis, University of Durham)

Brown, P.R.C. (1967) 'The diffusion of Manichaeism in the Roman Empire', *Journal of Roman Studies* 59: 92–103

—— (1978) *The Making of Late Antiquity*, Cambridge, Mass.

Brunt, P.A. (1971) *Italian Manpower 225 B.C.-A.D. 14*, Oxford

Bulliet, R.W. (1975) *The Camel and the Wheel*, Cambridge, Mass.

—— (1979) *Conversion to Islam in the Medieval Period: An Essay in Quantitative History*, Cambridge, Mass. and London

Burckhardt, J.L. (1822) *Travels in Syria and the Holy Land*, ed. W. Martin Leake, London

Burns, N. (1951) 'Development projects in Iraq: the Dujaylah Land Settlement', *Middle East Journal* 5: 362–6

Burton, R.F. (1893) *Personal Narrative of a Pilgrimage to Al-Madinah and Meccah*, memorial edn, London

—— and C.T. Tyrwhitt-Drake (1871) *Unexplored Syria: Visits to the Libanus, the Tulúl el Safá, the Anti-Lebanus, the Northern Lebanon, and the 'Aláh*, London

Butler, A.J. (1978) *The Arab Conquest of Egypt*, 2nd edn, Oxford

Butler, H.C. (1920) 'Desert Syria: the land of lost civilisation', *Geographical Review* 9: 77–108

Butzer, K.W. (1957) 'Der Umweltfaktor in der grossen arabischen expansion', *Saeculum* 8: 359–71
—— (1970) 'Physical conditions in eastern Europe, western Asia and Egypt before the period of agricultural and urban settlement' in *The Cambridge Ancient History*, 3rd edn, vol. 1, pt 1, Cambridge, 35–69
—— (1972) *Environment and Archaeology*, 2nd edn, London
—— (1974) 'Accelerated soil erosion: a problem of man-land relationships' in I.R. Manners and M.W. Mikesell (eds.), *Perspectives on Environment*, Association of American Geographers, Washington DC, 57–78
—— (1976) *Early Hydraulic Civilisation in Egypt: A Study in Cultural Ecology*, Chicago and London
Cadoux, H.W. (1906) 'Recent changes in the course of the lower Euphrates', *Geographical Journal* 28: 266–76
Cahen, C. (1940) *La Syrie du Nord à l'Epoque des Croisades et la Principauté franque d'Antioch*, Institute français de Damas, Paris; reprinted 1977
—— (1950–51) 'Notes sur l'histoire des croisades et de l'Orient Latin II: le régime rurale syrien au temps de la domination franque', *Bulletin de la Faculté des Lettres de Strasbourg* 24: 286–310
—— (1968) *Pre-Ottoman Turkey*, London
—— (1970) 'Quelques notes sur le déclin commerciale du monde musulman à la fin du Moyen Age' in M.A. Cook (ed.), *Studies in the Economic History of the Middle East*, London, New York and Toronto, 31–6
—— (1975) 'Tribes, cities and social organisation' in *The Cambridge History of Iran*, vol. 4, Cambridge, 305–28
*Cambridge Ancient History, The* 3rd edn: (1970) vol. 1, pt 1, *Prolegomena and Prehistory*, ed. I.E.S. Edwards, C.J. Gadd and N.G.L. Hammond, Cambridge
—— (1971) vol. 1, pt 2, *Early History of the Middle East*, ed. I.E.S. Edwards, C.J. Gadd and N.G.L. Hammond, Cambridge
—— (1973) vol. 2, pt 1, *History of the Middle East and the Aegean Region c.1800–1380 B.C.*, ed. I.E.S. Edwards, C.J. Gadd, N.G.L. Hammond and E. Sollberger, Cambridge
—— (1975) vol. 2, pt 2, *History of the Middle East and the Aegean Region c.1380–1000 B.C.*, ed. I.E.S. Edwards, C.J. Gadd, N.G.L. Hammond and E. Sollberger, Cambridge
*Cambridge History of Iran, The* (1983), vol. 2, *The Median and Achaemenian Periods*, ed. I. Gershevitch, Cambridge
—— (1982), vol. 3, *The Seleucid, Parthian and Sasanian Periods*, ed. E. Yar-Shater, Cambridge
—— (1975), vol. 4, *The Period from the Arab Invasion to the Saljuqs*, ed. R.N. Frye, Cambridge
Canard, M. (1959) 'Le riz dans le Proche Orient aux premiers siècles de l'Islam', *Arabica* 6: 113–31
Caponera, D.A. (1954) *Water Laws in Moslem Countries*, FAO Development Paper no. 43, Rome
Carmel, A. (1975) 'The German settlers in Palestine and their relations with the local Arab population and the Jewish community 1868–1918' in M. Ma'oz (ed.), *Studies in Palestine during the Ottoman Period*, Jerusalem, 442–65
Carneiro, R.L. and D.F. Hilse (1966) 'On determining the probable rate of population growth during the Neolithic', *American Anthropologist* 68: 178–80
Carruthers, D. (1928) *The Desert Route to India, being the Journals of Four Travellers by the Great Desert Caravan Route between Aleppo and Basra 1745–1751*, Hakluyt

Society, 2nd series, 63, London

Cary, M. and E.H. Warmington (1963) *The Ancient Explorers*, revised edn, Harmondsworth

Catling, H.W. (1973) 'Cyprus in the middle Bronze Age' in *The Cambridge Ancient History*, 3rd edn, vol. 2, pt. 1, Cambridge, 165–76

Cauvin, J. (1978) *Les Premiers Villages de Syrie-Palestine du IXème au VIIème millénaire avant J.C.*, Lyon

Chapman, R.W. (1971) 'Climatic changes and the evolution of landforms in the Eastern Province of Saudi Arabia', *Bulletin of the Geological Society of America* 82: 2,713–28

Chappell, J. (1970) 'Climatic change reconsidered: another look at "The Pulse of Asia" ', *Geographical Review* 60: 347–73

Charanis, P. (1961) 'The transfer of population as a policy in the Byzantine Empire', *Comparative Studies in Society and History* 3: 140–54

Chardin, J. (1735) *Voyage de Chevallier Chardin en Perse et autres Lieux de l'Orient*, 2, Paris

Charlesworth, M.P. (1926) *Trade-routes and Commerce of the Roman Empire*, 2nd edn, reprinted 1970, New York

Chayanov, A.V. (1966) *The Theory of Peasant Economy*, ed. D. Thorner, B. Kerblay and R.E.F. Smith, Homewood, Ill.

Chevallier, D. (1968) 'Western development and eastern crisis in the mid-nineteenth century: Syria confronted with the European economy' in W.R. Polk and R.L. Chambers (eds.), *Beginnings of Modernisation in the Middle East: The Nineteenth Century*, Chicago and London, 205–22

——— (1970) 'Les cadres sociaux de l'économie agraire dans le Proche-Orient au début du XIXe siècle: le cas de Mont Liban' in M. Cook (ed.), *Studies in the Economic History of the Middle East*, London, 333–45

Childe, V.G. (1950) 'The urban revolution', *Town Planning Review* 21: 9–16

——— (1952) *New Light on the Most Ancient East*, London

Chirol, V. (1903) *The Middle East Question, or Some Political Problems of Indian Defence*, London. (Originally published as a series of articles in *The Times*, beginning 14 Oct. 1902)

Chitty, D.J. (1977) *The Desert a City. An Introduction to the Study of Egyptian and Palestinian Monasticism under the Christian Empire*, London and Oxford

Christodoulou, D. (1959) *The Evolution of the Rural Land Use Pattern in Cyprus*, World Land Use Survey Monograph 2, Bude

Cicero, 'Pro Lege Manilla' in *Cicero: The Speeches*, transl. H.G. Hodges, Loeb Classical Library, London and New York, 1927

Clark, B.D. (1972) 'Iran's changing population patterns' in J.I. Clarke and W.B. Fisher (eds.), *Populations of the Middle East and North Africa: A Geographical Approach*, London, 68–96

Clarke, E.C. (1974) 'The Ottoman industrial revolution', *International Journal of Middle East Studies* 5: 65–76

Clark, G. (1944) 'Presidential address', *Geographical Journal* 104: 4–5

Clark, J.D. (1971) 'A re-examination of the evidence for agricultural origins in the Nile valley', *Proceedings of the Prehistoric Society*, new series, 37 (pt. 2): 34–79

Clarke, J.I. (1972) 'Introduction' to J.I. Clarke and W.B. Fisher (eds.), *Populations of the Middle East and North Africa: A Geographical Approach*, London, 15–39

Cobham, C.D. (1908) *Excerpta Cypriaca: Materials for a History of Cyprus*, Cambridge

Cohen, A. (1973) *Palestine in the Eighteenth Century*, Jerusalem

——— and B. Lewis (1978) *Population and Revenue in the Towns of Palestine in the Sixteenth Century*, Princeton

Cohen, E. (1977) 'Recent anthropological studies of Middle Eastern communities and ethnic groups', *Annual Review of Anthropology* 6: 315–47

Cohen, E.J. (1967) *Turkish Forest Products*, USAID, Ankara

Cohen, G.M. (1978) *The Seleucid Colonies. Studies in Founding, Administration and Organisation*, Historia-Einzelschriften 30, Wiesbaden

Cohen, H.R. and O. Erol (1969) 'Aspects of the palaeogeography of central Anatolia', *Geographical Journal* 135: 388–98

Colledge, M.A.R. (1967) *The Parthians*, New York

Collinder, P. (1954) *A History of Marine Navigation*, London

*Constantinople and the Scenary of the Seven Churches of Asia Minor illustrated . . . by Thomas Allom with . . . descriptions of the Plates by the Rev. Robert Walsh*, London 1838

Cook, M. (1983) *Muhammad*, Past Masters, Oxford and New York

—— (1972) *Population Pressure in Rural Anatolia, 1460–1600*, London Oriental Series 27, London

Cortesão, A. (1944) *The Summa Oriental of Tomé Pires*, vol. 1, Hakluyt Society, 2nd series, 89, London

Costello, V.J. (1973) 'The industrial structure of a traditional Iranian city', *Tijdshrift voor Economische en Sociale Geografie* 64: 108–20

—— (1977) *Urbanization in the Middle East*, Cambridge

Cotton, E. (1949) *East Indiamen: The East India Company's Maritime Service*, London

Crawford, D.J. (1979) 'Food: tradition and change in Hellenistic Egypt', *World Archaeology* 11: 136–46

Cressey, G.B. (1960) *Crossroads: Land and Life in Southwest Asia*, Chicago, New York and Philadelphia

Cresswell, K.A.C. (1958) *A Short Account of Early Muslim Architecture*, Harmondsworth

Cuinet, V. (1890–94) *Le Turquie d'Asie. Géographie administrative, statistique, descriptive et raisonée de chaque Provence de l'Asie Mineure*, 4 vols., Paris

Culican, W. (1966) *The First Merchant Venturers. The Ancient Levant in History and Commerce*, London

Curwen, E.C. and G. Hatt (1953) *Plough and Pasture. The Early History of Farming*, New York

Curzon, G. (1892) *Persia and the Persian Question*, London

Cvetkova, B. (1970) 'Les celep et leur rôle dans le vie économique des Balkans a l'époque ottomane (XVe–XVIIIe siècles)' in M. Cook (ed.), *Studies in the Economic History of the Middle East*, London, 172–92

Daniel, N. (1966) *Islam, Europe and Empire*, Edinburgh

Darlington, C.D. (1969) *The Evolution of Man and Society*, London

Davis, R. (1967) *Aleppo and Devonshire Square: English Traders in the Eighteenth Century*, London

Davison, R.H. (1960) 'Where is the Middle East?' *Foreign Affairs* 38: 665–75

Dawood, N.J. (transl.) (1959) *The Koran*, revised edn, Harmondsworth

Derry, T.K. and T. Williams (1970) *A Short History of Technology*, Oxford

Dewdney, J.C. (1971) *Turkey*, London

Dickinson, H.R.P (1951) *The Arab of the Desert*, London

Diester-Haas, L. (1973) 'Holocene climate in the Persian Gulf as deduced from grain-size and pteropod distribution', *Marine Geology* 14: 207–23

Dimbleby, G.W. (1977) 'Climate, soil and man' in J. Hutchinson, G. Clark, E.M. Jope and R. Riley (eds.), *The Early History of Agriculture*, Oxford, 197–208

Dio Cassius, *Roman History*, transl. E. Cary, Loeb Classical Library, London and New York 1924

*Diodorus Siculus*, transl. C.H. Oldfather, Loeb Classical Library, London and Cambridge, Mass. 1946

Doe, B. (1972) *Southern Arabia*, London

Dols, M.W. (1977) *The Black Death in the Middle East*, Princeton

Donaldson, W.J. (1981) 'Fisheries of the Arabian peninsula' in J.I. Clarke and H. Bowen-Jones (eds.), *Change and Development in the Middle East. Essays in Honour of W.B. Fisher*, London, 189–98

Donner, F.M. (1977) 'Mecca's food supplies and Muhammad's boycott', *Journal of the Economic and Social History of the Orient* 20: 249–66

—— (1979) 'Muhammad's political consolidation in Arabia up to the conquest of Mecca: a reassessment?', *Muslim World* 69: 229–47

—— (1980) ''The Bakr Wa'il tribes and politics in north eastern Arabia on the eve of Islam', *Studia Islamica* 51: 5–38

Dostal, W. (1959) 'The evolution of bedouin life' in F. Gabrielli (ed.), *L'Antica Società Bedouina*, Studi Semitici 2, Rome, 11–34

Downey, G. (1961) *A History of Antioch in Syria*, Princeton

Drower, M.S. (1971) 'Syria before 2200 B.C.' in *The Cambridge Ancient History*, 3rd edn, vol. 1, pt 2, Cambridge, 315–62

Duckham, A.N. and G. Masefield (1970) *Farming Systems of the World*, London

Durand, J.R. (1977) 'Historical estimates of world population: an evaluation', *Population and Development Review* 3: 253–96

During Caspers, E.C.L. (1977) 'Sumer, coastal Arabia and the Indus valley in Protoliterate and Early Dynastic eras', *Journal of the Economic and Society History of the Orient* 22: 122–35

Dyson-Hudson, N. (1972) 'The study of nomads', *Journal of Asian and African Studies* 7: 2–29

Earle, P. (1970) *Corsairs of Malta and Barbary*, London

Ehlers, E. (1975) 'Die Stadt Bam und ihr Oasen-umland/Zentraliran', *Erdkunde* 29: 38–52

Eickelman, D.F. (1967) 'Musaylima: an approach to the social anthropology of seventh century Arabia', *Journal of the Economic and Social History of the Orient* 10: 17–52

—— (1974) 'Is there an Islamic city? The making of a quarter in a Moroccan town', *International Journal of Middle East Studies* 4: 274–94

Eisma, D. (1962) 'Beach ridges near Selcuk, Turkey', *Tijdschrift van het Koninklijk Nederlandsch Aardrijskundig Genootschap* 79: 234–46

—— (1978) 'Stream deposition and erosion by the eastern shore of the Aegean' in W.C. Brice (ed.), *The Environmental History of the Near and Middle East since the Last Ice Age*, London, New York and San Francisco, 67–81

Emery, W.B. (1961) *Archaic Egypt*, Harmondsworth

*Encyclopaedia Britannica*, 14th edn, Chicago, London, etc, 1973

English, P.W. (1966) *City and Village in Iran: Settlement and Economy in the Kirman Basin*, Madison, Milwaukee and London

—— (1967) 'Urbanities, peasants and nomads: the Middle Eastern ecological triology', *Journal of Geography* 61: 54–9

—— (1968) 'The origin and spread of qanats in the Old World', *Proceedings of the American Philosophical Society* 112: 170–81

—— (1973) 'Geographical perspectives on the Middle East: the passing of the ecological trilogy' in M.W. Mikesell (ed.), *Geographers Abroad*, Chicago, 134–64

Erder, L. and S. Faroqhi (1979) 'Population rise and fall in Anatolia 1550–1620', *Middle*

*Eastern Studies* 15: 322–45

—— and S. Faroqhi (1980) 'The development of the Anatolian urban network during the sixteenth century', *Journal of the Economic and Social History of the Orient* 23: 265–303

Erinç, S. (1950) 'Climatic types and the variation of moisture regions in Turkey', *Geographical Review* 40: 224–35

—— (1978) 'Changes in the physical environment in Turkey since the end of the last glacial' in W.C. Brice (ed.), *The Environmental History of the Near and Middle East since the Last Ice Age*, London, New York and San Francisco, 87–110

—— and M. Bener (1961) Türkiyede Uzun Süreli İki Yağış Rasadı İstanbul ve Tarsus', *Istanbul Üniversitesi Coğrafya Enstitüsü Dergisi* 12: 100–16

Erman, A. (1971) *Life in Ancient Egypt*, New York

Ernst, J.E.F. and H.R. Merrens, (1978) 'Praxis and theory in the writing of American historical geography', *Journal of Historical Geography* 4: 277–90

Erol, O. (1978) 'The Quaternary history of the lake basins of central and southern Anatolia' in W.C. Brice (ed.), *The Environmental History of the Near and Middle East since the Last Ice Age*, London, New York and San Francisco, 111–39

Eton, W. (1798) *A Survey of the Turkish Empire*, London

Evenari, M. *et al.* (1961) 'Ancient agriculture in the Negev: archaeological studies and experimental farms show how agriculture was possible in Israel's famous desert', *Science*, 133: 979–96

Faegre, T. (1979) *Tents: Architecture of the Nomads*, London

FAO (1972) *Production Yearbook 1971*, Rome

Faroqhi, S. (1979) 'Sixteenth-century periodic markets in various Anatolian *sancaks*' İcel, Hamid, Karahisar-i Sahib, Kütahya, Aydın, and Menteşe', *Journal of the Economic and Social History of the Orient* 22: 32–80

—— (1980) 'Taxation and urban activities in sixteenth-century Anatolia', *International Journal of Turkish Studies* 1: 19–53

Fawcett, C. and R. Burn (1947) *Travels of the Abbé Carré in India and the Near East 1672 to 1674*, Hakluyt Society, 2nd series, 95, London

Fenton, A. (1972) 'A fuel of necessity: animal manure' in E. Ennen and G. Wiegelmann (eds.), *Festschrift Matthias Zender. Studien zu Volkskultur, Sprache und Landesgeschichte*, vol. 2, Bonn, 722–34

Ferguson, J. (1978) 'China and Rome' in H. Temporini and W. Haase (eds.), *Austeig und Niedergang der römischen Welt*, Berlin and New York, II. 9, 581–603

Fernea, R.A. (1970) *Shaykh and Effendi. Changing Patterns of Authority among the El Shabana of Southern Iraq*, Cambridge, Mass.

Ferrier, R.W. (1973) 'The Armenians and the East India Company in the seventeenth and early eighteenth centuries', *Economic History Review*, 2nd series, 26: 38–62

ffrench Blake, R.L.V. (1971) *The Crimean War*, London

Field, M. (1973) 'Developing the Nile', *World Crops* 25: 11–15

Finegan, J. (1979) *Archaeological History of the Ancient Middle East*, Boulder and Folkestone

Firoozi, F. (1974) 'Teheran — a demographic and economic analysis', *Middle Eastern Studies* 10: 60–76

Fischel, W.J. (1956) 'The city in Islam', *Middle Eastern Affairs* 7: 227–32

Fisher, A.W. (1978) 'The sale of slaves in the Ottoman Empire: markets and state taxes on slave sales. Some preliminary considerations', *Boğaziçi Üniversitesi Dergisi* 6: 149–74

Fisher, S. (1919) *Ottoman Land Laws*, Oxford

Flannery, K.V. (1969) 'Origins and ecological effects of early domestication in Iran and the

Near East, in P.J. Ucko and G.W. Dimbleby (eds.), *The Domestication and Exploitation of Plants and Animals*, London, 73–100

—— (1972) 'The origins of the village as a settlement type in Mesoamerica and the Near East: a comparative study' in P. Ucko, R. Tringham and G.W. Dimbleby (eds.), *Man, Settlement and Urbanism*, London, 23–501

Floor, W.M. (1976) 'The merchants (tujjār) in Qājār Iran', *Zeitschrift der Deutschen Mörganlandischen Gesellschaft* 126: 101–35

FO 78/24, Lord Elgin to Lord Grenville, 28 Dec. 1799, Public Record Office, London

Forbes, H.A., H.A. Koster and L. Foxhall (n.d.) 'Terrace agriculture and erosion: environmental effects of population instability in the Mediterranean' (unpublished paper)

Forbes, R.J. (1965) 'Irrigation and drainage' in his *Studies in Ancient Technology*, vol. 2, Leiden, 1–130

Foss, C. (1972) 'Byzantine cities of western Asia Minor' (unpublished doctoral dissertation, Harvard University)

—— (1975) 'The Persians in Asia Minor and the end of Antiquity', *English Historical Review* 90: 721–42

—— (1976) *Byzantine and Turkish Sardis*, Cambridge, Mass. and London

Foster, W. (1949) *The Red Sea and Adjacent Countries at the Close of the Seventeenth Century as described by Joseph Pitts, William Daniel and Charles Jacques Poncet*, Hakluyt Society, 2nd series, 100, London

Frankenstein, S. (1979) 'The Phoenicians in the far West: a function of Neo-Assyrian imperialism' in M.T. Larsen (ed.), *Power and Propaganda: A Symposium on Ancient Empires*, Copenhagen Studies in Assyriology 7, Copenhagen, 263–94

Frantz-Murphy, G. (1981) 'A new interpretation of the economic history of medieval Egypt. The role of the textile industry 254–567/868–1171', *Journal of the Economic and Social History of the Orient* 24: 274–97

Frézoub, E. (1980) 'Les fonctions du Moyen-Euphrate à l'époque romaine' in J.C. Margueron (ed.), *Le Moyen-Euphrate: Zone de Conflict et d'Echanges. Actes du Colloque de Strasbourg 10–12 mars 1977*, Leiden, 355–86

Frye, R.N. (1966) *The Heritage of Persia*, New York, Toronto and London

G/29/25, 'Report on the British Trade with Persia and Arabia, 1709', East India Company Archives, *Persia*, vol. 21, India Office Library, London

Gabrieli, F. (1968) *Muhammad and the Conquests of Islam*, London

Gadd, C.J. (1971) 'The cities of Babylonia' in *The Cambridge Ancient History*, 3rd edn, vol. 1, pt 2, Cambridge, 93–144

George, C.J. (1972) 'The role of the Aswan High Dam in changing the fisheries of the southeastern Mediterranean' in M.T. Farvar and J.P. Milton (eds.), *The Careless Technology: Ecology and International Development*, Garden City, NY, 159–78

Gerber, H. (1979) 'The population of Syria and Palestine in the nineteenth century', *Asian and African Studies* 13: 58–80

Ghirshman, R. (1954) *Iran from the Earliest Times to the Islamic Conquest*, Harmondsworth

Gibb, H.A.R. (1963) *Arabic Literature: An Introduction*, 2nd edn, Oxford

—— and H. Bowen (1950) *Islamic Society and the West*, vol. 1, pt 1, London, New York and Toronto

Gibson, M. (1974) 'Violation of fallow and engineering disaster in Mesopotamian civilisation' in T.E. Downing and M. Gibson (eds.), *Irrigation's Impact on Society*, Anthropological Papers of the University of Arizona 25, Tucson, 7–19

Gilbar, G.G. (1976) 'Demographic developments in later Qājār Persia, 1870–1906', *Asian and African Studies* 11: 125–56

—— (1978) 'Persian agriculture in the late Qājār period, 1860–1906: some economic and social aspects', *Asian and African Studies* 12: 312–65

—— (1979) 'The Persian economy in the mid-19th century', *Die Welt des Islam* 19: 172–211

Gilbert, M. (1974) *The Arab–Israeli Conflict: Its History in Maps*, London

Glacken, C.T. (1956) 'Changing ideas of the habitable world' in W.L. Thomas (ed.), *Man's Role in Changing the Face of the Earth*, Chicago, 70–92

Glesinger, E. (1960) 'The Mediterranean Project', *Scientific American* 204: 86, 1–20

Globb, P.V. and T.G. Bibby (1960) 'A forgotten civilisation of the Persian Gulf', *Scientific American* 203: 62–71

Glubb, J.B. (1963) *The Great Arab Conquests*, London

—— (1973) *Soldiers of Fortune: The Story of the Mamlukes*, London

Goblot, H. (1963) 'Dans l'ancien Iran, les techniques de l'eau et la grand histoire', *Annales: Economies, Sociétés, Civilisations* 18: 500–20

Goiten, S.D. (1967) *A Mediterranean Society. The Jewish Communities of the Arab World as Portrayed in the Documents of the Cairo Geniza*, vol. 1, Berkeley and Los Angeles

—— (1973) *Letters of Medieval Jewish Traders*, Princeton

Goodwin, A. (1971) *A History of Ottoman Architecture*, London

—— (1977) *Islamic Architecture: Ottoman Turkey*, London

Gould, A.G. (1973) 'Pashas and brigands: Ottoman provincial reform and its impact on the nomadic tribes of Southern Anatolia, 1840–1885' (unpublished doctoral dissertation, University of California, Los Angeles)

Göyünç, N. (1969) *XVI Yüzyılda Mardin Sancağı*, Istanbul Üniversitesi Edebiyat Fakultesi Yayınları, no. 1458, Istanbul

Grabar, O. (1969) 'The architecture of the Middle Eastern city' in I.M. Lapidus (ed.), *Middle Eastern Cities*, Berkeley and Los Angeles, 26–46

Gray, E.W. (1970) Review of J.I. Miller, 'The Spice Trade of the Roman Empire', *Journal of Roman Studies* 60: 222–4

Griswold, W.J. (1977) 'The Little Ice Age: its effects on Ottoman history 1585–1625' (unpublished paper, Fort Collins, Color.)

Groom, N. (1981) *Frankincense and Myrrh: A Study of the Arabian Incense Trade*, London and New York

Grunebaum, G.E. von (1955a) *Islam: Essays in the Nature and Growth of a Cultural Tradition*, London

—— (1955b) 'The Muslim town and the Hellenistic town', *Scientia* 90: 364–70

Güçer, L. (1953) 'Le commerce intérieur des céréales dans l'Empire Ottoman pendant la seconde moitié du XVIème siècle', *Revue de la Faculté des Sciences Economiques de l'Université d'Istanbul* 11: 1–26

Gurney, O.R. (1961) *The Hittites*, revised edn, Harmondsworth

Hambly, G. (1964) 'An introduction to the economic organisation of early Qājār Iran', *Iran* 2: 69–81

Hamdan, G. (1961) 'Evolution of irrigation agriculture in Egypt' in L.D. Stamp (ed.), *A History of Land Use in Arid Regions*, Arid Zone Research 17, Paris, 119–42

—— (1962) 'The pattern of medieval urbanisation in the Arab world', *Geography* 47: 121–34

Hansman, J.F. (1978) 'The Mesopotamian delta in the first millenium B.C.', *Geographical Journal* 144: 49–61

Hanson, W.S. and L.J.E. Keppie (1980) *Roman Frontier Studies 1979. Papers Presented to the 12th International Congress of Roman Frontier Studies*, British Archaeological Reports, International Series 71, Oxford

Harlan, J.R. (1967) 'A wild wheat harvest in Turkey', *Archaeology* 20: 197–201

——— and D. Zohary (1966) 'Distribution of wild wheats and barley', *Science* 153: 1,074–80

Harris, D.R. (1977) 'Alternative pathways toward agriculture' in C.A. Reed (ed.), *Origins of Agriculture*, The Hague and Paris, 179–243

Harvey, P. (ed.) (1955) *The Oxford Companion to Classical Literature*, Oxford

Harvey, P.D.A. (1980) *The History of Topographical Maps*, London

Hasan, M.S. (1958) 'Growth and structure of Iraq's population, 1867–1947', *Bulletin of the Oxford Institute of Statistics* 20: 339–523. Reprinted in C. Issawi (ed.), *The Economic History of the Middle East, 1800–1914*, Chicago and London 1966, 155–62

——— (1970) 'The role of foreign trade in the economic development of Iraq, 1864–1964: a study in the growth of a dependent economy' in M. Cook (ed.), *Studies in the Economic History of the Middle East*, London, 346–72

Hassan, F.A. (1977) 'The dynamics of agricultural origins in Palestine: a theoretical model' in C.A. Reed (ed.), *Origins of Agriculture*, The Hague and Paris, 589–609

Haudricourt, A.G. and M.J.-B. Delamarre (1955) *L'Homme et la Carrue à travers le Monde*, 2nd edn, Paris

Hayes, W.C. (1973) 'Internal affairs from Tuthmosis I to the death of Amenophis III' in *The Cambridge Ancient History*, 3rd edn, vol. 2, pt 1, Cambridge, 313–407

Healy, J.F. (1977) *Mining and Metallurgy in the Greek and Roman Worlds*, London

Hegazi, M. (1972) 'Land use in Egypt after the Aswan High Dam', *Reading Geographer* 1: 24–30

Heichelheim, F.M. (1938) 'Roman Syria' in T. Frank (ed.), *An Economic Survey of Rome*, vol.´4, Baltimore, 121–257

——— (1956) 'Effects of Classical Antiquity on the land' in W.L. Thomas (ed.), *Man's Role in Changing the Face of the Earth*, Chicago, 165–82

——— (1970) *An Ancient Economic History*, vol. 3, Leiden

Helbaek, N. (1959) 'How farming began in the Old World', *Archaeology* 12: 183–9

Herodotus, *The Histories*, Scriptorum Classicorum Bibliotheca Oxoniensis, Oxford 1918

Hershlag, Z.Y. (1964) *Introduction to the Modern Economic History of the Middle East*, Leiden

Herzfeld, E. (1948) *Geschichte der Stadt Samarra*, Berlin

Hess, A.C. (1970) 'The creation of the Ottoman seaborne empire in the age of the oceanic discoveries, 1453–1525', *American Historical Review* 75: 892–919

——— (1973) 'The Ottoman conquest of Egypt (1517) and the beginning of the sixteenth century world war', *International Journal of Middle East Studies* 4: 55–76

Hicks, J. (1969) *A Theory of Economic History*, Oxford

Hill, D.R. (1975) 'The role of the camel and the horse in the early Arab conquests' in V.J. Parry and M.E. Yapp (eds.), *War, Technology and Society in the Middle East*, London, New York and Toronto, 32–43

Hitti, P.K. (1964) *History of the Arabs*, 8th edn, London

Hobsbawm, E.J.E. (1969) *Bandits*, London

Hodgson, M.G.S. (1960) 'A comparison of Islam and Christianity as framework for religious life', *Diogenes* 32: 49–74

Hoffmann, K. (1927) *Oelpolitik und anglsächsischer Imperialismus*, Berlin. Translated extracts in *The Economic History of the Middle East, 1800–1914*, edited by C. Issawi, Chicago and London 1966, 199–200

Hogarth, D.G. (1902) *The Nearer East*, London

Holm, D.A. (1960) 'Desert geomorphology in the Arabian peninsula', *Science* 132: 1,369–79

Holt, P.M. (1966) *Egypt and the Fertile Crescent 1516–1922*, London

Holz, R.K. (1968) 'The Aswan High Dam', *Professional Geographer* 20: 230–7

Hookham, H. (1962) *Tamburlaine the Conqueror*, London

Hornell, J. (1942) 'A tentative classification of Arab seacraft', *Mariners' Mirror* 28: 11–40

Horowitz, A. (1971) 'Climate and vegetational developments in Israel during Upper Pleistocene–Holocene times', *Pollen et Spores* 13: 255–78

Hourani, A.H. (1957) 'The changing face of the Fertile Crecent in the eighteenth century', *Studia Islamica* 8: 89–122

—— and S.M. Stern (eds.) (1969) *The Islamic City*, Oxford

Hourani, G.F. (1963) *Arab Seafaring in the Indian Ocean in Ancient and Early Medieval Times*, 2nd edn, Beirut

Hours, F. (1982) 'Prehistoric settlement patterns in the Levant in relation to environmental conditions' in J.L. Bintliff and W. van Zeist (eds.), *Palaeoclimates, Palaeoenvironments and Human Communities in the Eastern Mediterranean Region in Later Prehistory*, British Archaeological Reports, International Series 133, Oxford, 419–48

Housego, J. (1973) 'The 19th century Persian carpet boom', *Oriental Art* 19: 169–71

Huntington, E. (1911) *Palestine and its Transformation*, Boston

Hütteroth, W-D. (1968) *Ländliche Siedlungen im südlichen Inneranatolien in den letzten Vierhundert Jahren*, Göttinger Geographische Abhandlungen 46, Göttingen

—— (1973) 'Zum Kenntnisstand über Verbreitung und Typen von Bergnomadismus und Halbnomadismus in den Gebirgs und Plateaulandschaften Südwest Asiens' in C. Rathjens, C. Troll and H. Uhlig (eds.), *Vergleichende Kulturgeographie der Hochgebirge des Südlichen Asien*, Erdwissenschaftliche Forschung 5, Wiesbaden, 146–56

—— (1974) 'The influence of social structure on land division and settlement in inner Anatolia' in P. Benedict, E. Tümertekin and F. Mansur (eds.), *Turkey: Geographic and Social Perspectives*, Leiden, 19–47

—— (1980) 'The demographic and economic organisation of the southern Syrian sancaks in the late 16th century' in O. Okyar and H. İnalcik, *Social and Economic History of Turkey (1071–1920)*, Ankara, 35–47

—— and K. Abdulfattah (1977) *Historical Geography of Palestine, Transjordan and Southern Syria in the late 16th Century*, Erlangen Geographische Arbeiten 5, Erlangen

Ibn Khaldūn, *The Muqaddimah*, transl. F. Rosenthal, ed. N.J. Dawood, London 1967

Ibrahim, S.E.M. (1975) 'Over-urbanisation and under-urbanisation: the case of the Arab World', *International Journal of Middle East Studies* 6: 29–45

İnalcık, H. (1954) 'Ottoman methods of conquest', *Studia Islamica* 20: 103–29

—— (1955) 'Land problems in Turkish history', *Muslim World* 45: 221–8

—— (1960) 'Bursa and the commerce of the Levant', *Journal of the Economic and Social History of the Orient* 3: 131–42

—— (1973) *The Ottoman Empire: The Classical Age 1300–1600*, London

—— (1975) 'The socio-political effects of the diffusion of fire-arms in the Middle East' in V.J. Parry and M.E. Yapp (eds.), *War, Technology and Society in the Middle East*, London, 195–217

Ingrams, H. (1966) *Arabia and the Isles*, 3rd edn, London

Ionides, M.G. (1937) *The Regime of the Rivers Euphrates and Tigris*, London

*Iranian Studies* (special edn on Esfahān) 7: 1974

Isaac, E. (1970) *Geography of Domestication*, Englewood Cliffs, NJ

Islamoğlu, H. and S. Faroqhi (1979) 'Crop patterns and agricultural production trends in sixteenth-century Anatolia', *Review* 2: 401–36

Ismail, A.A. (1972) 'Origin, ideology and physical patterns of Arab urbanisation', *Ekistics* 33: 113–23

Issawi, B. (1976) 'An introduction to the physiography of the Nile Valley' in F. Wendorf and R. Schild (eds.), *Prehistory of the Nile Valley*, New York, San Francisco and London, 3–22

Issawi, C. (ed.) (1966) *The Economic History of the Middle East, 1800–1914*, Chicago and London

—— (1969) 'Economic change and urbanisation in the Middle East' in I.M. Lapidus (ed.), *Middle Eastern Cities*, Berkeley and Los Angeles, 102–8

—— (1970) 'The Tabriz–Trazon trade 1830–1900: rise and decline of a route', *International Journal of Middle East Studies* 1: 18–27

—— (ed.) (1971) *The Economic History of Iran, 1800–1914*, Chicago and London

—— (1977) 'British trade and the rise of Beirut, 1830–1860', *International Journal of Middle East Studies* 8: 91–101

—— (1980) *The Economic History of Turkey, 1800–1914*, Chicago and London

Jacobsen, T. and R.M. Adams (1958) 'Salt and silt in ancient Mesopotamian agriculture', *Science* 128: 1,251–8

Jameson, M.H. (1977–78) 'Agriculture and slavery in classical Athens', *Classical Journal* 73: 122–45

Jardé, A. (1925) *Les Céréales dans l'Antiquité grecque*, Bibliothèque des Ecoles françaises d'Athènes et de Rome 130, Paris

Jennings, R.C. (1976) 'Urban population in Anatolia in the sixteenth century: a study of Kayseri, Karaman, Amasya, Trabzon and Erzurum', *International Journal of Middle East Studies* 7: 21–57

—— (1978) 'Sakaltutan four centuries ago', *International Journal of Middle East Studies* 9: 89–98

Jesus, P.S. de (1980) *The Development of Prehistoric Mining and Metallurgy in Anatolia*, British Archaeological Reports, International Series 74, Oxford

Johnson, A.C. (1936) 'Roman Egypt' in T. Frank (ed.), *An Economic Survey of Rome*, vol. 2, Baltimore

—— and L.C. West (1949) *Byzantine Egypt: Economic Studies*, Princeton

Johnson, D.L. (1969) *The Nature of Nomadism: A Comparative Study of Pastoral Migrations in Southwestern Asia and Northern Africa*, Department of Geography Research Paper 118, Chicago

Johnson, G.A. (1972) 'A test of the utility of central place theory in archaeology' in P.J. Ucko, R. Tringham and G.W. Dimbleby (eds.), *Man, Settlement and Urbanism*, London, 769–85

Johnstone, T.M. and J. Muir (1964) 'Some nautical terms in the Kuwaiti district of Arabia', *Bulletin of the School of Oriental and African Studies* 27: 299–32

Jones, A.H.M. (1953a) 'Census records of the later Roman Empire', *Journal of Roman Studies* 43: 49–64

—— (1953b) 'Inflation under the Roman Empire', *Economic History Review*, 2nd series, 5: 293–318

—— (1964) *The Later Roman Empire*, Oxford

—— (1966a) *The Greek City from Alexander to Justinian*, 2nd edn, London

—— (1966b) *The Decline of the Ancient World*, London

—— (1970) 'Asian trade in Antiquity' in R. Richards (ed.), *Islam and the Trade of Asia*, Oxford

Jones, D.R.W. (1963) 'Apple production in the Lebanon: a study of agricultural development in an underdeveloped area', *Economic Geography* 43: 245–57

Jones, L.W. (1969) 'Rapid population growth in Baghdad and Amman', *Middle East Journal* 23: 209–15

Josephus, *The Jewish Wars*, transl. H.St.J. Thackery, Loeb Classical Library, London and New York 1927

Jullian, P. (1977) *The Orientalists: European Painters of Eastern Scenes*, Oxford

Kark, R. (1983) 'Millenarism and agricultural settlement in the Holy Land in the nineteenth century', *Journal of Historical Geography* 9: 47–62

Karpat, K.H. (1978) 'Ottoman population records and the census of 1881/82–1893', *International Journal of Middle East Studies* 9: 237–74

Kassas, M. (1972) 'Impact of river control schemes on the shoreline of the Nile Delta' in M.T. Farvar and J.P. Milton (eds.), *The Careless Technology: Ecology and International Development*, Garden City, NY, 179–88

Kedar, Y. (1957) 'Water and soil from the desert: some ancient agricultural achievements in the central Negev', *Geographical Journal* 123: 179–88

Keddie, N.R. (1973) 'Is there a Middle East?', *International Journal of Middle East Studies* 4: 255–71

Keen, M. (1969) *The Pelican History of Medieval Europe*, Harmondsworth

Kees, H. (1977) *Ancient Egypt: A Cultural Topography*, Chicago and London

Kelley, K.B. (1976) 'Dendritic central-place systems and the regional organisation of Navajo trading posts' in C.A. Smith (ed.), *Regional Analysis* vol. 1, *Economic Systems*, New York, London and San Francisco, 219–54

Kelman, J. (1908) *From Damascus to Palmyra*, London

Kemal, Y. (1963) *The Wind from the Plain*, London

Kemp, B.J. (1977) 'The early development of towns in Egypt', *Antiquity* 51: 185–200

Kemp, W.R. (1971) 'The flow of energy in a hunting society', *Scientific American* 225: 105–15

Kennedy, D.L. (ed.) (1982) *Explorations on the Roman Frontier in North-eastern Jordan*, British Archaeological Reports, International Series 134, Oxford

Khachikian, L. (1967) 'Le registre d'un marchand armenien en Perse, en Inde et au Tibet (1682–93)', *Annales: Economies, Sociétés, Civilisations* 22: 231–78

Kinglake, W. (1844) *Eothen, or Traces of Travel brought Home from the East*, London

Kinnier, M. (1813) *A Geographical Memoir of the Persian Empire*, London

Kirk, G.E. (1964) *A Short History of the Middle East from the Rise of Islam to Modern Times*, 7th edn, London

Kohl, P.L. (1978) 'The balance of trade in southwestern Asia in the mid-third millennium B.C.', *Current Anthropology* 19: 463–75

Kolars, J. (1974) 'Systems of change in Turkish village agriculture' in P. Benedict, E. Tümertekin and F. Mansur (eds.), *Turkey: Geographic and Social Perspectives*, Leiden, 204–33

Koppes, C.R. (1976) 'Captain Mahan, General Gordon and the origin of the term "Middle East" ', *Middle Eastern Studies* 12: 95–8

Krader, L. (1959) 'The ecology of nomadic pastoralism', *International Social Science Journal* 11: 499–510

Kramers, J.H. and G. Wiet (1964) *Ibn Hauqal Configuration de la Terre (Kitab Surai al-Ard)*, Beirut and Paris

Kraybill, N. (1977) 'Pre-agricultural tools for the preparation of foods in the Old World' in C.A. Reed (ed.), *Origins of Agriculture*, The Hague and Paris, 485–521

Krupper, J.R. (1957) *Les Nomades en Mesopotamie au temps des Rois de Mari*, Paris

—— (1959) 'Le rôle des nomades dans l'histoire de la Mesopotamie ancienne', *Journal of the Economic and Social History of the Orient* 2: 113–27

Kurmuş, O. (1974) 'The role of British capital in the economic development of Western Anatolia, 1850–1913' (unpublished PhD thesis, University of London)

Lafont, F. and H-L. Rabino (1910) *L'Industrie séricole en Perse*, Montpellier. Extracts published in C. Issawi (ed.), *The Economic History of Iran, 1800–1914*, Chicago and London 1971, 235–8

Lamb, H.H. (1968) 'Climatic background to the birth of civilisation', *Advancement of Science* 25: 103–20

—— (1977) *Climate: Present, Past and Future*, London

Lamberg-Karlovsky, C.C. (1973) 'Urban interaction on the Iranian plateau: excavations at Tepe Yahya 1967–1973', *Proceedings of the British Academy* 59: 283–319

Lampl, P. (1968) *Cities and Planning in the Ancient Near East*, London

Landay, S. (1971) 'The ecology of Islamic cities: the case for the ethnocity', *Economic Geography* 49: 303–13

Landström, B. (1969) *Sailing Ships*, London

Lane, E.W. (1842) *An Account of the Manners and Customs of the Modern Egyptians*, 3rd edn, London

Lane, F.C. (1940) 'The Mediterranean spice trade', *American Historical Review* 45: 581–90

Lapidus, I.M. (1967) *Muslim Cities in the Later Middle Ages*, Cambridge, Mass.

—— (1969) 'Muslim cities and Islamic societies' in I.M. Lapidus (ed.), *Middle Eastern Cities*, Berkeley and Los Angeles, 47–79

—— (1972) 'The conversion of Egypt to Islam', *Israel Oriental Studies* 2: 248–62

—— (1973) 'The evolution of Muslim urban society', *Comparative Studies in Society and History* 15: 21–50

—— (1981) 'Arab settlement and economic development of Iraq and Iran in the age of the Umayyad and early Abbasid caliphs' in A. Udovitch (ed.), *The Islamic Middle East, 700–1900: Studies in Economic and Social History*, Princeton, 177–208

Larsen, C.E. (1975) 'The Mesopotamian delta region: a reconsideration of Lees and Falcon', *Journal of the American Oriental Society* 95: 43–57

—— and G. Evans (1978) 'The Holocene geological history of the Tigris–Euphrates–Karun delta' in W.C. Brice (ed.), *The Environmental History of the Near and Middle East since the Last Ice Age*, London, New York and San Francisco, 227–43

Larsen, M.T. (1976) *The Old Assyrian City-state and its Colonies, Mesopotamia*, Copenhagen Studies in Assyriology 4, Copenhagen

Lassner, J. (1963a) 'Notes on the topography of Baghdad: the systematic description of the city and the Khaṭīb al-Baghdad', *Journal of the American Oriental Society* 83: 458–69

—— (1963b) 'The *Habl* of Baghdad and the dimensions of the city: a metrological note', *Journal of the Economic and Social History of the Orient* 6: 228–9

—— (1966) 'Massignon and Baghdad: the complexities of growth in an imperial city', *Journal of the Economic and Social History of the Orient* 9: 1–27

—— (1969) 'The Caliph's personal domain: the city plan of Baghdad re-examined' in A.H. Hourani and S.M. Stern (eds.), *The Islamic City*, Oxford, 103–18

Latham, R. (transl.) (1958) *The Travels of Marco Polo*, Harmondsworth

Lattimore, O. (1951) *Inner Asian Frontiers of China*, American Geographical Society Research Series no. 21, 2nd edn, New York

Law, C.M. (1967) 'The growth of urban population in England and Wales, 1801–1911', *Transactions of the Institute of British Geographers* 41: 125–44

Lawrence, T.E. (1962) *Seven Pillars of Wisdom: A Triumph*, Harmondsworth

Lawton, R. (1978) 'Population and society 1730–1900' in R.A. Dodgshon and R.A. Butlin (eds.), *An Historical Geography of England and Wales*, London, New York and San

Francisco, 313–66

Leake, W.M. (1802) Notebook: *Cerigo to England 1802.* 3. Museum of Classical Archaeology, Cambridge

—— (1824) *Journal of a Tour in Asia Minor*, London

Lebon, J.H.G. (1955) 'The new irrigation era in Iraq', *Economic Geography* 31: 47–59

Leemans, W.F. (1960) *Foreign Trade in the Old Babylonian Period as Revealed by Texts from Southern Mesopotamia*, Leiden

—— (1968) 'Old Babylonian letters and economic history. A review article with a digression on trade', *Journal of the Economic and Social History of the Orient* 11: 171–225

Lees, G.M. and N.L. Falcon (1952) 'The geographical history of the Mesopotamian plains', *Geographical Journal* 118: 24–39

Lerner, J.A. (1976) 'British travellers' accounts as a source for Qājār Shiraz', *Bulletin of the Asia Institute of Pahlavi University* 1–4: 205–72

Le Strange, G. (1895) 'Description of Mesopotamia and Baghdad, written about the Year 900 A.D. by Ibn Serapion', *Journal of the Royal Asiatic Society*, 33–76, 255–315

—— (1896) *Description of Syria, including Palestine, by Mukaddasi*, Palestine Pilgrims' Text Society 3, London, reprinted New York 1971

—— (1900) *Baghdad during the Abbasid Caliphate*, Oxford

—— (1919) *The Geographical Part of the Nuzhat-al-Qulūab composed by Hamd-Allāh Mustawfī of Qazwīn in 740 (1340)*, Leiden and London

*Letters of Gertrude Bell*, ed. Lady Bell, London 1927

Lewis, B. (1958) 'Some reflections on the decline of the Ottoman Empire', *Studia Islamica* 9: 111–27

—— (1962) 'Ottoman observers of Ottoman decline', *Islamic Studies* 1: 71–87

—— (1963) *Istanbul and the Civilisation of the Ottoman Empire*, Norman

—— (1964a) *The Middle East and the West*, London

—— (1964b) *The Arabs in History*, 3rd edn, London

—— (1979) 'Ottoman land tenure and taxation in Syria', *Studia Islamica* 50: 27–36

Lewis, G. (transl.) (1974) *The Book of Dede Korkut*, Harmondsworth

Lewis, N.N. (1949) 'Malaria, irrigation and soil erosion in central Syria', *Geographical Review* 39: 278–90

—— (1955) 'The frontier of settlement in Syria, 1800–1950', *International Affairs* 31: 48–60. Reprinted in C. Issawi (ed.), *The Economic History of the Middle East, 1800–1914*, Chicago and London 1966, 259–68

—— and M. Reinhold (eds.) (1966) *Roman Civilisation. Sourcebook II: The Empire*, New York

Little, T. (1965) *High Dam at Aswan*, New York

Lloyd, S. (1956) *Early Anatolia: The Archaeology of Asia Minor before the Greeks*, Harmondsworth

Lockhart, L. (1938) *Nadir Shah*, London

Lombard, M. (1969) 'Un problème cartographié: le bois dans la Méditerranée mussulmane (VII–XIe siècles)', *Annales: Economies, Sociétés, Civilisations* 14: 234–54

—— (1975) *The Golden Age of Islam*, Amsterdam and Oxford

Longrigg, S.H. (1925) *Four Centuries of Modern Iraq*, Oxford

Lopez, R., H. Miskimin and A. Udovitch (1970) 'England to Egypt, 1350–1500: long-term trends and long-distance trade' in M. Cook (ed.), *Studies in the Economic History of the Middle East*, London, 93–128

Lorraine, P. (1943) 'Perspectives of the Near East', *Geographical Journal* 102: 6–12

Luke, J.T. (1965) 'Pastoralism and politics in the Mari period' (unpublished doctoral dissertation, University of Chicago)

Lukenbill, D. (1927) *Ancient Records of Assyria*, vol. 2, Chicago

McClure, H.A. (1976) 'Radiocarbon chronology of the late Quaternary lakes in the Arabian desert', *Nature* 263: 755–6

McDougall, T. (1956) 'Climate in Roman times' (unpublished PhD thesis, University of London)

Mcfadyen, W. and C. Vita-Finzi (1978) 'Mesopotamia: the Tigris–Euphrates delta and its Holocene Hammar fauna', *Geological Magazine* 115: 287–300

Macmunn, G. and C. Falls (1928) *History of the Great War. Military Operations: Egypt and Palestine*, vol. 1, London

McNicoll, A. (1972) 'The development of urban defences in Hellenistic Asia Minor' in P.J. Ucko, R. Tringham and G.W. Dimbleby (eds.), *Man, Settlement and Urbanism*, London, 787–91

Mabro, R. (1974) *The Egyptian Economy, 1952–1972*, Oxford

Magalhães-Godinho, V. (1969) *L'Economie de l'Empire Portugais aux XVe et XVIe siècles*, Paris

Mahan, A.T. (1902) 'The Persian Gulf and international relations', *National Review* 40: 39–45. Reprinted in his *Retrospect and Prospect*, London 1903

Malamat, A. (1971) 'Syro-Palestinian destinations in a Mari tin inventory', *Israel Exploration Journal* 21: 31–8

Malcolm, J. (1815) *History of Persia*, London

—— (1830) *The Melville Papers*, ed. A.T. Wilson, London

Mantran, R. (1962) 'Baghdad à 1 'époque ottomane', *Arabica* 9: 311–24

—— (1965) *La Vie quotidienne à Constantinople au Temps de Suleiman le Magnifique et ses Successeurs (XVIe et XVIIe siècles)*, Paris

Ma'oz, M. (1968) *Ottoman Reform in Syria and Palestine, 1840–1861*, Oxford

Marchese, R.T. (1976) 'A history of urban organisation in the Lower Meander River Valley: regional settlement patterns in the second century A.D.' (unpublished doctoral dissertation, University of New York)

Markham, C.R. (1859) *Narrative of the Embassy of Ruy Gonzalez de Clavijo to the Court of Timour at Samarkand A.D. 1403–06*, Hakluyt Society, 1st series, 26, London

Martin, L. (1944) 'The miscalled Middle East', *Geographical Review* 34: 335 (see also editor's note, pp. 335–6)

Marx, E. (1966) *Bedouin of the Negev*, Manchester

Masterman, E.W.G. (1908) 'The fisheries of Galilee', *Palestine Exploration Fund*, 40: 40–51

Mayerson, P. (1956) 'Arid zone farming in antiquity: a study in ancient agricultural and related hydrological practices in Southern Palestine' (unpublished doctoral dissertation, University of New York)

Meeker, M.E. (1979) *Literature and Violence in North Arabia*, Cambridge

Mejcher, H. (1970) 'Oil and British policy towards Mesopotamia, 1914–18', *Middle Eastern Studies* 8: 377–92

Mellaart, J. (1967) *Çatal Hüyük: A Neolithic Town in Anatolia*, London

—— (1970) 'The earliest settlements in western Asia from the ninth to the end of the fifth millennium B.C.' in *The Cambridge Ancient History*, 3rd edn, vol. 1, pt 1, Cambridge, 248–326

—— (1971) 'Anatolia c.4000–2300 B.C.' in *The Cambridge Ancient History*, 3rd edn, vol. 1, pt 2, Cambridge, 263–410

—— (1975) *The Neolithic of the Near East*, London

Mellars, P.A. (1976) 'Fire ecology, animal populations and man: a study of some ecological relationships in prehistory', *Proceedings of the Prehistoric Society* 42: 15–45

Meuleau, M. (1965) 'Mesopotamia under Persia rule' in H. Bengtson (ed.), *The Greeks and the Persians from the Sixth to the Fourth Centuries*, London, 354–85

*Middle East and North Africa, The* (1960) Oxford Regional Economic Atlas, London

Mikesell, M.W. (1955) 'Notes on the dispersal of the dromedary', *Southwestern Journal of Anthropology* 11: 231–45

—— (1969) 'The deforestation of Mount Lebanon', *Geographical Review* 59: 1–28

Miles, G.C. (1964) 'Byzantium and the Arabs: relations in Crete and the Aegean area', *Dumbarton Oaks Papers* 18: 1–31

Miller, J.T. (1969) *The Spice Trade of the Roman Empire 29 B.C. to A.D. 641*, Cambridge

Miller, R. (1980) 'Water use in Syria and Palestine from the Neolithic to the Bronze Age', *World Archaeology* 11: 331–41

Mitchell, R.C. (1958) 'Instability of the Mesopotamian plains', *Bulletin de la Société de Geographie d'Egypte* 31: 129

Mitchell, S. (ed.) (1983) *Armies and Frontiers in Roman and Byzantine Anatolia*, British Archaeological Reports, International Series 156, Oxford

Mitchell, W.A. (1971) 'Turkish villages in interior Anatolia and von Thünen's "Isolated State": a comparative analysis', *Middle East Journal* 25: 355–69

Moore, N. (1835) *Commercial Report, 16 November, 1835*, Public Record Office FO 78/264, London

Morier, J. (1818) *A Second Journey through Persia, Armenia, and Asia Minor, to Constantinople, between the Years 1810 and 1816*, London

—— (1824) *The Adventures of Hajji Baba of Ispahan*, London

Morony, M.G. (1976) 'The effects of the Muslim conquest on the Persian population of Iraq', *Iran* 14: 41–59

Mumford, L. (1966) *The City in History*, Harmondsworth

Murray, G.W. (1951) 'The Egyptian climate: an historical outline', *Geographical Journal* 117: 422–34

—— (1967) 'Trogodytica: the Red Sea littoral in Ptolemaic times', *Geographical Review* 133: 24–33

Murvar, V. (1966) 'Some tentative modifications of Weber's typology: occidental versus oriental city', *Social Forces* 44: 381–9. Reprinted in P. Meadows and E.H. Mizrnchi (eds.), *Urbanism, Urbanization and Change. Comparative Perspectives*, Reading, Mass. 1969, 51–63

Naval Intelligence Division (1943) *Turkey*, Geographical Handbook Series, London

—— (1944) *Iraq and the Persian Gulf*, Geographical Handbook Series, London

—— (1946) *Western Anatolia and the Red Sea*, Geographical Handbook Series, London

Neely, J.A. (1974) 'Sas'anian and early Islamic water-control and irrigation systems on the Deh Luran Plain, Iran' in T.E. Downing and M. Gibson (eds.), *Irrigation's Impact on Society*, Anthropological Papers of the University of Arizona 25, Tucson, 21–42

Nelson, C. (ed.) (1973) *The Desert and the Sown: Nomads in the Wider Society*, Berkeley

Netting, R.M. (1974) 'Agrarian ecology', *Annual Review of Anthropology* 3: 21–56

Newberie, J. (1905) 'Two voyages of Master John Newberie' in S. Purchas (ed.), *Purchas His Pilgrimes*, vol. 8, Glasgow

Nibbi, A. (1979) 'Some remarks on the assumption of ancient Egyptian sea-going', *Mariners' Mirror* 65: 201–8

Niebuhr, C. (1792) *Travels through Arabia and Other Countries of the East*, Edinburgh

Noel, E. (1944) 'Qanāts', *Royal Central Asian Journal* 31: 191–202

Nour el Din, N. (1968) 'The High Dam and land reclamation in Egypt', *Mediterranea* 20: 262–9

Nützel, W. (1978) 'To which depth are prehistoric civilisations to be found beneath the present alluvial plain of Mesopotamia', *Sumer* 34: 17–26

Oates, J. (1973) 'The background and development of early farming communities in Mesopotamia and the Zagros', *Proceedings of the Prehistoric Society*, new series, 39: 147–81

—— (1982) 'Archaeological evidence for settlement patterns in Mesopotamia and Eastern Arabia in relation to possible environmental conditions' in J.L. Bintliff and W. van Zeist (eds.), *Palaeoclimates, Palaeoenvironments and Human Communities in the Eastern Mediterranean Region in Later Prehistory*, British Archaeological Reports, International Series 133, Oxford, 359–93

—— *et al.* (1977) 'Seafaring merchants of Ur?', *Antiquity* 51: 211–34

Oded, B. (1979) *Mass Deportations and Deportees in the Neo-Assyrian Empire*, Wiesbaden

Ökçün, A.G. (1970) *Osmanlı Sanayii 1913, 1915 Yıllari Sanayi İstatistiki*, Ankara

Oliphant, L. (1886) *Haifa, or Life in Modern Palestine*, London

Olmstead, A.T. (1948) *History of the Persian Empire*, Chicago

Oppenheim, A.L. (1954) 'The seafaring merchants of Ur', *Journal of the American Oriental Society* 74: 6–17

—— (1970) *Ancient Mesopotamia: Portrait of a Dead Civilisation*, Chicago and London

Orgels, B. (1963) *Contribution à l'Etude des Problèmes agricoles de la Syrie*, Brussels

Orhonlu, G. (1969) *Osmanlı Imparatorluğunda Aşiretleri Iskân Teşebbüsü, 1691–1696*, Istanbul Üniversitesi Edebiyat Fakültesi Yayınları no. 998, Istanbul

Ormerod, H.A. (1924) *Piracy in the Ancient World: An Essay in Mediterranean History*, Liverpool

Ostrogorsky, G. (1962) 'La commune rurale byzantine — Loi agraire — Traité fiscal — Cadastre de Thèbes', *Byzantion* 32: 139–66

Ouseley, W. (1819) *Travels in Various Countries of the East; more particularly Persia*, London

Owen, R. (1969) *Cotton and the Egyptian Economy 1820–1914*, Oxford

—— (1976) 'The Middle East in the eighteenth century — an "Islamic" society in decline? A critique of Gibb and Bowen's *Islamic Society and the West*', *Bulletin of the British Society for Middle Eastern Studies* 3: 110–17

—— (1981) *The Middle East in the World Economy, 1800–1914*, London

Owen, W.F. (1964) 'Land and water use in the Egyptian High Dam era', *Land Economics* 40: 277–93

Palestine Exploration Fund (1881–88) *The Survey of Western Palestine*, 9 vols, London

Pamuk, S. (1978) 'Foreign trade, foreign capital and the peripheralization of the Ottoman Empire, 1830–1913' (unpublished doctoral dissertation, University of California, Berkeley)

—— (1982) *Commodity Production for Export and Changing Relations of Production in Ottoman Agriculture, 1840–1913*, GETA Centre for Development Studies and Social Research, Discussion Papers no. 2, Ankara

Panzac, D. (1973) 'La peste à Smyrne au XVIIIe siècle', *Annales: Economies, Sociétés, Civilisations* 28: 1,071–93

*Parliamentary Debates, House of Commons* (1951) 5th series, vol. 1, London, pp. 448–9

*Parliamentary Debates, House of Lords* (1911) 5th series, vol. 7, London, col. 575

*Parliamentary Papers 1831–32*, vol. 10, pt 2, London

*Parliamentary Papers 1844*, vol. 46, London

*Parliamentary Papers 1867–68*, vol. 68, London

Parrot, A. (1938) *Mari: Une Ville perdue*, Paris

Parry, V.J. (1970) 'Materials of war in the Ottoman Empire' in M. Cook (ed.), *Studies in the Economic History of the Middle East*, London, 219–39

—— (1975) 'Le manière de combattre' in V.J. Parry and M.E. Yapp (eds.), *War, Technology and Society in the Middle East*, London, 218–56

Patai, R. (1951) 'Nomadism: Middle Eastern and Central Asian', *Southwestern Journal of Anthropology* 7: 401–14

—— (1952) 'The Middle East as a culture area', *Middle East Journal* 6: 1–21. Reprinted in A.M. Lutfiyya and C.W. Churchill (eds.), *Readings in Arab Middle East Society and Culture*, Paris and The Hague 1970, 187–204

Pearcy, G.E. (1959) 'The Middle East — an indefinable region', *Department of State Bulletin* 40: 407–16

Pentzopoulos, D. (1962) *The Balkan Exchange of Minorities and its Impact upon Greece*, Paris and The Hague

*Periplus of the Erythraean Sea, The*, ed. G.W.B. Huntingford, Hakluyt Society, n.s., 151, London 1981

Perrin de Brichambaut, G. and C.C. Wallén (1963) *A Study of Agroclimatology in Semi-Arid and Arid Zones of the Near East*, World Meteorological Organisation Technical Note no. 56, Geneva

Perry, J.R. (1979) *Karim Khan Zand. A History of Iran 1747–1779*, Publications of the Centre for Middle Eastern Studies 12, Chicago

Peters, E.L. (1977) 'Local history in two Arab communities', *Bulletin of the British Society for Middle East Studies* 4: 71–81

Peters, F.E. (1970) *Harvest of Hellenism: History of the Near East from Alexander the Great to the Triumph of Christianity*, New York

—— (1978) 'Roman and bedouin in southern Syria', *Journal of Near Eastern Studies* 37: 315–26

Petrushevsky, I.P. (1968) 'The socio-economic condition of Iran under the Il-Khāns' in *The Cambridge History of Iran*, vol. 5, *The Saljuq and Mongol Periods*, ed. J.A. Boyle, 483–537

Philips, C.H. (1940) *The East India Company, 1784–1834*, Manchester

Pigulevskaya, N. (1963) *Les Villes de l'Etat Iranien aux époques Parthe et Sassanide*, Ecole Pratique des Hautes Etudes, Sorbonne, VI Section, Documents et Recherches 6, Paris and The Hague

Pinkerton, J. (1802) *Modern Geography*, London

Pirenne, H. (1939) *Mohammed and Charlemagne*, London

Planhol, X. de (1950) 'Estivage et exploitation des montagnes en Pisidie', *Bulletin de l'Association de Géographes français*, 81–8

—— (1959) *The World of Islam*, Ithaca

—— (1961, 1962, 1964, 1965, 1967) 'Nomades et pasteurs', *Revue de Géographie de l'Est* 2: 291–320; 3: 295–318; 4: 315–28; 5: 355–74; 6: 373–4

—— (1963) 'A travers les chaines pontiques: planations côtières et vie montagnarde', *Bulletin de l'Association de Géographes français* 311–12: 2–12

—— (1966) 'La signification géographique du Livre de Dede Korkut', *Journal Asiatique* 254: 225–44

Plato, *Platonis Opera*, ed. J. Burnet, Scriptorum Classicorum Bibliotheca Oxoniensis, vol. 5, Oxford 1915

Playfair, I.S.O. (1954) *The Mediterranean and the Middle East*, vol. 1, *History of the Second World War*, United Kingdom Military Series, Campaigns, London

Pleiner, R. and J.K. Bjorkman (1974) 'The Assyrian Iron Age: the history of iron in the

Assyrian civilisation', *Proceedings of the American Philosophical Society* 118: 283–313

Pliny the Elder, *Natural History*, transl. H. Rackham *et al.*, Loeb Classical Library, London and Cambridge, Mass. 1938–62

Plutarch, *Lives*, transl. B. Perrin, Loeb Classical Library, Cambridge, Mass. 1949

Pococke, R. (1745) *A Description of the East and Some Other Countries*, London

Poidebard, A. (1934) *La Trace de Rome dans le Déserts de Syrie. Recherches aériennes*, Paris

Poirier, J.P. (1980) 'Large historical earthquakes and seismic risk in northwest Syria', *Nature* 285: 217–20

Poliak, A.N. (1938) 'The demographic evolution of the Middle East: population trends since 1348', *Palestine and the Middle East* 10: 201–5

Polk, W.R. (1962) 'Rural Syria in 1845', *Middle East Journal* 16: 508–14

Porter, R.K. (1821) *Travels in Georgia, Persia, etc. during the years 1817–1820*, London

Postgate, J.N. (1974) 'Some remarks on conditions in the Assyrian countryside', *Journal of the Economic and Social History of the Orient* 17: 225–43

Potter, D. (1955) 'The bazaar merchant' in J.N. Fisher (ed.), *Social Forces in the Middle East*, Ithaca, 99–115

Pounds, N.J.G. (1974) *An Historical Geography of Europe, 1500–1840*, Cambridge

*Power and Propaganda: A Symposium on Ancient Empires*, ed. M.T. Larsen, Copenhagen Studies in Assyriology 7, Copenhagen 1979

Prawer, J. (1972) *The Latin Kingdom of the Crusaders*, London

Preussner, A.H. (1920) 'Date culture in ancient Babylonia', *American Journal of Semitic Languages and Literature* 36: 212–32

Pritchard, J.B. (ed.) (1955) *Ancient Near Eastern Texts relating to the Old Testament*, 2nd edn, Princeton

Procopius, *History of the Wars*, transl. H.B. Dewing, Loeb Classical Library, London and New York 1914–35

Ptolemy, Claudius, *Geography*, transl. and ed. by E.L. Stevenson, New York 1932

Quataert, D. (1975) 'Dilemma of development: the Agricultural Bank and agricultural reform in Ottoman Turkey, 1888–1908', *International Journal of Middle East Studies* 6: 210–27

—— (1977) 'Limited revolution: the impact of the Anatolian Railway on Turkish transportation and the provisioning of Istanbul, 1890–1908', *Business History Review* 51: 139–60

—— (1980) 'The commercialisation of agriculture in Ottoman Turkey, 1800–1914', *International Journal of Turkish Studies* 1: 38–55

—— (1981) 'Agricultural trends and government policy in Ottoman Anatolia, 1800–1914', *Asian and African Studies* 15: 69–84

Rafeq, A-K. (1966) *The Province of Damascus, 1723–1783*, Beirut

Raisz, E. (1951) *Landforms of the Near East*, Cambridge, Mass.

Rappaport, R.A. (1971) 'The flow of energy in an agricultural society', *Scientific American* 225: 117–32

Raschke, M.G. (1978) 'New studies in Roman commerce with the East' in H. Temporini and W. Haase (eds.), *Austeig und Niedergang der römischen Welt*, Berlin and New York, II.9, 605–1,361

Raymond, A. (1973, 1974) *Artisans et Commerçants au Caire au XVIIIe siècle*, Damascus

—— (1979) 'La conqûete ottomane et le développement des grandes villes Arabes: le cas de Caire, de Damas et d'Alep', *Revue de l'Occident Mussulman et de la Méditerranée* 27: 114–34

Reade, J. (1975) 'Studies in Assyrian geography', *Revue d'Assyriologie* 72: 47–72, 157–80

Redman, C.L. (1977) 'Man, domestication and culture in southwestern Asia' in C.A. Reed (ed.), *Origins of Agriculture*, The Hague and Paris, 523–41

—— (1978) *The Rise of Civilisation: From Early Farmers to Urban Society in the Ancient Near East*, San Francisco

Reed, C.A. (1974) 'The beginnings of animal domestication' in H.H. Cole and M. Ronning (eds.), *Animal Agriculture: The Biology of Domestic Animals and their Use by Man*, San Francisco, 5–17

– (1977) 'A model for the origin of agriculture in the Near East' in C.A. Reed (ed.), *Origins of Agriculture*, The Hague and Paris, 543–67

Reiss, R.O. (1969) ' "The Middle East": the problem of geographic terminology', *Journal of Geography* 68: 34–40

Renfrew, A.C. (1975) 'Trade as action at a distance: questions of integration and communication' in J.A. Sabloff and C.C. Lamberg-Karlovsky (eds.), *Ancient Civilisation and Trade*, Albuquerque, 3–59

Renfrew, J.M. (1973) *Palaeoethnobotany. The Prehistoric Food Plants of the Near East and Europe*, London

Rennell of Rodd, Lord (1946) 'Presidential address', *Geographical Journal* 107: 85–6

Richard, J. (1978) *The Latin Kingdom of Jerusalem*, revised edn, London

Richards, A. (1978) 'Technical and social change in Egyptian agriculture: 1890–1914', *Economic Development and Cultural Change* 26: 725–46

Ricks, T.M. (1973) 'Towards a social and economic history of eighteenth-century Iran', *Iranian Studies* 6: 110–26

—— (1974) 'Politics and trade in Southern Iran and the Gulf, 1745–1765' (unpublished doctoral dissertation, University of Indiana)

Riising, A. (1952) 'The fate of Henri Pirenne's thesis on the consequences of the Islamic expansion', *Classica et Mediaevalia* 13: 87–130

Roberts, N. (1982a) 'Lake levels as an indicator of Near Eastern palaeo-climates: a preliminary appraisal' in J.L. Bintliff and W. van Zeist (eds.), *Palaeoclimates, Palaeoenvironments and Human Communities in the Eastern Mediterranean Region in Later Prehistory*, British Archaeological Reports, International Series 133, Oxford, 235–67

—— (1982b) 'Forest re-advance and the Anatolian Neolithic' in M. Bell and S. Limbrey (eds.), *Archaeological Aspects of Woodland Ecology*, British Archaeological Reports, International Series 146, Oxford, 231–46

—— *et al.* (1979) 'Radiocarbon chronology of the late Pleistocene Konya lake, Turkey', *Nature* 281: 662–4

Rogers, J.M. (1969) 'Samarra: a study in medieval town-planning' in A.H. Hourani and S.M. Stern (eds.), *The Islamic City*, Oxford, 119–56

—— (1976) *The Spread of Islam*, London

Ron, Z. (1969) 'Agricultural terraces in the Judaean mountains', *Israel Exploration Journal* 16: 33–49, 111–22

Rosenan, N. (1955) 'One hundred years of rainfall in Jerusalem', *Israel Exploration Journal* 5: 137–53

Rostovtsev, M.I. (1926) *Social and Economic History of the Roman Empire*, Oxford

—— (1932) *Caravan Cities*, Oxford

—— (1941) *The Social and Economic History of the Hellenistic World*, Oxford

Rousseau, J-B. (1809) *Description du Pachalik de Baghdad*, Paris. Reproduced in C. Issawi (ed.), *The Economic History of the Middle East, 1800–1914*, Chicago and London, 1966, 135–6

Roux, G. (1966) *Ancient Iraq*, Harmondsworth

Rowton, M.B. (1967) 'The woodlands of ancient Asia', *Journal of Near Eastern Studies* 26: 261–77

—— (1973) 'Autonomy and nomadism in western Asia', *Orientalia* n.s. 42: 247–58

Runciman, S. (1961) *Byzantine Civilisation*, London

—— (1965) *A History of the Crusades*, Harmondsworth

Russell, A. (1734) *The Natural History of Aleppo*, 2nd edn, London

Russell, J.C. (1958) 'Late ancient and medieval population', *Transactions of the American Philosophical Society*, new series 48, pt 3: 1–132

—— (1968) 'That earlier plague', *Demography* 5: 174–84

Rycaut, P. (1686) *The History of the Present State of the Ottoman Empire*, 6th edn, London

Sahillioğlu, H. (1965) 'Dördüncü Murad'ın Bağdat seferi Menzilnamesi (Bağdat seferi Harp Jurnali)', *Belgeler* 2: 1–35

St John, O. (1876) *Eastern Persia: An Account of the Journey of the Persian Boundary Commission*, London

Salim, S.M. (1962) *Marsh Dwellers of the Euphrates Delta*, Monographs on Social Anthropology 23, London

Sandars, N.K. (1972) *The Epic of Gilgamesh*, revised edn, Harmondsworth

Sarç, O.C. (1941) 'Tanzimat ve Sanayimiz' in *Tanzimat*, Istanbul, 423–40. Transl. in C. Issawi (ed.), *The Economic History of the Middle East, 1800–1914*, Chicago and London 1966, 48–59

Sarnthein, M. (1972) 'Sediments and history of the postglacial transgression in the Persian Gulf and NW Gulf of Oman', *Marine Geology* 12: 245–66

Saunders, J.J. (1971) *The History of the Mongol Conquests*, London

Sauvaget, J. (1948) *Relation de la Chine et de l'Inde, rédigée en 851*, Paris

Savory, R. (1980) *Iran under the Safavids*, Cambridge

Schalie, H. Van der (1974) 'Aswan Dam revisited', *Environment* 16: 18–20

Schede, M. (1964) *Die Ruinen von Priene*, Berlin

Schölch, A. (1981) 'The economic development of Palestine, 1856–1892', *Journal of Palestine Studies* 10: 35–58

Schumm, M. (1977) *The Fluvial System*, New York and London

Schweizer, G. (1972) 'Tabriz (Nordwest-Iran) und der Tabrizer Bazar', *Erdkunde* 26: 32–46

Searight, S. (1969) *The British in the Middle East*, London

Semple, E.C. (1916) 'Pirate coasts of the Mediterranean Sea', *Geographical Review* 2: 134–51. Reprinted in her *The Geography of the Mediterranean Region: Its Relation to Ancient History*, London 1932, 638–59

Serjeant, R.B. (1963) *The Portuguese off the South Arabian Coast*, Oxford

—— (1964) 'Some irrigation systems in Hadramaut', *Bulletin of the School of Oriental and African Studies* 27: 33–76

—— (1968) 'Fisher folk and fish-traps in al-Bahrain', *Bulletin of the School of Oriental and African Studies* 31: 486–514

—— (1974) 'The cultivation of cereals in medieval Yemen', *Arabian Studies* 1: 25–75

Setton, K.M. (1953) 'On the importance of land tenure and agrarian taxation in the Byzantine Empire', *American Journal of Philology* 74: 225–59

—— (ed.) (1962–77) *A History of the Crusades*, Philadelphia and Madison

Shaban, M.A. (1976) *Islamic History: A New Interpretation*, London

Shanin, T. (ed.) (1971) *Peasants and Peasant Societies*, Harmondsworth

Sharon, M. (1975) 'The political role of the bedouins in Palestine in the sixteenth and seventeenth centuries' in M. Ma'oz (ed.), *Studies on Palestine during the Ottoman Period*, Jerusalem, 11–30

Shaw, S.J. (1962) *The Financial and Administrative Organisation and Development of Ottoman Egypt, 1517–1798*, Princeton

—— (1964) *Ottoman Egypt in the Age of the French Revolution by Huseyn Efendi*, Harvard Middle Eastern Monographs 11, Cambridge, Mass.
—— (1976) *History of the Ottoman Empire and Modern Turkey*, vol. 1, Cambridge, London and New York
—— and E.K. Shaw (1977) *History of the Ottoman Empire and Modern Turkey*, vol. 2, Cambridge
Siroux, M. (1974) 'Les caravanserais routiers Safavids', *Iranian Studies* 7: 348–75
Slofstra, J. (1974) 'Ecological and systems approaches to prehistoric archaeology: a summary of recent discussions', *Helinium* 14: 163–73
Smith, C.G. (1968) 'The emergence of the Middle East', *Journal of Contemporary History* 3: 3–17
—— (1975) 'The geography and natural resources of Palestine as seen by British writers in the nineteenth and early twentieth centuries' in M. Ma'oz (ed.), *Studies on Palestine during the Ottoman Period*, Jerusalem, 87–100
Smith, P.E.L. (1975) 'Ganji Dareh Tepe', *Iran* 13: 178–80
—— and T. Cuyler Young (1972) 'The evolution of early agriculture and culture in greater Mesopotamia: a trial model' in B. Spooner (ed.), *Population Growth: Anthropological Implications*, Cambridge, Mass.
Snodgrass, A.M. (1977) *Archaeology and the Rise of the Greek State*, Cambridge
Solecki, R. (1974) 'The Old World Paleolithic' in R. Stigler, R. Holloway, R. Solecki, D. Perkins and P. Daly (eds.), *The Old World. Early Man to the Development of Agriculture*, New York, 45–70
Sourdel-Thomine, J. (1954) Le peuplement de le région des "villes mortes" (Syrie du Nord) à l'époque ayyūbide', *Arabica* 1: 187–200
Southern, R.W. (1962) *Western Views of Islam in the Middle Ages*, Cambridge, Mass.
Southgate, H. (1840) *Narrative of a Tour through Armenia, Koordistan, Persia, Mesopotamia . . .*, New York
Spender, H. (1911) 'Introduction' to the Everyman Edition of Kinglake's *Eothen*, London
Sperber, D. (1974) 'Drought, famine and pestilence in Amoraic Palestine', *Journal of the Economic and Social History of the Orient* 17: 272–98
Stager, L.E. (1976) 'Farming in the Judaen desert during the Iron Age', *Bulletin of the American Schools of Oriental Research* 221: 145–58
Stanislawski, D. (1975) 'Dionysus westwards: early religion and the economic geography of wine', *Geographical Review* 65: 427–44
Stauffer, T.R. (1965) 'The economics of nomadism in Iran', *Middle East Journal* 22: 284–302
Stavrianos, D.S. (1958) *The Balkans since 1453*, New York
Steadman, J.M. (1969) *The Myth of Asia*, New York
Steensberg, A. (1977) 'The husbandry of food production' in J. Hutchinson, G. Clark, E.M. Jope and R. Riley (eds.), *The Early History of Agriculture*, Oxford, 43–54
Steensgaard, N. (1974) *The Asian Trade Revolution of the Seventeenth Century. The East India Companies and the Decline of the Caravan Trade*, Chicago and London
Stigler, R. (1974) 'The later Neolithic in the Near East and the rise of civilisation' in R. Stigler, R. Holloway, R. Solecki, D. Perkins and P. Daly (eds.), *The Old World. Early Man to the Development of Agriculture*, New York, 98–126
Stoianovich, T. and G.C. Haupt (1962) 'Le maïs arrive dans les Balkans', *Annales: Economies, Sociétés, Civilisations* 17: 84–92
Strabo, *The Geography of Strabo*, transl. H.L. Jones, Loeb Classical Library, London and Princeton 1954
Sweet, L. (1965) 'Camel raiding of north Arabian bedouin: a mechanism of ecological

adaptation', *American Anthropologist* 67: 1,132–50

—— (1968) 'The Arabian Peninsula: an annotated bibliography' in *The Central Middle East. A Handbook of Anthropology and Published Research*, vol. 2, New Haven, Conn., 287–8

Swidler, W.W. (1973) 'Adaptive processes regulating nomad-sedentary interaction in the Middle East' in C. Nelson (ed.), *The Desert and the Sown: Nomads in the Wider Society*, Berkeley, 23–41

Sykes, M. (1915) *The Caliph's Last Heritage*, London

*Syria, The Holy Land, Asia Minor etc., Illustrated by W.H. Bartlett, William Purser, etc., with Descriptions of the Plates by John Carne, Esq.*, London, Paris and New York, 1837–38

Tavernier, J.B. (1684) *Collections of Travels through Turky [sic] into Persia and the East Indies*, London

Taylor, E.G.R. and M.W. Richey (1962) *The Geometrical Seaman*, London

Tchalenko, G. (1953) *Villages antiques de la Syrie du Nord*, Paris

Thesiger, W. (1959) *Arabian Sands*, London

Thévenot, J. de (1727) *Voyages de Monsieur de Thévenot en Europe, Asie et Afrique*, 3rd edn, Amsterdam

Thirgood, J.V. (1981) *Man and the Mediterranean Forest: A History of Resource Depletion*, London

Thiriet, F. (1959) *Le Romanie vénitienne au Moyen Age*, Paris

Thomas, B. (1938) *Arabia Felix: Across the Empty Quarter of Arabia*, London

Thorner, D. (1964) 'Peasant economy as a category in economic history' in *Second International Conference of Economic History, Aix-en-Provence, 1962*, Congrès et Colloques VIII, Ecole Pratique des Hautes Etudes — Sixième Section: Sciences économiques et sociales, Paris and The Hague, 287–300

Tibbetts, G.R. (1971) *Arab Navigation in the Indian Ocean before the Coming of the Portuguese*, London

Toussoun, O. (1924) *Mémoire sur les Finances de l'Egypte*, Cairo

Toynbee, A.J. (1959) *Hellenism: The History of a Civilisation*, London

*Travels of Abbé Carré in India and the Near East 1672 to 1674*, ed. C. Fawcett and R. Burn, Hakluyt Society, 2nd series, 95 (I) 1947; 96 (II) 1947; 97 (III) 1948, London

Trécourt, J.B. (1942) *Mémoires sur l'Egypte année 1791*, ed. and annotated by G. Wiet, Cairo

Trigger, B.G. (1983) 'The rise of Egyptian civilisation' in B.G. Trigger, B.J. Kemp, D. O'Connor and A.B. Lloyd (eds.), *Ancient Egypt: A Social History*, Cambridge, 1–70

—— B.J. Kemp, D. O'Connor and A.B. Lloyd (1983) *Ancient Egypt: A Social History*, Cambridge

Tümertekin, E. (1955) 'Time relationships between the wheat growing period and dry months in Turkey', *Review of the Geographical Institute of the University of Istanbul*, International Edition 2: 73–84

Turkowski, L. (1969) 'Peasant agriculture in the Judean hills', *Palestine Exploration Quarterly* 101: 21–33, 101–12

Tute, R.C. (1927) *The Ottoman Land Laws*, Jerusalem

Ubicini, M.A. (1856) *Letters on Turkey: An Account of the Religious, Political, Social and Commercial Condition of the Ottoman Empire*, London

Ülker, N. (1974) 'The rise of Izmir, 1688–1740' (unpublished doctoral dissertation, University of Michigan)

UN (1983a) *Demographic Yearbook 1981*, New York

—— (1983b) *Statistical Yearbook 1981*, New York

Unger, E. (1935) 'Ancient Babylonian maps and plans', *Antiquity* 9: 311–22

Urquhart, D. (1833) *Turkey and its Resources*, London

Vance, J.E. (1970) *The Merchants' World*, Englewood Cliffs, N.J.

Vasiliev, A.A. (1933) 'On the question of Byzantine feudalism', *Byzantion* 8: 584–604

Vaumas, E. de (1965) 'L'écolement des eaux en Mésopotamie et la provenance des eaux de Tello', *Iraq* 27: 81–99

Vaux, R. de (1971) 'Palestine in the early Bronze Age', in *The Cambridge Ancient History*, 3rd edn, vol. 1, pt 2, Cambridge, 208–37

Veinstein, G. (1975) 'Ayan de la région d'Izmir et le commerce du Levant (deuxième moitié du XVIIIe siècle)', *Revue de l'Occident Mussulman et de la Méditerranée* 20: 131–46

Vesey-Fitzgerald, D.F. (1955) 'Vegetation of the Red Sea coast of Jedda, Saudi Arabia', *Journal of Ecology* 43: 477–89

—— (1957a) 'The vegetation of the sea coast north of Jedda, Saudi Arabia', *Journal of Ecology* 45: 547–62

—— (1957b) 'The vegetation of central and eastern Arabia', *Journal of Ecology* 45: 779–98

Vidal de la Blache, P. (1911) 'Les genres de vie dans la géographie humaine', *Annales de Géographie* 20: 193–214, 289–304

—— (1926) *Principles of Human Geography*, London

Villiers, A. (1940) *Sons of Sinbad*, London

—— (1948) 'Some aspects of the Arab dhow trade', *Middle East Journal* 2: 399–416

Vita-Finzi, C. (1969a) *The Mediterranean Valleys: Geological Changes in Historical Times*, Cambridge

—— (1969b) 'Late Quaternary continental deposits of central and western Turkey', *Man* 4: 605–19

—— (1969c) 'Late Quaternary alluvial chronology of Iran', *Geologie Rundschau* 58: 951–73

—— (1978) 'Recent alluvial history in the catchment of the Arabo-Persian Gulf' in W.C. Brice (ed.), *The Environmental History of the Near and Middle East since the Last Ice Age*, London, New York and San Francisco, 255–61

Volney, C.F. (1786) *Voyage en Syrie et en Egypte pendant les années 1783, 1784 et 1785*, Paris

—— (1788) *Travels through Syria and Egypt in the Years 1783, 1784 and 1785*, London

Vryonis, S. (1971) *The Decline of Medieval Hellenism in Asia Minor and the Process of Islamization from the Eleventh through the Fifteenth Century*, Berkeley, Los Angeles and London

—— (1975) 'Nomadization and Islamization in Asia Minor', *Dumbarton Oaks Papers* 29: 41–71

Waber, J.G. (1969) 'A study of selected economic factors and their contribution to the understanding of the history of Palestine during the Hellenistic period' (unpublished doctoral dissertation, Duke University)

Wagstaff, J.M. (1981) 'Buried assumptions: some problems in the interpretation of the "Younger Fill" raised by recent data from Greece', *Journal of Archaeological Science* 8: 247–64

Waldbaum, J.C. (1978) *From Bronze to Iron: The Transition from the Bronze Age to the Iron Age in the Eastern Mediterranean*, Studies in Mediterranean Archaeology 54, Göteburg

Wallén, C.C. (1969) 'Arid zone meteorology' in E.S. Hills (ed.), *Arid Lands*, London, 31–51

Waring, E.S. (1807) *A Tour to Sheeraz . . .*, London

Warriner, D. (1962) *Land Reform and Development in the Middle East. A Study of Egypt, Syria and Iraq*, 2nd edn, London

Watson, A.M. (1974) 'The Arab agricultural revolution and its diffusion, 700–1100', *Journal of Economic History* 34: 8–35

—— (1983) *Agricultural Innovation in the Early Islamic World*, Cambridge

Watt, W.M. (1961) *Muhammad: Prophet and Statesman*, London

Weinbaum, W.G. (1977) 'Agricultural policy and development politics in Iran', *Middle East Journal* 31: 434–50

Wells, D. (1970) 'The water requirements of particular stock producing systems' in J.A. Taylor (ed.), *The Role of Water in Agriculture*, Oxford, 121–32

Wenke, R.J. (1975–76) 'Imperial investments and agricultural developments in Parthian and Sasanian Khuzestan: 150 B.C. to A.D. 640', *Mesopotamia* 10–11

Wertime, T.A. (1973) 'The beginnings of metallurgy: a new look', *Science* 182: 875–87

West, L.C. (1924) 'Commercial Syria under the Roman Empire', *Transactions and Proceedings of the American Philological Association* 54: 159–89

Weulersse, J. (1946) *Paysans de Syrie et du Proche Orient*, Paris

Wheatley, P. (1972) 'The concept of urbanism in P.J. Ucko, R. Tringham and G.W. Dimbleby (eds.), *Man, Settlement and Urbanism*, London, 601–37

Whittaker, C.R. (1976) 'Agri Deserti' in M.I. Finley (ed.), *Studies in Roman Property*, Cambridge, 137–65

—— (1980) 'Inflation and economy in the fourth century A.D.' in *Imperial Revenue, Expenditure and Monetary Policy in the Fourth Century A.D.*, British Archaeological Reports, International Series 76, Oxford, 1–22

Wiet, G. (1937) *Ya'kubi Les Pays*, Cairo

—— (1962) 'Le traité des famines de Maqrīzī', *Journal of the Economic and Social History of the Orient* 5: 1–90

Wightman, G.B.H. and A.Y. al-Udhari (1975) *Birds through a Ceiling of Alabaster: Three Abbasid Poets*, Harmondsworth

Wilkinson, J.C. (1974) *The Organisation of the Falaj Irrigation System in Oman*, School of Geography Research Papers 10, Oxford

—— (1977) *Water and Settlement in South-east Arabia. A Study of the Aflaj of Oman*, Oxford

Williams, A. (1968) *Britain and France in the Middle East and North Africa, 1914–1967*, London and New York

Willcox, G. (1974) 'A history of deforestation as indicated by charcoal analysis of four sites in eastern Anatolia', *Anatolian Studies* 24: 117–33

Wilson, C.T. (1906) *Peasant Life in the Holy Land*, London

Wirth, E. (1966) 'Damaskus-Aleppo-Beirut, ein geographischer Vergleich dreir nahöstlicker Städte im Spiegel inhrer sozial und wirtschaftlich tonangebenden Schichten', *Die Erde* 97: 96–137

—— (1968) 'Strukturwandlungen und Entwicklungstendzen der Orientalischen Stadt', *Erdkunde* 22: 101–28

Wittek, P. (1938) *The Rise of the Ottoman Empire*, Royal Asiatic Society Monographs 23, London (reprinted 1965)

Wittfogel, K.A. (1957) *Oriental Despotism: A Comparative Study of Total Power*, New Haven, Conn. and London

Wittman, W. (1803) *Travels in Turkey, Asia Minor, Syria and Across the Desert into Egypt During the Years 1799, 1800, and 1801 . . .*, London

Wolf, E.R. (1966) *Peasants*, Englewood Cliffs, N.J.

Woolley, C.L. and T.E. Lawrence (1915) *The Wilderness of Zin*, London

*World Weather Records*, ed. H.H. Clayton *et al.*, Smithsonian Miscellaneous Collections, vol. 79, Washington D.C.

Worthington, E.B. (1972) 'The Nile catchment — technological change and aquatic biology' in M.T. Farvar and J.P. Milton (eds.), *The Careless Technology: Ecology and International Development*, Garden City, N.Y., 189–205

Wright, H.E. (1961) 'Late pleistocene soil development, glaciation and cultural change in the eastern Mediterranean region', *Annals of the New York Academy of Sciences* 95

—— (1969) *The Administration of Production in an Early Mesopotamian Town*, Anthropological Papers, Museum of Archaeology 38, Ann Arbor

—— (1977a) 'Recent research on the origin of the state', *Annual Review of Anthropology* 6: 379–97

—— (1977b) 'Environmental change and the origins of agriculture in the Old and New Worlds' in C.A. Reed (ed.), *Origins of Agriculture*, The Hague and Paris, 281–318

—— and G.A. Johnson (1975) 'Population, exchange and early state formation in southwestern Iran', *American Anthropological* 77: 267–89

——, J.A. Neely, G.A. Johnson and J. Speth (1975) 'Fourth millennium developments in southwestern Iran', *Iran* 13: 129–47

Wulff, H.E. (1966) *The Traditional Crafts of Persia*, Cambridge, Mass. and London

Xenophon, *The Anabasis*, Scriptorum Classicorum Bibliotheca Oxoniensis, Oxford

Yapp, M.E. (1980) *Strategies of British India: Britain, Iran and Afghanistan, 1798–1850*, Oxford

Young, T.C. (1972) 'Population densities and early Mesopotamian urbanism' in P.J. Ucko, R. Tringham and G.W. Dimbleby (eds.), *Man, Settlement and Urbanism*, London, 827–42

Zarins, J. (1978) 'The camel in ancient Arabia: a further note', *Antiquity* 52: 44–6

Zeist, W. van (1977) 'Plant foods in southwestern Asia' in J. Hutchinson, G. Clark, E.M. Jope and R. Riley (eds.), *The Early History of Agriculture*, Oxford, 27–41

—— and S. Bottema (1977) 'Palynalogical investigations in western Iran', *Palaeohistoria* 19: 19–85

—— and S. Bottema (1982) 'Vegetational history of the eastern Mediterranean and the Near East during the last 20,000 years', in J.L. Bintliff and W. van Zeist (eds.), *Palaeoclimates, Palaeoenvironment and Human Communities in the Eastern Mediterranean Region in Later Prehistory*, British Archaeological Reports, International Series 133, Oxford, 277–321

—— et al. (1975) 'Late Quaternary vegetation and climate of south-western Turkey', *Palaeohistoria* 17; 55–143

Zeuner, F.E. (1963) *A History of Domesticated Animals*, London

Ziraî Istatistik Özetleri 1941–1962, Türkiye Cumhuriyeti Devlet Istatistik Enstitüsü, Yayın 447, Ankara 1963

Zohary, D. (1969) 'The progenitors of wheat and barley in relation to domestication and agricultural dispersal in the Old World' in P. Ucko and G.W. Dimbleby (eds.), *The Domestication and Exploitation of Plants and Animals*, London, 47–66

—— and M. Hopf (1973) 'Domestication of pulses in the Old World', *Science* 182: 887–94

Zohary, Z. and P. Spiegel-Roy (1975) 'The beginnings of fruit growing in the Old World', *Science* 187: 319–27

Zubrow, E.B.W. (1975) *Prehistoric Carrying Capacity: A Model*, Menlo Park, Calif.

Zuckermann, B. (1971) 'Das Euphratprojekt in der Syrischen Arabischen Republik und sein Einfluss auf die Territorialstruktur der syrischen Volkswirtschaft', *Petermanns Geographische Mitteilungen* 115: 98–101

# INDEX